SURVEY METHODS FOR ECOSYSTEM MANAGEMENT

SURVEY METHODS FOR ECOSYSTEM MANAGEMENT

Wayne L. Myers, Associate Professor

School of Forest Resources
Pennsylvania State University

Ronald L. Shelton, Associate Professor

Department of Resource Development
Michigan State University

A Wiley-Interscience Publication

JOHN WILEY & SONS

New York • Chichester • Brisbane • Toronto

Library of Congress Cataloging in Publication Data:

Myers, Wayne L. 1942-
Survey methods for ecosystem management.

"A Wiley-Interscience publication."
Includes bibliographies and index.
1. Ecological surveys—Methodology.
2. Information storage and retrieval systems—Ecology. I. Shelton, Ronald L., joint author. II. Title. III. Title: Ecosystem management.
QH541.15.S95M93 574.5′028 79-25404
ISBN 0-471-62735-6

Printed in the United States of America

10 9 8 7 6 5 4 3 2 1

To

Scientific Management of Ecosystems

PREFACE

Expanding population and technology are increasing the pressure on our ecosystems. Rising demands, pollution, and conflicts between interest groups make it essential that ecosystems be managed scientifically under the guidance of interdisciplinary teams. Such integrated management requires comprehensive information on the status of ecosystems along with a broad base of ecological understanding among all those involved in the management effort.

This book is concerned directly with building information bases for scientific management and indirectly with contributing to a common base of ecological understanding. Basic survey methods for collecting information on major components of terrestrial ecosystems are covered in detail. Potential applications of space age technology still under development are covered in a more general way to impart an appreciation for the promise that they hold. The rapidly expanding technology of remote sensing is one example of these research frontiers. All of these techniques are presented in the context of integrated information systems.

The growth of scientific knowledge through research in combination with the need for broader social and ecological awareness has posed a real quandary for educators in disciplines dealing with the management of environmental resources. More material must be covered in a shorter time. Basic techniques of surveying and statistics are being condensed into discipline oriented courses on survey methods at the same time that the scope of these courses is broadened to cover more environmental resources. We have attempted to provide a text that will meet the needs of these courses and will also serve to broaden the perspective of those who are professionally involved in the management of environmental resources. When combined with case studies and problems selected by instructors, this book should serve a broad range of educational purposes.

WAYNE L. MYERS
RONALD L. SHELTON

University Park, Pennsylvania
East Lansing, Michigan
January 1980

CONTENTS

CHAPTER 1–IMPORTANCE OF SURVEY INFORMATION IN ECOSYSTEM MANAGEMENT

Natural resources management in developing countries typically has progressed from a simple use and exploitation of the resources at hand to the rudimentary management of extensive area units in which nature does most of the work with only a little help from man, followed by a long-term trend toward more intensive management of successively smaller area units for increased production as the resources become more scarce relative to population numbers and their values increase. Noneconomic resources such as air and water (except in arid regions) are often taken for granted in the beginning, with concern developing only as the degradation of quality becomes serious. The transitions, however, are not necessarily smooth. Especially in societies without central planning, the shifts in management often take place sporadically in response to crisis or near crisis situations.

So has it been in the United States. In the first 400 years—between the founding of the Plymouth Colony and the closing of the frontier—natural resources of most types were plentiful, and the management philosophy (or lack thereof) was "let daylight into the swamp." One who worked with natural resources needed only to know how to harvest nature's bounty. In those days the only practical need for the measurement of forests, for example, was to decide whether a timber stand offered a lucrative logging chance or not.

Exploitation of natural resources during this period fueled explosive economic growth. Large forest areas of the Northeast and the Michigan pineries were sacrificed for farmland and building materials to support the development of the rest of the Midwest. Likewise large deposits of high grade mineral ores and coal were mined with the primary objective of removing the material as fast as possible rather than maximizing recovery or preserving other resource values. The historical novel "Holy Old Mackinaw" by Stewart H. Holbrook provides a colorful documentary of logging in this era.

1.1 THE INCREASING COMPLEXITY OF ECOSYSTEM MANAGEMENT

As the wholesale harvest proceeded apace, it became evident to some that the supply of natural resources might not be infinite under such continued exploitation. The conservation movement of the late nineteenth century was the long overdue offspring to which this realization gave birth. The keynotes of the conservation movement were preservation and protection of the remaining resources coupled with restoration through such measures as reforestation and stabilization of soils eroded by wind and water. The game warden and forest fire lookout became symbols of the conservation movement embraced by the public.

The conservation movement marked a shift from exploitation to extensive-area low-intensity management of environmental resources. The inspired leadership of Theodore Roosevelt and Gifford Pinchot gave us the noble goals of management for the greatest good for the greatest number in the long run, which have evolved into the present-day concepts of sustained yield and multiple use.

Philosophical goals, however, tend to be overshadowed by economic reality as reflected in the relative prices of raw materials, commodities, and services. Management inputs must be commensurate with the economic value of the resource. The economic climate in the first half of the twentieth century United States was characterized by relatively low unit values for most natural resources as determined by supply and demand. This did not permit small-area high-intensity management of natural resources like that practiced in many European countries. Consequently multiple use remained more of a philosophy than a reality, even for public agencies. In practice most areas were managed for a dominant use such as production of timber or forage for cattle. Other environmental values received only secondary consideration and were often completely ignored in the private sector.

Management in the extensive-area dominant-use context was found to be best effected by placing each area under the direction of a professional or a staff of professionals whose specialty was the dominant resource. The manager required detailed survey information only on the dominant resource, since his or her decisions were keyed to the production of that resource under the set of constraints that applied to the particular area. Information on other components of the ecosystem was needed only to the degree necessary to determine the constraints. Furthermore long experience with the dominant resource and the local area allowed a veteran manager to operate effectively on an intuitive basis, even when detailed data on the resource or area were lacking. Institutions training professionals in the several areas of natural resource management have been largely oriented toward the production of such discipline specialists.

A multitude of factors are operating, however, to shift the management of ecosystems away from the extensive-area low-intensity mode. The most pervasive of these forces are population growth, industrialization, environmental awareness, and improved communications at both the national and the international levels. Population growth has increased demands on domestic resources at the same time that attendant urbanization has reduced the area that can be managed for the production of these resources. Expansion of industrialization has increased the per capita consumption of natural resources at the same time that it has increased the capacity for damage to ecosystems through air pollution, water pollution, pesticides, and detrimental techniques of extracting raw materials. Environmental awareness has produced conflicting demands from various interest groups for setting priorities on the use of environmental resources as well as an outcry against the destruction of less economic environmental values. Improved communications have both enhanced the growing environmental awareness and expanded markets for resources. Underdeveloped countries are now capable of demanding compensation for the use of their resources and are using these resources to bargain in the international market for those resources needed to raise their own standard of living.

The results of these pressures are predictable and, as a matter of fact, experts have been predicting them for some time. We have already experienced some of these shortages and environmental crises. More important than short term effects, however, are the long term implications for the future. As demand increases relative to supply, prices of natural resources will probably continue to rise to the point that intensive management becomes not only feasible but mandatory. This trend toward increased prices relative to other commodities is already evident for timber products and is all too obvious for petroleum products and other forms of energy.

The pattern of land use will have to be planned to protect the natural resource base from degradation and encroaching urbanization. Each area has the capacity for producing a series of product mixes (timber, water, wildlife, recreation, and so on) under different kinds of management. The product mixes that are feasible without a degradation of inherent environmental values must be analyzed carefully and management objectives set accordingly.

The analysis will show some areas to be "prime lands" for the production of one type of commodity such as agricultural crops or timber. These areas will usually have relatively uniform terrain and an absence of sensitive ecosystem components that might be seriously damaged by intensive management for a single product. These prime lands should be zoned for maximizing yield of the product to which they are uniquely suited. Zoning may take the form of legal restrictions on other uses, incentives toward the prime use, or a combination of the two. Such zoning need not, however, preclude other uses of the same area which are compatible with its prime purpose. Hunting on prime timber land would be an example of a compatible secondary use.

Even on prime lands, however, there must be adequate environmental safeguards to ensure that management activities associated with a prime product do not affect surrounding ecosystems adversely. Examples of these adverse effects which can take place are water pollution from agricultural fertilization and careless logging or drift of pesticides. The best avenue to these safeguards is a carefully done and thoroughly reviewed environmental impact statement for each change in the management plan that controls the area.

At the other extreme from prime lands, areas will be found to contain components that are ecologically sensitive to many types of human activities. Development of these areas must be severely restricted, and it may even be necessary to designate natural areas in which human alterations are prohibited.

The majority of areas, however, will be neither prime for a single use nor ultrasensitive to any of a wide range of uses. It will be necessary to manage these latter areas under an intensive, truly multiple use system.

Interdisciplinary teams of resource specialists with a broad base of common ecological understanding will be needed for the land use planning process and to appraise managers of alternatives for meeting the planned objectives under environmental, social, and budgetary constraints.

The complexity of both ecology and economics involved will force the use of scientific as opposed to intuitive methods of analyzing management alternatives. This, in turn, will require readily accessible (probably by computer) resource information banks containing current survey data on the status of all components of the ecosystem.

In order to prepare discipline specialists to function more effectively as members of interdisciplinary teams, curricula must be broadened to provide a larger degree of common understanding which serves as the basis for communication and cooperative effort. Specialization is even more essential than before because there is far too much technical material on any one component of an ecosystem to allow comprehensive coverage in a 4 year program of study, and the amount of technical material in each field is expanding rapidly through research. Each specialist, however, must have a good general understanding of the factors that affect the management and use of other resources as well as the kinds of information that are needed to support management efforts and how such information can be obtained. This will give each member of the team an appreciation of the problems and importance of other members so that each may fulfill a cooperative role.

1.2 PURPOSE AND SCOPE

The purpose of this text is to provide an integrated view of the way in which the effectiveness of ecosystem management depends on the survey information available, and to provide a basic understanding of how the necessary information can be obtained for each component of the ecosystem. Emphasis is placed on an appreciation for the extent to which combined survey operations and data banks are feasible and desirable. At the same time special problems associated with surveying particular components of ecosystems are also discussed.

The material following this introductory chapter is divided into two major parts. Part I contains a discussion of the *fundamentals* of survey systems. Unless the reader is familiar with the materials presented in this part, we feel that his or her perception of the materials in the rest of the book would be superficial and lack perspective.

The amount of time that needs to be spent on this first part will vary with the background of the reader. Those who have prior knowledge of survey systems can treat it as a quick review, but those with no prior background should give it careful attention. Chapter 2 is the first of the two chapters in Part I, and it contains an introduction to survey information systems and survey planning. The next chapter (3) is devoted to the basics of quantification, geographic referencing, navigation in the field, and methods of measuring areas, heights, depths, and slopes.

Part II consists of four chapters which deal with the technical aspects of gathering and interpreting information on components of ecosystems. Given the material in Part I as common background, each of the chapters in Part II is essentially an independent unit dealing with a major ecosystem segment. This type of organization should facilitate the use of the book as a supplement in discipline oriented courses where it might not be appropriate as the prime text, and should generally enhance its value as a reference. Coverage in Part II includes physical factors of ecosystems, vegetation, and patterns of human interaction with the physical and biological environment.

Constraints on the length of the book have precluded a specific consideration of survey methods for animal populations. This omission is justified on the grounds that environmental data banks do not usually include information on the numbers of animals present in an area. Rather, an area is rated with respect to its suitability as a habitat for various species of wildlife. Such a rating is based primarily on physical factors, vegetation, and land use which are covered in Part II. Furthermore animal populations often change rapidly over time, and available census techniques do not yield very accurate estimates. Data banks should and often do contain information on the known presence of rare and endangered species. However, the detection of these rare and endangered species is more a matter of diligent search than of routine survey. Likewise the specialized nature of survey techniques used for mineral exploration has made it necessary to exclude this fascinating but highly technical aspect of environmental surveys.

Throughout Part II the emphasis in terms of techniques is on basic and relatively inexpensive methods for monitoring the status of ecosystems. Brief descriptions of sophisticated instrumentation and analytical techniques are included to give an appreciation for capabilities. Although one may not be able to operate them, the quantitative output of such systems is often available for routine use in the form of reports or data banks of governmental agencies.

The final portion of the book is an appendix devoted primarily to statistics. Our readers will inevitably have very diverse backgrounds with respect to mathematics and statistics. We do not want to burden the body of the text with numerous digressions into these fields, but statistical and mathematical concepts are so fundamental to surveys that they cannot be avoided. Appendixes offer a convenient solution to the problem.

In the balance of this chapter we pursue a further discussion of the part that survey information plays in ecosystem management, and we also attempt to establish a more complete perspective on our purpose and mode of presentation.

1.3 SURVEY INFORMATION AS A ROAD MAP FOR PLANNERS AND MANAGERS

In many ways the importance of survey information to land use planners and resource managers is analogous to the role of road maps for the traveler. First of all the importance of road maps varies with the length of the journey. When one travels in a familiar area there is little need for reference to a map. For longer trips into unfamiliar territory, however, maps become essential.

So it is with land use planning and resource management. If a relatively small area is being managed for a dominant use, the experienced manager becomes so familiar with both the area and the use that he or she can do a very creditable job of management with a minimal amount of survey information. When larger areas are being managed for multiple uses, however, a single individual has difficulty becoming intimately familiar with all aspects of the problem and must rely heavily on both survey information and the inputs of other discipline specialists on the management team. In short, the importance of both survey information and the multidisciplinary approach increases with increasing complexity of the management problem. The growing pressures on our environment are creating situations in which both are of critical importance.

The second point of the analogy is that different stages of planning the journey call for maps with different levels of detail. In the early stages of planning one is only concerned with selecting the best of several major routes, and a rather general map with little detail will suffice to guide the planning through these early stages. Having selected a general route to follow, one then needs maps with more detail for making reservations for lodging, planning meal stops, and so on.

The process of drawing up management plans is quite parallel. The establishment of objectives calls for figures aggregated into averages and totals for rather large areas. The subsequent determination of constraints, formulation of action plans, and allocation of budget calls for much more detailed and location specific information.

The third point of the analogy is that there are some things that cannot be determined from the usual maps and in which the traveler must exercise judgment. The map is not likely to show such things as road repairs in progress, fuel prices, periods during which service stations are open, quality and prices of hotels and restaurants, and "now playing" information for entertainment. The traveler must take into account budgetary considerations, consult other sources of information, and combine them with personal experience in working out the final details of travel plans.

Likewise survey data provide the manager or team of managers with information on the quantities and conditions of the various ecosystem components. Given a set of management objectives, this information is sufficient for determining what needs to be accomplished and the magnitude of the task. The managers still face the all important task of deciding how the desired ends can best be accomplished, and it is here that the managers earn their money. They must analyze the funds, equipment, personnel, and time available to formulate and implement an operational strategy. Scientific management techniques, operations research, and personal judgment are the keys to success in this phase of the program.

Another analogy is to say that survey data provide the foundation on which managers must build. A good foundation is essential, but does not guarantee a good structure unless skilled fabricators use quality materials to build upon it.

1.4 PERSPECTIVE

The perception of the need for land use planning and environmental impact analysis is relatively new at the legislative level in the United States. Since special purpose as opposed to integrated resource surveys have been the pattern of the past, many of the professionals charged with implementing this new legislation are somewhat poorly prepared to meet the new requirements for comprehensive data collection on the several components of ecosystems. There seems to be a common lack of understanding of exactly what to measure, how to measure it economically and accurately, and how to structure information storage and retrieval systems that will make the data readily available for use in the planning process.

A frequently traveled route is for the agencies to approach consulting firms with rather loosely written contracts in hopes that they will be able to satisfy the informational requirements. All too often the consulting firms are little better prepared than the contracting agency to execute such integrated surveys, or they may be promoting a new but untried technique that is purported to be a panacea for all data collection problems. Consequently substantial amounts of such contract funds are often dissipated in pilot projects which fall far short of the intended result. In addition to serving as a text for college courses in curricula that bear on applied ecology, this book should be a source of guidance to those presently involved in integrated resource surveys.

We are convinced that an ecological approach is the best way to ensure an orderly development and effective use of environmental data systems. For this reason the ecological role of each ecosystem component is discussed briefly in the process of presenting methods by which it may be surveyed.

Although we recognize that much of the current environmental furor and concern centers around conflicting priorities for resource use, we also feel that a substantial amount of it stems from a lack of complete and reliable data bases. Without this common ground for agreement on the current status of the human ecosystem there can be little agreement on the manner in which the ecosystem will respond to an alteration. The more rapidly data bases can be improved, the sooner will debate focus more sharply on the real issue of priorities for resource use. These needs are now being recognized and addressed as evidenced by the Renewable Resources Planning Act (RPA) of 1974 and the January 1978 workshop on integrated inventories of renewable natural resources (Lund, LaBau, Ffolliott, and Robinson, 1978). We hope that this book will provide added impetus in that direction.

1.5 BIBLIOGRAPHY

Avery, T. E. *Natural Resources Measurements*, 2nd ed. New York: McGraw-Hill, 1975.

Clepper, H. *Origin of American Conservation*. New York: Ronald, 1966.

Curry-Lindahl, K. *Conservation for Survival: An Ecological Strategy*. New York: Morrow, 1972.

Dassman, R., J. Milton, and P. Freeman. *Ecological Principles for Economic Development*. New York: Wiley, 1973.

Hines, L. *Environmental Issues: Population, Pollution, Economics*. New York: Norton, 1973.

Holbrook, S. *Holy Old Mackinaw*. New York: Macmillan, 1961.

Landsberg, H., L. Fischman, and J. Fisher. *Resources in America's Future: Patterns of Requirements and Availabilities, 1960–2000*. Baltimore: Johns Hopkins University Press, for Resources for the Future, 1963.

Lund, H. G., V. LaBau, P. Ffolliott, and D. Robinson, Eds. *Integrated Inventories of Renewable Natural Resources: Proceedings of the Workshop, January 8–12, 1978, Tucson, Arizona*. Fort Collins, Colo.: U.S. Department of Agriculture, Forest Service, Rocky Mountain Forest and Range Experiment Station, General Technical Report RM-55, 1978.

Nash, R., Ed. *The American Environment: Readings in the History of Conservation*. Reading, Mass.: Addison-Wesley, 1968.

National Academy of Sciences and National Academy of Engineering. *Man, Materials, and Environment*. Cambridge, Mass.: M.I.T. Press, 1973.

Owen, S. *Natural Resource Conservation: An Ecological Approach*. New York: Macmillan, 1971.

Pinchot, G. *Breaking New Ground*. New York: Harcourt Brace, 1947.

Smith, G., Ed. *Conservation of Natural Resources*. New York: Wiley, 1971.

PART I–FUNDAMENTALS OF SURVEY SYSTEMS

CHAPTER 2-INFORMATION SYSTEMS AND SURVEY PLANNING

Our purpose in this chapter is twofold. The first is to convey a clear conception of the structure and function of survey information systems. A lack of such understanding or neglecting to keep it clearly in mind has been a major reason for outright failure of many applications of survey data and caused much frustration and waste of money and time in still more applications. The second purpose is to place before the reader the many considerations involved in survey planning and to offer some systematic ways of approaching the survey planning process.

2.1 INTRODUCTION TO SURVEY INFORMATION SYSTEMS

The importance of survey information was stressed in the road map analogy in the first chapter (Section 1.3). Particularly with larger areas of land and more complex management and planning decisions some sort of formal information system—manual or computer based—is likely to be established. These systems are identified in several ways, such as natural resources information systems, planning information systems, management information systems, regional information systems, and geographic or spatial information systems. The details of their organization vary along with their exact purposes, but the structure and general functions of each type of system are basically comparable and can be described in common terms. This chapter focuses on information systems which are appropriate for use with survey data collected using the methods described in this book and applied to the various problems of ecosystem management arising in land use planning, forestry, range management, watershed management, wildlife management, outdoor recreation, and so on.

Information systems have probably existed as long as humans had material objects that represented or recorded information and could be organized. Written materials such as books, and the libraries that contain them, form the basis for one type of information system. Maps are another, with map indexes and files being the core of the information system. During the past 20 years, with the development of modern computers, data recorded on cards, paper tape, magnetic tape, and other devices associated with computers have provided the basis for a new type of information system: emphasizing quantitative or numerical (rather than qualitative) data, permitting storage of large quantities of data with rapid retrieval and analysis, and stimulating the development of new data creation techniques (such as digitization of mapped information) and data analysis (such as simulation models of complex systems).

Paralleling the rapid development of computer technology have been the fields of cybernetics, systems science, information science, operations research, and the like. From these fields evolved the fundamental concepts of systems, information, and information systems.

Information systems for natural resources management, land use planning, and environmental impact assessment are relatively poorly developed. We know best how to design and implement information systems for accounting, inventory control, and other business management operations. Routine business and governmental operations involving numerical data are easily organized into a management information system. Less easy to accomplish are planning or regional or urban information systems intended to coordinate large quantities of diverse data for quantitative analysis. These may involve public census data, survey data gathered by public agencies, reports, projections, and so on, primarily intended for the generation of topical reports for the area being studied. Coordinating data from reports, tapes, cards, and so on in a variety of formats—perhaps obtained as products of other information systems—requires an effort of a different magnitude from that needed for operating a single purpose, internally controlled information system.

Information systems with a geographic, spatial, or areal dimension add a special constraint on structure of the system and an additional magnitude of effort. Data on various attributes of the area being studied have not only to be identified but also coded as to their *locations*. Data generated by the survey methods described in this book usually have important locational characteristics, and applications of the data to problems of ecosystem management usually require that these locational characteristics be known. Understanding the special requirements of information systems intended for handling survey data is essential both in conducting surveys and in the subsequent use of the data.

The point of view in the following sections is that of the technician, planner, analyst, or manager involved in providing survey information to an agency or set of decision makers. He or she may not only have to acquire the information but also to organize it in some systematic way so that it may be used as needed. For better or worse he or she will be using or creating an information system for acquiring, handling, and using survey data.

2.2 COMPONENTS OF A SURVEY INFORMATION SYSTEM

There are several approaches that one can take in understanding a survey information system. Since the present concern is with the total system rather than with any particular feature, a functional approach seems most appropriate. This provides a general framework within which any component or other approach that one may encounter should fit quite nicely.

Figure 2.1 is a flowchart showing the structure and function of a survey information system in terms of six component subsystems. This diagram provides the focal point of discussion for the rest of this chapter and, in fact, for the entire book.

2.2.1 User Component

Users of the system constitute the component at the top center of Figure 2.1 since this is the position that users must occupy if the survey information system is to be even moderately successful. If a survey system does not exist primarily to serve a fairly well-defined set of users, then there will be no one to sponsor its care, and it will most likely die on the vine. Like an author, the architects of a survey information system must consider their audience carefully at the very beginning, and this concern for users must carry through every phase of construction and operation of the system.

First, potential users must be aware of the system and its capabilities, or they will have no basis for interacting with it.

Second, the system must be easily accessible to potential users, or they will either be unable to use it or become so frustrated with red tape that they will ignore the system. Therefore continuing communication with users is of utmost importance to the success of the system.

Third, the system must provide users the information they want, when they want it, and in the form they want it. Users normally avail themselves of a survey information system because they must make a decision or prepare a

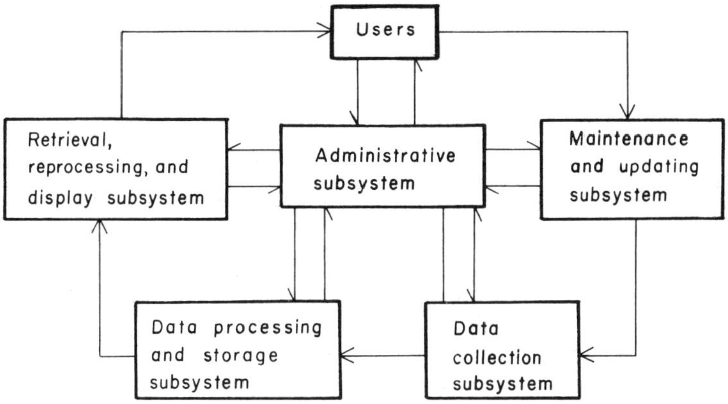

Figure 2.1 Flowchart showing structure and function of a survey information system.

plan that will control future decisions. Unless the user has a peculiar penchant for extraneous information, he or she is not likely to be too enthralled with the right answers to the wrong questions. Likewise, information from such a system does little good if it reaches the user after the deadline for making the decision has already passed. Furthermore the output of the system must be understandable to the users, else they either waste time learning how to interpret the output or run the risk of misinterpretation.

Fourth, the cost of the information to the user must be reasonable in terms of intended use. It makes little sense to spend more for information than the potential loss associated with a lack of the information. This means that some users will be prepared to pay more for the information than others, depending upon the values at stake in their decision making. Thus the nature of the users and their uses determines the financial resources available for constructing and maintaining the system itself.

We shall return to the matter of purpose in survey information systems in Section 2.3, but first let us follow the flow of information backwards from the user through the rest of the system.

2.2.2 Retrieval, Reprocessing, and Display Component

The retrieval, reprocessing, and display subsystem constitutes the interface between the user and the data bank, and its composition is determined by the specific characteristics of these two components which it links.

On the output side of the interface is the user with a specific set of needs for information. These information needs can be grouped into two broad categories. The first category encompasses various types of statistical summaries that can be displayed in either tabular or graphical format. The second category consists of spatially oriented information for which maps are the usual mode of display.

On the other side of the interface is the data bank containing the information currently resident in the system. The structure of the data bank may take many different forms, ranging from a simple library of document files to a highly sophisticated set of computer files. The type of storage plays a major role in determining the equipment and personnel needed for retrieving and reprocessing information into the display format required by the user.

In the case of the simple document library the retrieval, reprocessing, and display subsystem may consist of an indexing system, a librarian, copying facilities, and a work area. At the other extreme the computerized data bank may reside in a large computer system in some remote location. This large computer system probably also would handle most of the retrieval and reprocessing with a local computer terminal linked by phone line serving for display.

2.2.3 Data Processing and Storage Component

The data processing and storage subsystem must be geared to the amount of information involved, along with the speed and flexibility needed in retrieving and reprocessing the information after storage. As with many manufacturing systems, there are economies of scale involved with information systems.

Manual methods coupled with relatively simple and inexpensive devices such as typewriters, calculators, and drafting equipment often provide the most economical and manageable way of handling relatively small amounts of information which do not require recasting into a variety of output formats. As the amounts of information or the requirements for speed and flexibility of processing increase, these simple methods become inadequate, and more automated procedures involving microfilm systems, punched cards, and small computers become both necessary and economical. With a still further expansion of information quantities and requirements of processing, large computers, digitizers, automatic plotters, and storage on magnetic tape or disk are in order.

Although users provide the justification for the existence of a survey information system, the storage and processing subsystem coupled with the data collection subsystem make up the foundation on which the system is built. These two components are placed at the bottom of Figure 2.1 in an attempt to symbolize their basic role in supporting the other components.

Once a structure for the storage and processing component is implemented, it becomes the most difficult and expensive feature of the system to alter. Therefore the choices made in developing this subsystem are indeed crucial.

2.2.4 Data Collection Component

No amount of sophisticated processing or impressive displays can improve the basic information content of the data that go into the system. The best that can be done in processing poor information is to leave the user with a great deal of uncertainty. Careless processing of poor information can give it an unwarranted appearance of accuracy and easily leads to poor decisions as a result.

Since poor information is of little benefit and can be worse than no information at all, quality control in the data collection subsystem is very important. On the other hand the mere act of measuring something adds nothing to its inherent value, so information should not be collected unless there is a good chance that it will be needed for decision making, and the cost per unit of information collected should be kept as small as possible without compromising the requirements of accuracy.

One of the keys to obtaining reliable information at reasonable cost is to integrate the available sources of information so that each item of data comes from the most economical source.

Existing data banks external to the system constitute the lowest cost source of information available and should be drawn upon wherever possible. In particular many governmental agencies routinely collect information of value in ecosystem analysis and resource management.

The expanding technology of *remote sensing* provides a second source of relatively low cost information that is utilized much less than it should be. Aerial photography is the oldest and most familiar form of remote sensing, but several newer forms are also available.

Field observations made on site comprise the third major source of information and usually tend to be the most expensive. The effective use of statistical approaches to sample design helps to return the greatest amount of information per dollar spent on field observations.

2.2.5 Maintenance and Updating Component

Like photographs in an album, the information elements in the system become less representative of their real world counterparts with the passage of time. Surprisingly enough the successful use of a system often accelerates this aging process. A successful system furnishes its users the information they need to make progress toward their management objectives. The resulting management actions in turn bring about changes which move the real world away from the conditions reflected by the data in the system.

The natural sequence of events is for the users to return to the system for an assessment of progress achieved through management so that plans can be revised accordingly. Unless the information base of the system has been updated in the interim, there can be no satisfactory response to queries regarding change. If an information system is to be more than a history book, plans and procedures for updating must be an integral part of the system. Furthermore a simple replay of prior surveys may not be appropriate. Some information elements decay more rapidly than others, and there may even be major changes in the needs of users for information.

Still another aspect of change comes about through obsolescence of functional components of the system. A replacement of these obsolete parts by newer devices and/or techniques is quite likely to call for some restructuring in other parts of the system as well. Since change is a fact of life, the continuing process of maintenance and updating must be treated as an essential part of the system as a whole.

2.2.6 Administrative Component

The remaining component of a survey information system is the administrative subsystem. As its position in Figure 2.1 would suggest, the administrative subsystem can be likened to the hub of a wheel. Integrated operation of the several components as an effective system revolves around the direction and coordination provided by the administrative unit.

The essential ingredients of this unit are personnel and the framework of responsibility within which they operate. The size of the administrative unit will depend on the magnitude of the system as a whole and must be in balance with other components. Regardless of size, however, the administrative unit must serve several essential functions as indicated by the arrows in Figure 2.1 connecting it with the other components.

If the system is already functioning, much of the interaction with users will be taken up with making arrangements for access to the system and carrying out necessary accounting activities such as logging of use and billing. These interactions with experienced users are largely routine and can be handled by clerical staff.

More technically oriented personnel are required, however, to serve as the first line of contact with new users through personal response to inquiries, by conducting workshops and seminars, and by preparing and disseminating materials such as brochures, or some combination of these vehicles for communication. Those who show sufficient interest to run pilot tests of the system on their needs must be given directions in operating procedures, and even experienced users will occasionally run into problems that require the services of technically competent consultants.

There interactions with users will inevitably point out alterations, additions, or adaptations needed in the retrieval, reprocessing, and display subsystem to meet special needs or simply to give better overall service. These latter types of functions are represented by the arrows connecting the administrative unit to the retrieval, reprocessing, and display unit in Figure 2.1.

Likewise, interactions with users will also indicate needs for updating the information content of the system. This, in turn, requires the redesign and activation of procedures for data collection and perhaps also an expansion or revision of the data processing and storage subsystem to handle the new information. Major activities in the data collection or storage and processing subsystems are likely to require a temporary expansion of the administrative unit in the form of consultants in statistics, systems analysis, and computer programming as well as contracting for outside services such as acquisition of new aerial photography.

This completes the tour of survey information systems in general terms, and we shall focus next on specific aspects of survey planning and survey products which require examination at close range.

2.3 ANALYSIS OF INFORMATION NEEDS IN RELATION TO MANAGEMENT OBJECTIVES

Hopefully it has already been made quite clear that a survey information system must justify its existence in terms of user's needs for information.

Since quantification is the name of the game in survey systems, supervisors of these systems should be receptive to continuous monitoring of cost-effectiveness in meeting users' needs.

The first requirement for monitoring cost-effectiveness in meeting users' needs is to know who the users are. This is usually not a major problem since the system should not exist in the first place without a group of potential users providing the impetus for development. Keeping track of changes in the user population over time is simply a matter of record keeping. Often the survey system is set up in the context of a large organization which deals in resource management, with the decision makers in a well-defined hierarchy of responsibility being the prime users of the system. In other cases the survey system is set up by legislative mandate to serve a rather broad spectrum of users involved in public planning. In such cases the original mandate should describe a generalized user, and a brief period of evolution is usually sufficient to define the characteristics of regular and intermittent users as well as to determine which of these are really the primary users and which are more or less secondary.

The second prerequisite for evaluating the appropriateness of a survey system is to obtain from each user a clear specification of objectives and the information needed to satisfy these objectives. Unfortunately this specification of objectives and needs on the part of the user tends to receive far less attention than the more technically elegant considerations of statistical efficiency; but without it there is really no basis for evaluating the survey system. Optimization of survey systems for the wrong information can hardly be considered efficient.

Most survey analysts do subscribe to the need for specification of objectives and information needs, but the matter often suffers from benign neglect. The process of elaborating objectives and information needs, along with some sort of priority rating, requires a major input at the managerial level. Typically the request for this information is duly made, but the survey analyst is usually disappointed at the superficial nature of the response. In fact the disappointment with this response may even engender a feeling somewhat akin to rejection.

The reaction is often to assume the position, "O.K., the decision makers didn't have time to give this proper attention so I will go ahead and make whatever assumptions I deem reasonable." Unless the survey analyst is specifically instructed to proceed in this manner, doing so is very poor strategy. He or she must realize that the decision makers are busy making decisions, and the pressures of the moment may well mitigate against a time consuming digression into the philosophy of management objectives and the associated information needs. Therefore a superficial response to such an abstract question is to be expected. However, the likelihood of a vigorous response is much better if the decision makers are given something to which they can react.

A reasonable course of action is to draw up a provisional statement of

management objectives with information needs to satisfy each objective as the survey analyst perceives the situation. Each objective and each information need within an objective should also receive a priority rating of some sort. Furthermore the statement of information needs must include required limits of accuracy.

Various schemes of presentation and priority rating are possible. The scheme should be relatively simple to understand and easy to apply. The manager's time is already short, and a complicated, time consuming system which requires special mathematical or statistical savy will not likely be greeted with enthusiasm. The scheme should be flexible enough for application at several levels of detail, which will allow a focus on different problems by different persons at different times. While preserving simplicity, the system should provide a meaningful point of departure for a more sophisticated analysis of optimization and allocation problems by survey planners.

One approach to the problem is to use a tabular format for displaying the relationship between management objectives and information needed to achieve those objectives. A suggested structure for the objectives-needs table is shown in Table 2.1.

First list the management objectives in the first row of the table. Next view the relevant sphere of management activity as 100% and assign a percentage rating of importance to each objective. Place these percentage ratings of importance in the second row of the table below the respective objectives. Thus completion of these first two rows constitutes the important step of defining management objectives and assigning them priority ratings.

If the interaction between managers and survey planners takes place in several steps (as will normally be the case), the objectives can be stated in rather broad categories at the first cut with a subdivision of broad categories in later sessions. If the basis for assigning percentages seems too abstract,

Table 2.1 Generalized model of objectives–needs table for survey planning

	Management objectives			
	I	II	III	
% importance:	60	30	10	% importance of
Information need				information need
Need 1	50	5	20	33.5
Need 2	20	0	45	16.5
Need 3	20	40	0	24.0
Need 4	10	35	5	17.0
Need 5	0	20	30	9.0
	100	100	100	100.0

it may help to think in budgetary terms such as percentage allocation of money to be spent on the procurement of information to guide management decisions. If there seems to be a difficulty thinking in terms of management objectives, then think instead of problems to be solved or tasks to be accomplished during some planning period.

Now array the various information needs for satisfying the management objectives along the left margin of the table. Each information need thus becomes a row of the table. Work down for each objective, and assign a percentage importance to each information need in satisfying that objective. Of course some information needs will not apply to a particular objective and will receive zero entries in the column below that objective. When the needs for a given objective have been analyzed in this manner, the column below that objective should total 100% (excluding the initial importance rating given to the objective in the second row). Again, the information needs can be stated in rather broad terms at the first round, with an expansion in subsequent rounds.

When the table has been completed to this point, a final column can be computed which shows the overall percentage importance of each information need. The entry in this column for a given information need is obtained by multiplying each percentage in that row by the corresponding percentage in the second row and summing these products, and then dividing this total by 100. For example the percent importance of need 1 in Table 2.1 is figured as follows:

$$\frac{(60 \times 50) + (30 \times 5) + (10 \times 20)}{100} = 33.5\%$$

To provide a simple illustration of the process, let us assume that an agency with responsibilities in the area of wildlife management is planning to impound the waters of a stream to improve the habitat for waterfowl. The general area in which the proposed impoundment will be located has been selected on the basis of existing information. However, additional information is needed to develop the details of the construction plan and to prepare an environmental impact statement for the project.

The first step in preparing an objectives–needs table is to state the objectives of management and rate their importance on a percentage basis. In this case we have two management objectives: (1) to prepare the construction plan and (2) to prepare an environmental impact statement for the project. For present purposes we will assume that management places relative importance values on these two objectives of 80 to 20%, respectively. Therefore the first two lines of the objectives–needs table appears as shown in Table 2.2a.

The information needs for satisfying the objectives must be determined next. Preparation of the construction plan requires a topographic map of the

Table 2.2a First stage in preparation of objectives–needs table for impoundment project—objectives and priorities

	Objectives	
	Prepare construction plan	Prepare environmental impact statement
% importance	80	20

area with a contour interval of 2 ft and a set of soil map overlays showing texture and depth of A, B, and C horizons. A classification system for the soil maps has already been developed, and differences in classification larger than 0.1 acre (minimum type size or mapping area) are to be shown by delineations on the map. Areas smaller than 0.1 acre are to be lumped with the surrounding type and not delineated separately on the map. This set of map overlays is needed to determine the permeability of the soil, locate the construction materials for the earth dam, and help in determining the kinds of plants and animals that will inhabit the resulting pond.

In addition to information on soils and topography, the preparation of an environmental impact statement requires a series of three vegetation map overlays showing tree layer, shrub layer, and herbaceous layer and a census of animal life currently resident in the area to be flooded. The minimum type size for vegetation map overlays is to be 0.5 acre, except that any rare or endangered species are to be shown individually. The animal census is to include a listing of major species of mammals, birds, reptiles, amphibians, and fishes present with a rating of abundance and notes on special features of the habitat that might not be evident from other information collected. As with vegetation, the animal census is to include a complete count of individuals of rare or endangered species.

Furthermore it seems desirable to include in the environmental impact statement a set of aerial photographs before flooding.

The next step in preparing the objectives–needs table is to add these information needs along the left margin of the table as shown in Table 2.2b. Then comes an examination of each objective to assess the importance of each information need in satisfying the objective. The most critical need for the construction plan is a good topographic map from which planners can determine the maximum extent of flooding for any given height and the placement of the dam. The soil survey will be used primarily to determine the rate at which impounded water will drain away through the soil, but also to determine the best places within the impoundment area for digging earth materials to construct the dam. With these considerations in mind, the topographic map might be rated about 70% importance in the construction plan with

Table 2.2b Second stage in preparation of objectives–needs table for impoundment project—information needs

	Objectives	
	Prepare construction plan	Prepare environmental impact statement
% importance:	80	20
Information needs		
Topographic map		
Soil maps		
Vegetation maps		
Animal census		
Airphotos		

25% for the soil survey. The remaining 5% might be assigned to airphotos since the overview they provide will facilitate general planning efforts.

Information on vegetation and resident animal life in the flooding area is not expected to enter directly into planning efforts unless it reveals some very unusual conditions which had not been known previously.

For the second objective the topographic map might be rated about 30% importance for preparing the environmental impact statement since it will be used to determine the area to be flooded, the water depth, and seasonal fluctuations in both area and depth. Information on vegetation and resident animal life might be equally weighted at 25% for each, with the remaining 20% being split evenly between soil information and airphotos. When these hypothetical importance weightings for information needs are included, the developing objectives–needs table appears as shown in Table 2.2c.

The final phase in the preparation of the objectives–needs table is to combine the importance ratings for objectives and needs within objectives to obtain an overall index of importance for each of the respective information needs. To accomplish this, first multiply each entry in the line that pertains to the need by the corresponding percent importance for that objective in the second line, then add these products, and finally divide by 100. The importance index for the topographic map, for instance, is computed as follows:

$$\frac{(70 \times 80) + (30 \times 20)}{100} = \frac{5600 + 600}{100} = 62\%$$

The completed objectives–needs table for this illustration is given in Table 2.2d. Note that the importance indexes in the last column total 100, so they can be interpreted as percentages.

Table 2.2c Third stage in preparation of objectives–needs table for impoundment project—importance of needs within objectives

	Objectives	
	Prepare construction plan	Prepare environmental impact statement
% importance:	80	20
Information needs		
Topographic map	70	30
Soil maps	25	10
Vegetation maps	0	25
Animal census	0	25
Airphotos	5	10
	100	100

Table 2.2d Fourth stage in preparation of objectives–needs table for impoundment project—combined importance of information needs

	Objectives		
	Prepare construction plan	Prepare environmental impact statement	
% importance:	80	20	% importance of information needs
Information needs			
Topographic map	70	30	62
Soil maps	25	10	22
Vegetation maps	0	25	5
Animal census	0	25	5
Airphotos	5	10	6
	100	100	100

When the provisional objectives–needs table has been completed to the best of the survey analyst's ability, it is passed on to the decision makers and other members of the management team for review. They may restructure the table, append comments, or call a conference for restructuring the table. A series of such revisions should eventually lead to a table or a set of tables with a satisfactory degree of detail for planning new surveys or revising existing survey procedures. All memos, schedules, minutes of meetings, and other materials that document the evolution of the objectives-needs table should be retained in a file.

The objectives-needs table (or the series of tables if one becomes unwieldy) with the supporting materials provides decision makers and other members of the management team an effective vehicle for communicating information needs to survey analysts. Decision makers can keep a copy handy and make notations on it to record frustrations of the moment at a lack of information which otherwise might never reach those responsible for the survey system. It is one of the inputs to the budgetary planning process and provides a point of departure for the next level of survey planning. It also provides a solid basis for evaluating the performance of the survey information system in meeting the needs of users. A preparation of such tables should be a beneficial exercise for organizations with an ongoing survey program as well as for those planning new surveys. The objectives-needs table or its equivalent is especially helpful in providing a continuity for organizations with relatively rapid turnover in either management or survey personnel.

To summarize, the key features of the objectives-needs table are the following:

1. It forces the manager or team of managers to state objectives in a concise, written form.
2. It forces the manager or team of managers to rate the relative importance of objectives.
3. It requires careful consideration of information needs for meeting each objective and relative importance of each need for each objective.
4. It serves as a communication vehicle by allowing each member of the team to understand the thinking of other members.

As such, the tabulation of objectives and needs becomes a focal point of discussions regarding management objectives and the information needed to meet objectives, which should lead more rapidly to a concensus among team members.

There is nothing magic about the format suggested for the objectives-needs table, however, and any modification that will facilitate the underlying purpose of the table should be adopted without hesitation. In situations where the survey information system is part of the organizational hierarchy, objectives-needs analysis should proceed quite smoothly. If the survey information system is a more or less autonomous unit serving external users on a contract or public basis, however, some resistance to this approach may be encountered.

In particular, users may feel that there is some attempt to infringe on areas that should be confidential. Also the person who serves as a contact on the user's side may be working under a tight timeline or lack the kind of access to supervisors needed in objectives-needs analysis. This latter remoteness from ultimate users is really a further justification for insisting on some sort of objectives-needs analysis to ensure that the contact person is actually on the same wavelength as those whom he or she represents. A little persuasion should convince potential users that such an analysis is in the interests of both parties, although it may not be possible to carry the analysis as far as might be desirable. In such cases it may also help to disguise things a bit by talking in terms of problems that the information will help solve rather than in terms of "management objectives."

2.4 DATA SOURCES AND PROCUREMENT

The objectives-needs table or its equivalent becomes a point of departure for survey analysts in planning new surveys or modifying existing survey procedures. It provides a list of information needs along with the justification for and relative priority of each. In and of itself, however, the objectives-needs analysis does not determine either the survey methods that should be used or what level of survey budgeting is required to satisfy the respective information needs. These matters are primarily the province of survey specialists and will require an input of higher level management only at key points.

The first requirement for economical procurement is to know the sources of supply. Although the three major sources of information—existing data, remote sensing, and field measurements—have been mentioned previously, they were not really discussed in any detail and deserve more thorough introduction at this point.

2.4.1 Existing Data

If someone else has already collected all or part of the information needed, they are often willing to share this information for a nominal fee which helps them recover part of their original costs. In the case of public agencies specifically charged with survey operations, the information is usually available for the cost of reproduction and dissemination. If such existing information meets current needs, it is usually the most economical source of supply, and every effort should be made to take advantage of it.

There are many such sources of information, some being better known and more accessible than others. The ideal situation is where information is collected on a nationwide basis by a public agency and the nature of the data is such that its utility does not decay rapidly with time. The "quadrangle" series of topographic and geologic maps prepared by the U.S. Geological Survey (USGS) in the U.S. Department of the Interior (USDI) approaches this ideal. Neither geological features nor topography change very rapidly with time, although some other information included on these maps does change more rapidly. Although preparation of these maps takes place on a nationwide basis, there are gaps in the 1:24,000 scale topographic coverage, and coverage of detailed geologic maps is much less extensive.

The National Weather Service in the National Oceanographic and Atmospheric Administration (NOAA) is an example of an agency that operates nationwide to collect meteorological information which is available to the public, but in this case the utility of the information often decays very rapidly with time, so the combination of access and reporting intervals becomes very important. Demographic information collected by the Bureau of the Census in the U.S. Department of Commerce (USDC) occupies an intermediate position on the scale of decay rates between the extremes of geology and weather.

Soil information is relatively stable over time and is basic to most activities in planning and managing ecosystems. The National Cooperative Soil Survey Program produces soil maps through a joint effort of the Soil Conservation Service (SCS) in the U.S. Department of Agriculture (USDA) with counterpart agencies in the individual states.

Agencies that collect information related to land use run the gamut from federal to local with little coordination between units, and the characteristics of their data products vary correspondingly. Compatibility of classification and coding systems is an important consideration for interchanging materials between information systems and aggregating data for larger areas.

The scope of the data collection activities for renewable natural resources such as vegetation and wildlife is diverse and lacks coordination between the agencies involved. The USDA conducts widespread activities of this type in rural areas. In more localized situations, where forest and range lands are administered by the federal government, the U.S. Forest Service (USFS) in the USDA and the Bureau of Land Management (BLM) in the USDI conduct comprehensive natural resource surveys. The USFS is also responsible for a periodic assessment of the nation's timber resources in the form of the National Forest Survey. Many states have similar departments responsible for state lands, and they too conduct natural resource surveys necessary to support management activities in these areas as well as more general types of statewide surveys.

Several state universities also house various types of data centers, with the Resource Information Laboratory at Cornell University in New York being a good example. The files of agencies responsible for monitoring pollution such as the U.S. Environmental Protection Agency (EPA) constitute still another vast source of environmental information, especially for air and water resources. Several directories to environmental information may be helpful. Examples are:

The Federal Environmental Monitoring Directory
Council on Environmental Quality
Executive Office of the President
Washington, D.C., May 1973
GPO Stock No. 4111-0016

Environmental Information Centers and Environmental Research Laboratories in the Federal Government, Environmental Information System Directory
Environmental Information System Office
Oak Ridge National Laboratory
Oak Ridge, Tennessee, 1973

NFEC Directory of Environmental Information Sources
The National Foundation for Environmental Control, Inc.
Boston, Massachusetts, 1972

References to sources of information on particular components of ecosystems are given in the chapters that deal with those components. In summary, existing information is where you find it, and a little time spent foraging can often reduce survey costs as well as uncover possibilities for exchange of data with other groups.

2.4.2 Remote Sensing

The space age technology of remote sensing provides a second major source of information on many features of the environment. In its most basic sense the term remote sensing simply means collection of information at some distance from the origin of that information. Strictly speaking, visual observation qualifies as a form of remote sensing and can serve as a point of departure for a brief excursion into this realm.

The necessary elements of a remote sensing system are (1) a *sensor* to collect the information, (2) some form of *energy* to transmit information from *target* to sensor, (3) a way to establish the desired *remoteness* or distance between target and sensor, and (4) a way to *record* the information collected by the sensor. For a human observer the eye is the sensor, light is the energy, the brain is the recorder, and distance can be achieved in several ways.

Aerial Photography. Aerial photography is the classic, and still perhaps the most useful, form of remote sensing for surveys. A camera serves as the sensor in aerial photography. Depending on the application, the camera can range from a simple hand held 35 mm type to a precision instrument with gyrostabilized mount in which the lens alone costs thousands of dollars. Likewise, the airplane for establishing the distance between camera and target can range from a light pleasure craft operating at low altitudes to a sophisticated aircraft like the U-2 which routinely operates at altitudes in excess of 50,000 ft and was originally designed for military intelligence purposes.

Films are available in a similar variety, encompassing those that approximate the light sensitivity of the human eye such as black and white panchromatic or conventional color film as well as those that portray healthy vegetation in a brilliant magenta color using infrared (IR) radiation which is invisible to the human eye. Filters can also be used to give still more selectivity in highlighting or suppressing certain features in the scene.

One of the most fascinating things to the novice is the three-dimensional or *stereo* effect that can be achieved by simultaneously viewing through a stereoscope two photos of the same scene which were taken from different positions. This causes hills, trees, and so forth to stand in a relief which is usually quite exaggerated. Besides being an interesting phenomenon, this stereo effect is an aid to the photo interpreter and facilitates measurements of heights and slopes from the photos.

If one is primarily interested in having a visual record of the landscape for illustration, documentation, or examining the condition of objects in the scene, then a light plane and a hand-held camera are adequate. When the photos are used in mapping or making precise measurements, however, the long lens-to-subject distances require a stable aircraft coupled with a camera having a low distortion lens and high shutter speed. Furthermore photos taken with the camera pointed vertically downward are more useful than slant or *oblique* shots by virtue of the uniform scale in the former.

Although aerial photography is an amazingly versatile technique with its utility extending into most phases of ecosystem survey and management, it does have limitations. For one thing it is definitely limited by weather conditions. Precision photographic missions require crystal clear weather during daylight hours. Clouds, haze, air pollution, and similar atmospheric disturbances make it impossible to obtain satisfactory imagery for precise photogrammetric work. This sensitivity to atmospheric conditions is extremely limiting for many of the potential applications in ecosystem management.

Aerial photography would seem to be an ideal way to survey and monitor the vast expanses of tropical forests in equitorial regions like the Amazon where access by ground routes is difficult. However, completely clear weather is so infrequent or of such short duration in many of these areas that there is simply no opportunity to obtain adequate coverage with aerial photographs. Likewise many phenomena of interest in agriculture, natural resources, and other environmental studies are associated with inclement weather conditions. Where vegetational growth takes place in the rainy season there will be little chance to use airphotos to study phenology and energetics of the ecosystem.

Aerial observation is very useful in detecting wildfires, and it would seem that airphotos would be correspondingly useful in control operations. They do have some utility in this regard, but smoke usually obscures the advancing front of the fire which is the real focus of suppression efforts, and the time required to develop the photos reduces their usefulness.

Similarly airphotos have a potential for assessing damage caused by other natural catastrophes such as floods and large storm systems. If photos could be taken in real time (while the catastrophe is happening), they could be of great help in coordinating emergency operations and rescue efforts. Unfortunately floods and storms are caused by bad weather, which prohibits comprehensive coverage by aerial photography.

The time required to take the pictures is another limiting factor for many applications. Many natural phenomena occur over relatively large areas and take place in a short span of time. Airphotos taken of the entire area almost simultaneously and at repeated short intervals are needed. At the usual scales, however, several hundred photographs are required to cover even relatively small areas. The time involved in obtaining a single coverage is sufficiently long to make repeated coverage at daily or weekly intervals a practical impossibility with conventional aircraft.

Still another limitation lies in the range of sensitivity of photographic films. The sensitivity of these films is usually limited to the visible and near infrared portions of the electromagnetic spectrum. Many changes we seek to detect are best reflected in portions of the spectrum outside this range. Thermal infrared radiation (heat rays) emitted by features of the terrain is of particular interest. The usefulness of this type of radiation in the study of volcanic activity and thermal pollution should be apparent.

Although considerably faster than the equivalent ground reconnaissance, interpretation and mapping from photographs are still time consuming and expensive. These are largely manual operations with conventional photography, and opportunities for automation are often limited or awkward.

Taking pictures with cameras on photographic film from conventional aircraft is only one of the many ways that electromagnetic energy can be used in remote sensing to collect information about the environment. Other remote sensing systems that circumvent the limitations of conventional aerial photography have developed rapidly since World War II. The primary impetus for their development has come from their military potential, and the most sophisticated systems are still classified. However, the potential for applying such systems to environmental sciences is obvious. Different types of sensors replace the camera, magnetic tape replaces the film, other portions of the electromagnetic spectrum replace visible light, and the sensor may be mounted in spacecraft rather than conventional aircraft.

Radar and Microwave. Side looking airborne radar (SLAR) is operable under inclement weather conditions and at night. Radar is an *active* system which generates its own signals rather than passively sensing reflected radiation from the sun, thus making it independent of daylight or darkness. Furthermore the radar waves are considerably longer than light waves, so atmospheric moisture causes less interference. This helps alleviate a major drawback of conventional aerial photography. The known source of the signals in radar makes it possible to determine the distance of the target as well as its direction. In radar sensors a receiving antenna takes the place of the camera lens. Until recently the resolution obtainable with radar was limited by the size of the antenna. Synthetic aperture techniques have largely dispelled this difficulty. Radar has been used to survey forest resources of the Amazon region in the Brazilian "Project RADAM."

Passive microwave sensors are also receiving application. The passive systems share the interference-free characteristics of radar, but the signal is not controlled and therefore carries less information.

Thermal Infrared. Like radar, thermal infrared (heat) sensors are operable at night, but do not have the moisture penetrating capacity of radar. Thermal infrared systems are essentially temperature sensors. The hotter the feature the brighter it appears to the heat sensor, and this relationship can be quantified by building temperature standards into the sensor for comparison.

Monitoring of volcanic activity and thermal pollution are obvious applications of thermal infrared. Likewise it has proven useful for studying the dynamics of polar ice fields. The control of wildfires is still another fertile field for thermal infrared. The infrared sensor looks right through the smoke cover and shows the hot spots on a monitor from which polaroid photocopies can be prepared in flight. Such systems are even sensitive enough to detect the presence of warm-blooded animals when the body temperature is higher than that of the surrounding terrain. This capability has been used experimentally to estimate numbers of big game animals.

Since thermal infrared radiation cannot be transmitted through optical systems of the lens type, *scanners* based on reflecting optics are used instead. The scanner has a rotating mirror which sweeps a strip across the flight path at each scan. Each strip is segmented into a series of small *pixels* (picture elements). Radiation is sensed by special detectors on a pixel by pixel basis and stored on magnetic tape. The taped information is then processed by computer to remove distortions and filter out static. The rectified information can be plotted directly on paper by computer, displayed on cathode ray tubes resembling television screens, or transferred to photographic film.

Multispectral Scanners. Thermal infrared scanners are a special case of a relatively new class of sensors called *multichannel* or *multispectral scanners.* These sensors break their spectral range of sensitivity into bands (channels) with a separate recording being made of each portion. The records for the separate channels can then be integrated into any desired combination. Multichannel sensors hold a very high potential for discriminating between particular types of targets which are very difficult to differentiate with conventional photography. The *spectral signature* or reflectance of the various targets must be determined in order to take full advantage of this potential.

Remote Sensing from Spacecraft. Spacecraft show promise for resolving the dilemma of synoptic and repeated coverage. Satellite orbits are high enough that each frame of imagery covers a large area and thereby provides synoptic coverage. The satellite's orbit can be designed to return to the same point on the earth at regular intervals. This provides the means for repetitive coverage. The extreme case would be a stationary orbit in which the satellite's orbit is synchronous with the rotation of the earth, thus remaining over the same point for continuous monitoring.

Both multispectral scanners (MSS) and return beam vidicon (RBV) cameras similar to television are natural candidates for use with spacecraft which cannot be landed to retrieve data since their outputs can be telemetered to ground receiving stations. The effectiveness of telemetry was demonstrated vividly to the public during the Apollo missions to the moon.

Weather satellites have been around long enough that they must be considered part of routine atmospheric data collection rather than experimental curiosities. Likewise the National Aeronautics and Space Administration's (NASA) LANDSAT and SKYLAB satellite programs have placed earth resources surveys from spacecraft well on the way to operational status for surveying many large area phenomena.

Computer Processing of Remotely Sensed Data. There are several other advantages besides telemetry to switching away from conventional film as a recording medium, foremost among them being the computer compatible nature of magnetic tapes. Much of the information must be gleaned from photographs by human interpretation, which is more of an art than an exact science. If the spectral signature or reflectance of particular targets can be determined, data on magnetic tape can be processed by computers programmed to identify targets automatically. Computer recognition can range from a simple separation of signal levels in certain channels to a sophisticated statistical analysis for pattern recognition.

In summary, it is quite possible that remote sensing techniques now in the exploratory stage may generate a real revolution in the collection and processing of information regarding both natural and cultural resources.

Procurement of Remotely Sensed Imagery. The Agricultural Stabilization and Conservation Service (ASCS) in the USDA supervises the procurement and dissemination of black and white aerial photographs of much of the United States. ASCS photography is repeated on cycles ranging from 5 to 10 years. The earlier photos had a scale of 1:20,000, but more recent ones have a scale of 1:40,000. (A scale of 1:40,000 means that one unit of length on the photograph represents a condensation of 40,000 of the same units on the ground.)

NASA and the USDI maintain a data center at Sioux Falls, South Dakota, called the EROS Data Center. The EROS Data Center contains NASA satellite and aircraft imagery along with imagery from several other agencies, copies of which can be purchased by the public.

Many other governmental agencies involved in land management activities also have aerial photographs of regions with which they are concerned. These include several agencies such as the Forest Service at the federal level, state natural resources and construction departments, municipal construction and tax assessment divisions, and regional planning authorities.

Many large private landowners also have their own plane and camera or contract for aerial photographic services. Some of this photography is quite recent, and some of it is not so recent. Copies can often be purchased by contacting the organization responsible for the photography.

There is also a national index to aerial photography that can be obtained from:

National Cartographic Information Center
U.S. Geological Survey
507 National Center
Reston, Virginia 22092

The photographs indexed on this map must be obtained directly from the agencies holding them or through some other outlet such as the EROS Data Center.

The first step in obtaining airphotos of small areas is to learn what agencies have flown photographic missions in the vicinity. The next step is to purchase or borrow an *index mosaic* for each mission. An index mosaic shows exactly the area that each frame (print) covers. The final step is to order the desired prints. Mission code, roll number, and frame number within the roll as taken from the index mosaic must be specified in ordering. The cost per frame may decrease if large numbers of frames are ordered. One must also allow plenty of lead time in ordering since delivery times in excess of a month are common.

The considerations involved in writing contracts for new aerial photographic missions will not be reviewed here since they are adequately covered in several textbooks on aerial photography.

2.4.3 Field Surveys

Lest one get the impression that remote sensing is a cure for all ills, however, we must hasten to point out that it seldom constitutes a complete and independent data source. Many characteristics of ecosystems either cannot be measured at all with current remote sensing technology or cannot be measured accurately enough to serve the intended purpose. Even for those things best suited to remote sensing—like topography, drainage systems, landforms, transportation networks, and many aspects of land use—one normally has to go to the field to establish ground control or "ground truth" to guide the interpretation of imagery.

Thus we are led to the third and most comprehensive source of information, namely, observation or measurement of ecosystems in the field. This will be the primary subject of Part II of this book. Our goal at this point in the presentation is simply to introduce the many things that must be considered in organizing field data collection activities. The most important of these considerations can be stated in two words, *plan ahead*. Going to the field and then deciding what to do is a great way to waste time and money.

Importance of Locational Information. One consideration is the relative importance of *location* as compared to other characteristics measured. Locational information is intrinsic to most types of remotely sensed imagery. This is not necessarily true of data collected in field surveys. If the data are to be displayed in map form, positional information obviously must be sufficient to plot the data within the limits of accuracy set for the map.

If mapping is to be the sole form of display, the requirements of accuracy for measurements other than position probably will not be too stringent, and their quantification will most likely take the form of a classification. This arises from the limited variety of symbols, colors, shadings, and so forth that can be used to portray information without giving the map overlays a hopelessly cluttered and confusing appearance.

At the other extreme is the situation where initial interest centers primarily on a more accurate measurement of other characteristics such as volume of timber in trees. Since the main concern in the latter case is often with totals and averages for relatively large areas, and statistical summaries constitute the most concise display format for this sort of data, there would seem to be little need for positional referencing beyond that needed to find the features so that they can be measured in the first place. This is probably true if information is needed for only one point in time, and no use is to be made of the data beyond that which originally motivated the collection. However, we are looking now to combined data bases for several components of an ecosystem and to repeated uses for monitoring change over time. Since geographic referencing and dates of collection form the indexing threads for weaving the various data items into an information fabric, it is essential that they be part of every data collection effort.

This provides the means for plotting locations of data collection points for the several items of information in order to determine existing information densities in space and time. It also serves as the basis for sorting, matching, correlating, and otherwise studying the myriad interactions between components of ecosystems. These interactions often provide the key to successful planning, since an unexpected interplay between ecosystem components frequently makes well-intentioned treatments and legislation turn sour. Furthermore the most sensitive indications of change are obtained by repeated measurements on the same set of units as opposed to completely new surveys in which differences between present and previous units may confound estimates of change.

Choosing Measurements and Classifications on an Ecological Basis. A crucial choice which must be made early in the process of survey planning is the decision regarding exactly what is to be measured and how the results of the measurement are to be expressed. The importance of this step can hardly be overemphasized, since it has major implications for the utility of the survey data. Our focus on the compatibility of data bases leads us to stress the importance of considering a variety of possible uses for the survey data besides that which originally motivates the survey effort. The best route to achieving such compatibility lies in an ecological approach to deciding what should be measured and how it should be expressed. After all, ecological factors control the structure and function of ecosystems, and choices that make sense in terms of these underlying factors will also make sense for most applications in ecosystem analysis and resource planning.

To amplify this idea of ecologically based surveys, consider for a moment an area in which wetlands are prominent. A survey to provide information for land use planning might include mapping of *land cover* in the five classes: (1) water, (2) grass-herbaceous, (3) brush, (4) forest, and (5) manmade structures. Such a map conveys a clear picture to the layman and shows the extent of human development, but it is of little use as a guide to planning in the area. It tells little about the nature of the wetlands since the extent of open water fluctuates with rainfall and the season of the year. The character of the vegetation gives a much better picture of the dynamics of the wetlands, but the information on vegetation in the above map would not be of much help. There are both upland and lowland species among the forest, brush, and grass-herbaceous vegetational growth forms. So the land cover as structured above could not be used for further analysis of wetlands. Much more informative than the simple five class land cover map would be a set of overlays with one showing topography, drainage, and dry season water levels; a second showing vegetation by both form and habitat type; a third showing the extent and type of human development; and possibly a fourth showing soil types.

Another frequent tendency is to interject economic uses of the resources into the survey at such an early stage that it obscures a subsequent analysis and interpretation for other uses. A good example is the propensity of foresters to measure trees in terms of board feet and cords. A board foot is a unit of sawed lumber 1 ft square and 1 in. thick. A cord is a unit of stacked roundwood that occupies 128 cu ft of total space, including both wood and air spaces. Measuring standing trees by predicting the number of board feet that they will yield upon sawing only makes sense if the trees will actually be sawed into boards for lumber, which is only one of the possible economic uses for most trees. Furthermore, changes in the standards of utilization make such a prediction pretty uncertain anyhow. Since the cord includes both wood and the space between pieces of wood in a pile, the actual amount of wood in it depends on a number of factors such as average size of the pieces, straightness, care in stacking, and so on. It makes much more sense to measure wood in units of volume or weight, and then develop conversion factors for predicting the yield of particular products such as sawed lumber.

If the main focus is on the mapping of information which is largely qualitative in nature, like geological features, soils, broad vegetation types, and land use, then some sort of classification scheme will be in order. The categories must be meaningful in terms of ecology and economics and also reasonably easy to identify in the process of data collection. Rapid progress toward a comprehensive data base for ecosystem analysis can be made only if there is a serious attempt on the part of all involved in data collection activities to develop a more or less standard set of classification systems for the various components of ecosystems. This makes it possible to interchange and mesh data bases between different survey information systems. Special characteristics of particular uses and localities usually make it unrealistic to expect a complete standardization of classification systems. However, it is usually possible to confine these local adaptations to the lower levels of the classification so that the major categories at higher levels will remain compatible for quite diverse areas and applications. A conscientious attempt to accomplish this will go far toward solving many of the problems associated with aggregating data from various sources to cover larger areas.

A desirable precursor to any integrated survey effort is to draw up both ecological and economic overviews of the area in question based on all information available. The existence of these ecological and economic abstracts in written and/or diagrammatic form will help to ensure that the survey data speak directly to the most important considerations in both ecology and economics of the area. The results of the survey will provide the meat to fill out the bones of the abstracts. Ecological and economic roles of ecosystem components are discussed briefly in this book to encourage the preparation of such abstracts for areas to be surveyed.

Survey Equipment. The choice of survey equipment is often less critical than some of the other considerations because several different sets of instrumentation will often satisfy the same data collection purpose. However, such matters as cost, accuracy, speed, skill and training required for operation, versatility, and form of readout are not to be taken lightly since a series of poor choices can seriously hinder the survey effort. Even with the mapping of qualitative data, choices must be made with regard to the equipment for controlling the positional accuracy of the map product. Details of purpose, use, advantages, and limitations of the various survey devices are covered in Chapter 3 and Part II of this book.

Sampling. When the situation calls for a map, it is usually fairly obvious what area is in need of mapping. Furthermore the nature of a map is to cover the area of interest in its entirety. The case is not so clearcut, however, when one deals with quantitative characteristics that are more suited to representation via statistical summaries. In many of the latter situations it is economically impractical, if not physically impossible, to make observations or measurements on the whole population of interest. Note that the term *population* is used in a very broad sense here to mean the entire set of whatever it is that is under consideration, not just people or animals. Furthermore the measuring process may itself be destructive, as when dissections are required to determine the presence or absence of disease or pesticide residues. With destructive measurement it is obviously undesirable to do a complete census since this would eliminate the population under study and the data obtained would then be of historical interest only. Often we must be content to examine only a portion of the population in which our interest lies. The portion actually examined is a *sample.*

The measurements made on the sample are used to *estimate* the characteristics of the population. (The sampling sophisticate would call such a characteristic a *parameter*.) The estimates are called *statistics*, and the equations which are used to arrive at the statistics are called *estimation equations* or *estimators*. Since a sample contains only part of the population, such an estimate is likely to be somewhat different than the result that would be obtained from a complete census. This difference is called *sampling error*. Sampling is useful only if the sampling error can be held within acceptable limits.

The sampling error is affected by four major factors. The first of these is the *variability* inherent in the population. If all the units in the population were identical, measuring any one of them would be sufficient to describe the population without any sampling error.

The second factor is the *size* of the sample relative to the population. If the sample were to include the entire population, there would not be any sampling error, but then we would have a complete census rather than a sample.

The third factor is the *nature of the estimation equations* used in arriving at the statistics. Some estimators make better use of the information in the sample than others and thus produce smaller sampling errors.

The fourth factor is the method of *sample selection*. Since the method of sample selection is extremely important, we will examine briefly some of the alternatives and their merits. After considering the methods of selection we will return to the sample size, but we need a little more terminology before proceeding with either selection or sample size.

The individual unit in the population on which a measurement can be made is called an *elementary unit*. If one is trying to determine the volume of wood in a forest tract, the elementary unit will be the tree. If one undertakes to find the number of animals of a particular species in an area, the elementary unit will be the individual animal. If one is attempting to assess the extent of damage by leaf-eating insects, the elementary unit might be the individual leaf. Elementary units for measurement, then, vary according to the material under investigation and what it is that one wants to know about the material.

In order to reduce the number of choices that must be made in selecting the sample, elementary units may be grouped into *sampling units*. The actual sample is then assembled by selecting a certain number of sampling units. The essential feature of the sampling unit is that the elementary units within it are not considered for measurement unless that sampling unit is first selected. For example, one ordinarily does not begin by selecting plants individually for measurement in vegetation surveys. Instead the population is first subdivided into *plots* which serve as sampling units. If a particular plot is not selected for inclusion in the sample, the entire group of plants growing on the plot is automatically excluded from the sample. For determining damage by leaf-eating insects, the branch might be clipped and taken into the laboratory as a sampling unit. Formulation of sampling units can have a decided effect on the cost and accuracy of a survey.

The structure of the population in terms of elementary units and sampling units is called the *sampling frame.* With this as background we can go on to consider some of the possible approaches to sample selection. An ideal selection procedure should be efficient from an information standpoint, and yet not too costly in terms of time, travel, personnel, and so on. In other words, the goal is to obtain the most information per dollar spent in collecting and analyzing the sample.

Judgment Sampling. A natural approach to selecting a sample is to say, "Let's take a quick look at the population and select those sampling units that appear to be typical." This approach can be very economical since it requires only one experienced person working in the field to select the sample. The difficulty with judgment selection is that the quality of the sample depends entirely on the skill of the person making the selection. The experience and reputation of the sampler furnish the only basis for choosing between two different judgment samples taken from the same population, and this situation is usually unsatisfactory.

One individual may tend to choose units that are better than (greater than) average, whereas another individual may tend in the opposite direction. In the former case, estimates from a series of samples will average out too high. This is called *positive bias.* The term "positive" here only indicates the direction of the bias and does not imply that the bias is good. In the reverse case a series of estimates will average out too low, thus giving rise to a *negative bias.* If there is no consistent tendency toward either underestimation or overestimation, the procedure for arriving at the estimates is said to be *unbiased.* Since bias or lack thereof is an average property of the selection and estimation procedures, the concept is not applicable to a single estimate. For instance, it is quite possible that an estimator with a positive bias will produce an underestimate on any given try.

An unbiased estimator is usually preferable, but a slight bias is not serious if its direction and approximate amount are known so that a correction can be made. In the case of judgment sampling, however, this is a serious consideration because the direction and amount of bias tend to be quite unpredictable.

Convenience Sampling. Another approach that has similar drawbacks is to measure the units that happen to be most readily available on the assumption that they are representative of the population. This is known as *convenience* or *haphazard* selection. The assumption that the readily available units are also representative is often unfounded. The practice of examining roadkills in studies of wildlife populations is an example of convenience sampling. The areas along roads are usually the most accessible. However, there are often *edge effects* that make the flora and fauna along roads quite different from those situated at some distance from the road. Because of the high probability that the most convenient units will be atypical, this approach to selecting a sample will often lead to a bias of unknown magnitude.

Systematic Sampling. One simple approach that does not carry an inherent bias is to pick up units at regular intervals through the population. This is called *systematic sampling* because of the uniform spacing between sampling units. Equally spaced strips are often used for sampling vegetation or other features which can be associated with area, and equally spaced plots situated along equally spaced lines constitute another very common systematic layout. The spacing for the selection of sampling units can be keyed to other things than distance in the field as, for example, the use of a time sequence in sampling every *n*th automobile that passes through the entrance to a park or perhaps taking every *n*th name from a phonebook or other directory.

Systematic selection has certain advantages that make it very popular, especially for sampling area units in the field. One of these advantages is its simplicity in concept. Another is the fact that it can be used to estimate areas and collect data on area units without the aid of maps. A third advantage is that the uniform spacing gives convenient patterns for field travel. Still a fourth advantage is that the even distribution pattern of the data collection points tends to simplify the structure of the data base itself.

Systematic sampling is not without disadvantages, however, and it has been the subject of much statistical controversy. This controversy centers around two weaknesses. One is the possibility that the regular spacing of sample units might coincide with some cyclic fluctuation in the population being sampled. If there was a regular sequence of hills and valleys, for instance, the samples might all fall on hilltops in one survey of the tract and all in valleys in another. Neither of these surveys would be very representative of the tract as a whole. In practice, cyclic fluctuations of a serious nature will normally be evident on inspection so that the survey planner can take steps to avoid them. Such precautions may include running survey lines across the topography rather than along it and letting chance play a part in determining the starting point for the first line.

A more serious problem is the difficulty in estimating the variability of the population from systematic samples. This, in turn, makes it hard to determine how good the sample is likely to be. For the same reason previous sample results cannot be used to the best advantage in designing new surveys.

Simple Random Sampling and the Role of Probability. The theory of statistics dictates that chance ought to have the final say in deciding which units actually get into the sample, with the proviso that the chance of each unit showing up in the sample must be known. There must also be some degree of independence between the units that appear in the sample. The major defect of systematic sampling lies in the lack of this independence since the first unit in the sample completely determines the others that will appear with it. When chance plays such a role in determining the composition of the sample, we have one of the class of *probability selection methods.* The process by which chance is allowed to operate is called *randomization* and usually involves either the use of a random number table or programming a computer to generate numbers in a fashion that approximates a random draw.

The simplest way of meeting these statistical requirements is to give every possible *combination* of sampling units which could make up the desired sample size an equal chance of forming the sample actually selected. This is called *simple random sampling.* Simple random sampling may be either with or without replacement. Sampling with replacement is like catching a fish and then throwing it back before you try for the next one, since the same fish may bite again. In sampling with replacement a sampling unit is allowed to be measured more than once in the same sample. When sampling is without replacement, a sampling unit cannot be measured more than once in a sample. Since measuring the same unit twice would normally be a waste of time, simple random sampling is usually understood to be without replacement unless otherwise stated. This type of selection is achieved by giving every sampling unit the same chance of showing up *each time a new unit is drawn* until the sample is complete. If sampling is without replacement, one just rejects any unit showing up that is already in the sample and draws another.

In order to draw a simple random sample one must be in prior possession of a list or map of sampling units to serve as the basis of selection. Simple random sampling is a reasonable procedure to use when one has no prior knowledge about the population beyond the list or map that will serve as the basis for selection.

Stratified Random Sampling. Should one happen to have some prior knowledge about the population, however, one would not usually want to use simple random sampling. Simple random sampling does not allow one to use previous knowledge to improve the sample, except with respect to defining sampling units and determining how many sampling units are needed. In many cases one will be able to divide the population ahead of time into groups of sampling units that are already known to be similar. What is needed is a probability selection method that permits one to take advantage of one's ability to subdivide the population into relatively homogeneous parts. In a forestry setting one might have a forest composed of several stands that have different species, ages, site qualities, or stockings. The stands are much more uniform within themselves than the forest as a whole. The fact that the stands are different is obvious from a cursory examination of airphotos before a survey is started. It would be desirable to use the known differences between stands to improve the survey being planned. The selection method known as *stratified random sampling* is useful in these circumstances.

In sampling jargon the groups of relatively uniform sampling units are called *strata.* The general idea is to take separate samples in the respective strata, and then pool the information from the individual strata that make up the population. In this way the rather large differences between strata do not get mixed up in the sampling error. Stratification leaves only the relatively small variations within the strata to be reflected in the sampling error.

Stratification thus reduces the sampling error if one is successful in setting up the strata so that each contains quite uniform sampling units. However, stratification does nothing to reduce the travel time involved in taking the measurements since each stratum must be visited and sampled individually. Furthermore stratified sampling carries the same stringent requirements for a prior existence of lists and maps as does simple random sampling.

Multistage Sampling. One may wish to subdivide the population with an eye toward reducing travel costs or the need for prior listing/mapping rather than reducing the sampling error. One probability selection method for accomplishing this is known as *multistage sampling* or *subsampling*. The idea is to divide the population into contiguous groups of sampling units, and then randomly select only a sample of these groups for further attention.

The groupings of sampling units in this case are called *primary sampling units* or *first-stage sampling units* rather than strata. Political subdivisions such as counties or townships often provide a convenient basis for setting up primary units, since the boundaries can be readily identified in the field. To the extent that the need for travel is reduced to visiting only a subset of primary units and restricting operations to take place within these primary units, the costs of the survey are reduced accordingly. Likewise there is no need for a listing or mapping of sampling units in those primary units that are not to be visited.

There is no reason to stop at the primary level. The process of grouping can be continued in hierarchical or nested fashion by forming *secondary* or *second-stage* units within the first-stage units, *third-stage* units within second-stage units, and so on. This puts one in the position of taking samples from samples. One draws a sample of first-stage units; from the first-stage units that show up in the sample one draws a sample of second-stage units; from the second-stage units that show up in this sample one draws a sample of third-stage units; and so on down until one finally gets to the level where one has selected elementary units on which measurements are to take place. Note that the higher level stages of sample selection do not involve actual measurements. The measurements occur only at the final stage.

Unfortunately at the same time that it reduces costs, a multistage approach also tends to reduce the accuracy of estimates as compared to an equal number of measurements taken under a simple random sampling scheme. The impact of this tradeoff can be minimized, however, by making each of the groups at any given stage as much like the other groups in that stage as possible without sacrificing cost advantages. This is the opposite grouping strategy of that used in stratified sampling where the goal is the difference between groups and similarity within groups.

Multiphase Sampling. Still another technique which can be used to improve efficiency is known as *multiphase sampling*. This bears a superficial resemblence to multistage sampling in as much as the sampling is done in several steps. However, the situation in multiphase sampling is really quite different since measurements of one kind or another are required in all steps. The early phases involve low cost measurements on a large number of sampling units. Successive phases involve progressively more difficult or expensive measurements on smaller samples.

So-called *double sampling with stratification* is one variety of multiphase sampling, and an example of its possible use in assessing damage to forests by foliage-eating insects may help to illustrate the concept. The feeding activities of such insects often cause a thinning or discoloration of foliage that is quite noticeable from the air, and the infestations tend to be patchy with heavy damage in some areas and virtually no damage in others. The first phase of sampling, then, might involve aerial photography of randomly selected sample blocks and photointerpretation to separate the infested blocks from those that are not infested. Thus the photointerpretation serves to *stratify* photo sample blocks into infested and noninfested categories without the necessity of expensive ground visits to blocks that are not infested. This "stratification" is actually a type of measurement on the photo sample blocks and is not to be confused with conventional stratified sampling in which the categorization covers the entire population. The next phase of sampling would involve ground surveys of a subsample of photo blocks which were placed in the infested category by photointerpreters. These ground surveys would serve to determine the extent of damage in infested blocks.

By shifting the context only slightly, we can create a setting for another variety of multiphase sampling. In this case suppose that the forest is healthy, but we take airphotos as before to help in surveying the amount of wood present. By counting the number of trees and making measurements from the photos of height, crown size, and similar items, we can make predictions of the wood volume on the photo plots. A subsample of the photo plots would then be visited in the second phase, and measurements would be made more accurately on the ground. The ratios of ground measured volume to photo measured volume on plots measured both ways would be used to correct estimates from plots where measurements took place only from the photos. This would be called *double sampling with ratio estimation.*

Still a third variety of multiphase sampling might involve the use of a small sample to establish relationships between characteristics that are difficult to measure and other characteristics that are easier to measure, with the easier measurements also being made on a much larger sample. The relationship derived from the small sample is then used to make predictions for the larger sample. The equation describing the relationship is called a *regression equation*, and the sampling system would be called *double sampling with regression.* The "double" part of the term implies two phases of measurement, "triple" would imply three phases, and so on.

Varying Probabilities of Selection. As a final note on sample selection, some parts of the population may be more important than others. If so, it may be desirable to give these crucial units a greater chance of appearing in the sample than those that are less important. It is permissible to weight units in this way as long as the weighting is also taken into account when estimates are made from the sample data.

In practice, most situations call for a blend of these various sampling strategies in order to take advantage of the different kinds of economics that they offer. In fact, the same need for integration applies to the various sources of information, with a balance between existing data, remote sensing, and field work being preferable to complete reliance on a single source. A word of caution though is that the composite survey design can become rather complicated. If the consulting services of statistical experts are not available to ensure that things fit together properly, it is sometimes better to accept a little inefficiency while preserving simplicity rather than to risk getting caught in a statistical web and ending up with a bunch of data that is uninterpretable.

Sample Size. Before we leave the matter of sampling for the time being, the sample size (number of sampling units in the sample) needed to meet specified limits of precision deserves a little further comment. In particular there are two common sources of confusion that must be addressed early so that they do not cloud the interpretation of later material.

The first is a tendency to feel that the sampling error should decrease at the same rate as the sample size increases. On the contrary, it is only a slight oversimplification to say that the sampling error varies inversely with the square root of the sample size in randomized designs, which means that the sample size must be quadrupled in order to cut the average sampling error in half. Although the two situations are not completely analogous, it may make this seem more reasonable to think in terms of the relationship between the number of cells in a grid and the fineness of spacing between the mesh lines that form the grid. If the spacing of the grid mesh (sampling error) is cut in half, the number of cells (samples) in the grid is multiplied by a factor of four. The implication is that gains to be achieved by increasing the sample size are fairly good when the samples are composed of only a few sampling units, but there is little to be had by further increases in the size of samples that are already large.

The second point of confusion is an erroneous impression that the sample size should be directly proportional to the *size of the population.* The population size does have some effect on the size of the sample needed, as can be seen from the fact that measuring the entire population eliminates the sampling error completely. Much more important than the population size, however, is the inherent variability or diversity of the sampling units that make up the population. As pointed out earlier, when all the units in the population are exactly alike, then an examination of any one of the units is sufficient to describe the entire population regardless of how many units it may contain. Natural populations, however, are seldom if ever uniform. Unless the population contains very few sampling units, a safe rule of thumb is to say that samples consisting of less than 30 sampling units are small and therefore subject to a great deal of variation. At the other extreme, probability based samples composed of several hundred sampling units can be considered large and should provide good estimates. Further generalization is dependent on a knowledge of the variation and structure of the population.

When one has considerable knowledge of the population structure gained through sampling experience, it is sometimes possible to use a *sequential sampling* procedure in which the sample is examined continuously after each new unit is measured so that the sample size is held to a minimum. On the other hand there is little prior information available in setting up entirely new surveys, so pilot surveys may be necessary to avoid excessive guesswork.

Further details of statistical approaches to the sample size problem are presented in the Appendix, along with a more complete discussion of other statistical aspects of sampling and estimation. Concentrating the core of basic statistical material in an appendix has been chosen for at least three reasons. First of all, it facilitates reference. Second, it preserves the integrity of the individual chapters in Part II. Third, it eases the problem of accommodating heterogeneity in the background of readers. Those who are relative newcomers to the whole field of statistical analysis will need to spend some time going over the basics covered in the Appendix. It will probably be necessary for most readers to refer back to the Appendix now and then as they continue through the chapters on surveys of ecosystem components in Part II.

Nonsampling Errors and Survey Logistics. Sampling considerations are by no means the only sources of errors and inefficiency in field surveys. A complete census can easily contain major errors that creep into the processes of measuring, recording, and data reduction. Each of these phases can also contribute unduly to costs if survey planners are lax in their preparations. Poorly adjusted or improperly used instruments can generate a bias that is almost undetectable, as well as a great deal of frustration on the part of those who use them; so the best advice is to identify possible sources of difficulty ahead of time and be on guard against them. Instruments must undergo regular inspection and cleaning to ensure that they are properly calibrated and in good working order. Careful initial training of technicians followed by in-service training sessions will give the best insurance that the instruments are properly used and maintained. These well-trained personnel must also be provided good working conditions, orderly travel arrangements, and even sufficient opportunities for rest and relaxation during extended trips into the field. Emphasis on safety in all operations is essential.

Despite great vigilance, however, some errors are bound to crop up. A sufficient number of field checks and remeasurements by experienced supervisors is necessary to ensure quality control throughout the operation. Such mundane things as poorly structured recording forms or illegible handwriting can create real hassles. Likewise, data processing after return from the field can become a bottleneck if the mechanics of data reduction and storage are not planned and pretested before they are needed, then followed with careful editing and crosschecks. In fact it all boils back down to the two-word message with which we opened this discussion of field surveys—*plan ahead*—carefully and comprehensively.

2.5 SELECTION OF SURVEY METHODS ACCORDING TO INFORMATION NEEDS

A tabulation similar to that suggested for the objectives–needs analysis may help to guide this phase of survey planning. Since this tabulation focuses on the linkage between information needs and survey methods, it will be called a *needs–methods* table. It is somewhat premature at this point to discuss details of survey methods which are presented in later chapters, so letters will be used in this discussion to designate alternative survey methods which might be used to satisfy the information needs for the example of the impoundment project presented in Section 2.3. Method A, for instance, might involve the preparation of a topographic map from a field survey, whereas method B could make use of aerial photographs taken during the leafless season for this purpose. Perhaps method D, then, would be aerial photography during the growing season to show relationships between plant communities.

The first step in preparing a needs–methods table is to array the information needs from the objectives–needs analysis along the left margin (rows) with their respective priorities. The vertical columns will contain feasible sets of survey techniques which provide all or part of the needed information. This is illustrated for the example of the impoundment project in Table 2.3a. The information needs and associated priorities from Table 2.2d are: topographic map with 62% importance; set of soil maps with 22% importance; set of vegetation maps with 5% importance; animal census with 5% importance;

Table 2.3a First stage in development of needs–methods table for impoundment project example—entry of information needs and priorities

		Survey methods			
Information needs	Priority of need	A*	B*	C*	D*
Topographic map	62				
Soil maps	22				
Vegetation maps	5				
Animal census	5				
Airphotos	6				

*Since survey methods have not yet been considered, letters are used to designate possible alternative methods.

and airphotos with 6% importance. These information needs with their associated priorities comprise the first two columns of the needs-methods table.

If a particular survey method is adequate for satisfying a given information need, enter the priority rating of that need (as taken from the objectives-needs table) at the intersection of the need row and the method column. For example, method C might produce soil maps, vegetation maps, and animal census, but not a topographic map or airphotos. The three needs satisfied by this method have priority ratings of 22, 5, and 5%, respectively. These priority ratings are therefore listed in Table 2.3b in the column for method C. As in this case, a given survey method may satisfy several information needs. If a given need is only partially satisfied by a survey method, however, the need should be broken down further so that each one of the components can be associated with a single survey method. The total of the priority ratings satisfied by a survey method provides a measure of its *effectiveness* in meeting overall information needs. For instance, the effectiveness rating for method C is

$$22 + 5 + 5 = 32\%$$

The hypothetical analysis carried to this stage is shown in Table 2.3b.

The choice between possible survey methods will depend on cost as well as on effectiveness. The cost element can be included in the needs-methods table by adding two additional lines. The first of these two lines contains the estimated costs of the respective survey methods. The second line contains the *ratios of cost to effectiveness.* The *lower* the cost/effectiveness ratio, the more desirable the method. Hypothetical costs are included for the illustration in

Table 2.3b Second stage in development of needs-methods table for impoundment project example—entry of effectiveness ratings

		Survey methods			
Information needs	Priority of need	A*	B*	C*	D*
Topographic map	62	62	62		
Soil maps	22	22		22	
Vegetation maps	5	5		5	
Animal census	5	5		5	
Airphotos	6	—	—	—	6
Effectiveness		94	62	32	6

*Since survey methods have not yet been considered, letters are used to designate possible alternative methods.

Table 2.3c. Since method C has a cost of $500 and an effectiveness rating of 32, its cost/effectiveness ratio is

$$\frac{\$500}{32} = 15.6$$

In this hypothetical case the table shows that although method A has a high effectiveness index, it also has a high cost/effectiveness ratio. It will usually be the case that all information needed cannot be obtained through a single survey method. It is apparent from Table 2.3c that complete survey systems could be assembled by combining either methods A and D or methods B, C, and D. The combination of A, C, and D would also provide the needed information, but is not a likely choice because of the overlap between methods A and C. Combinations of methods such as B, C, and D can be evaluated in another column to the right of the table. Effectiveness entries for the combination are recorded in the first part of the column, then totaled at the bottom. The priority or effectiveness entry for a given need should be recorded only once in the combination column, even if it is satisfied by two or more methods in the combination. For purposes of illustration, the cost entry for the combination of B, C, and D will be taken as the sum of the costs for the methods included in the combination. In practice, however, combined travel or other shared costs will often make the cost of the combination less than the sum of the separate costs. Table 2.3d shows the addition of a column for evaluating the combination of methods B, C, and D.

As with the format suggested for objectives-needs analysis, any modification of the needs-methods table format that makes it more convenient for

Table 2.3c Third stage in development of needs-methods table for impoundment project example—entry of costs and cost/effectiveness ratios

Information needs	Priority of need	Survey methods A*	B*	C*	D*
Topographic map	62	62	62		
Soil maps	22	22		22	
Vegetation maps	5	5		5	
Animal census	5	5		5	
Airphotos	6				6
Effectiveness		94	62	32	6
Cost		$2000	$1000	$500	$50
Cost/effectiveness ratio		21.3	16.1	15.6	8.3

*Since survey methods have not yet been considered, letters are used to designate possible alternative methods.

Table 2.3d Fourth stage in development of needs–methods table for impoundment project example—analysis of combination of survey methods

Information needs	Priority of need	Survey methods A*	B*	C*	D*	B, C, & D
Topographic map	62	62	62			62
Soil maps	22	22		22		22
Vegetation maps	5	5		5		5
Animal census	5	5		5		5
Airphotos	6	—	—	—	6	6
Effectiveness		94	62	32	6	100
Cost		$2000	$1000	$500	$50	$1550
Cost/effectiveness ratio		21.3	16.1	15.6	8.3	15.5

*Since survey methods have not yet been considered, letters are used to designate possible alternative methods.

individual circumstances is all to the better. For the sake of completeness in perspective it should be pointed out that the problem of optimizing survey methods for effectiveness relative to costs can also be approached in a much more sophisticated manner through mathematical programming. Mathematical programming, however, is beyond the scope of this text. Furthermore because of the complexities of optimizing for several objectives simultaneously, the mathematical programming approach to planning integrated surveys has not progressed much beyond the exploratory stage.

2.6 OUTPUT PRODUCTS OF A SURVEY SYSTEM

Users are reluctant to accept survey information in a form that they do not understand, and rightly so since it is likely to be misinterpreted. One alternative that may appeal to the administrators of the system is to use a fixed format and educate the users in the interpretation of that format. This approach must be used with some caution since it is human nature for a person to dislike showing, much less admitting, ignorance. A rigidity in output formats is likely to repel potential users, with the result that the survey information system falls far short of its potential usefulness. It is much better to offer the users a choice of output formats, and better still to allow them a say in determining what these choices will be. In any case, clearly written guidebooks to the interpretation of available products are essential. In this section we take a quick look at some of the more commonly accepted recipes for serving information to users.

The discourse on ways of presenting data tends to become a little abstract unless accompanied by examples that have a strong geometric flavor. The surface configuration of ten small lakes on a hypothetical tract provides a simple example of the properties to be illustrated and is easy to visualize. In view of their hypothetical nature, however, this set of lakes will be called the "Mirage Lakes."

2.6.1 Map Products

If the spatial relationships of the features are important components of the information to be transmitted, then a map will be the natural choice for display. The map sheet represents some selected surface in the environment. The surface most frequently represented is the interface of the air with the land or water. In addition to the selected surface, the map will usually show the locations of certain other features of interest. Thus the map is a geometric abstraction or model of the real world, and the goal should be to make it clear and concise so that the reader may understand the information it was designed to transmit.

Map Projections. The first problem encountered in preparing a map is that the map sheet is flat, whereas the relevant environmental surface usually is not. How then is the environmental surface to be contorted so that it will correspond with a flat sheet? This is accomplished by means of a *map projection.* In reconnaissance mapping of small tracts the general curvature of the earth is slight. Since the intrinsic accuracy of reconnaissance maps is rather low anyhow, this slight curvature can be ignored without adversely affecting the usefulness of the map.

Over somewhat larger areas the curvature is still not very pronounced, and a simple *orthographic* projection works quite well. To visualize an orthographic projection, first imagine a very large flat sheet (plane) that would cover the area to be mapped. Mentally balance the large sheet at the center of the area to be mapped, so that it is perpendicular to a line through the center of the earth and the balance point. Now project each terrain point up to the sheet along a line that is perpendicular to the sheet. An orthographic projection is good near the center of the area, but deteriorates quickly as you move away from the center.

For larger areas on the order of a county or state in size the type of projection becomes more important. *Mercator* and *conic* projections are often used, but a further discussion of geographic referencing will be deferred to the next chapter.

Scale. When the environmental surface has been projected onto a plane so that it conforms with a flat sheet, the lifesize plane must still be shrunk to the desired size of the map sheet. The amount of shrinkage that takes place determines the *scale* of the map. In other words, scale is the relationship between map size and actual size. This may be stated in several ways.

Perhaps the most familiar form is to state map distance in terms of actual distance without any attempt to use the same units in both. For example, a scale might be stated as 1 inch to the mile or 1 in. = 1 mi. This means that each inch of distance on the map represents 1 mi on the ground (horizontal distance, not slope distance). Since a mile contains 5280 × 12 = 63,360 in., a 63,360-fold shrinkage has taken place in reducing the environmental surface to the map representation.

Another way of stating this is to use ratio or fractional forms with both map and ground distance in the same units. The equivalent *scale ratio* for 1 inch to the mile would be 1:63,360 and the *scale fraction* would be 1/63,360. Note that it is customary to simplify scale ratios and fractions so that the number on the left or in the numerator is unity. As long as the map and actual units are the same, scale ratios and fractions hold true for any unit of linear measure that one might choose.

Still another approach is to draw a *scale bar* on the map and mark off actual ground equivalents along the graphic scale. Each form has certain advantages.

The fractional form simplifies the calculation of distance and the conversion of units, but no longer holds true if the map undergoes photographic enlargement or reduction after initial preparation. On the other hand a graphic scale bar is awkward to interpret numerically but remains accurate under enlargement and reduction.

No matter what size of area is to be mapped, the choice of scale is very important. The closer the map distance approaches actual distance, the *larger* the scale. The scale must be large enough to accommodate the required detail without a crowded and messy appearance, but beyond this nothing is gained by going to a larger scale except unless bulk.

Map Contents. In addition to the projection and scale, a map has three other basic ingredients: title block, legend, and detail. The *title block* provides potential users information they need to decide whether the map meets their needs before doing a more careful examination of the material portrayed. Such things as description of the tract, purpose of the map, agency responsible for its preparation, and date of preparation help the user to determine whether the map is likely to be of interest. The title block may also contain other general information that the user might find helpful, but there should not be so much extraneous information that it gets confusing.

The *legend* tells the user how to interpret the map. A complete yet succinct explanation of the classification and coding systems characterizes a good legend. The importance of standardizing classifications and symbology wherever possible has already been stressed. A *north arrow* should be shown on every map, whether as part of the legend or in some other convenient place that will give the map a more balanced appearance. Since a compasss needle usually does not point true north, it is also highly desirable to show the direction in which the compass needle actually points in the given locality (called magnetic north). A full arrowhead is usually used for true north, with a half arrowhead for magnetic north. Custom also has it that north should lie at the top of the map. Sometimes the configuration of the area to be mapped will make this inconvenient, but departure from the "north up" rule is inadvisable and should be used with caution.

Details of maps are as variable as motives for making maps. Examples of a different emphasis in details are topographic maps, vegetation cover or type maps, soil maps, geologic maps, hydrographic maps, highway and street maps, weather maps, and land use maps. One way of breaking down the types of data to be mapped is by *geometric properties*. There are four broad categories of data in this respect.

The first category consists of *point* data. The task in mapping point data is to show the correct location of some feature that occupies an insignificant area relative to the map scale. This is not to imply that the feature itself is unimportant, but rather that the ground equivalent size of a dot on the map covers an area larger than the actual feature. Point data are relatively easy to handle in mapping since the problem boils down to locating the feature and choosing an appropriate symbol to represent it on the map.

The second type of data is *line* data typified by roads, railroads, small water courses, and so on. As with point data, the width of the line on the map is wider than the feature itself when the line width is expanded to ground equivalent distance according to map scale. This is not true for the length of the feature, however, and tracing the correct path on the map complicates the representation somewhat. When line data are being mapped, the best that can be done is to show all deviations in the path that are larger than the ground equivalent width of the map line. Thus both line width and locational accuracy must be considered.

The third category of mapped data consists of *area* data. In this case both the length and the width of the actual area are usually larger than the ground equivalent width of map lines. This does not necessarily mean, however, that all the variations on the ground can be shown on the map. For one thing there is no point in separating out an area on the map unless there is room on the map to designate the classification for that area. Colors can be used to mark smaller areas than do letter codes or shading patterns, but even here one encounters a practical limit. One of the necessary choices in mapping area data, then, is to decide the smallest area that will be broken out and categorized separately on the map. This is usually called *minimum type size* or *minimum mapping area.* Areas smaller than the minimum mapping area are lumped in with the surrounding category. One compromise that can be made is to choose a point symbol to flag such inclusions without giving further information as to the nature of the inclusion.

Problems may also be encountered in determining just where boundaries between mapped categories should be drawn. Manmade or *cultural* features usually have fairly sharp boundaries and present relatively few problems if the classification system is really adequate to describe the features without ambiguity. Some natural features also have distinct boundaries, but transition belts called *ecotones* between adjacent vegetation types, soil types, and so on often make it difficult to decide exactly where the boundary should be placed. Decisions on how to handle ecotones must be made before the mapping can proceed, and it may even be appropriate to explain these in the legend in some cases.

The fourth type of mapped data could be considered a special case of line data, but it is important enough to be discussed separately. *Isoquant* (also called *isopleth*) lines show the location of points having equal levels of some variable quantity being mapped. A common use is to represent the third dimension, *elevation* or height above an established point. On a *topographic* map of this type the lines are called *contour lines.* The difference in elevation between adjacent contour lines is called the *contour interval,* and this difference is constant throughout the map. Thus closely spaced contour lines indicate steep slopes, and widely spaced lines indicate gentle slopes. Topographic maps are treated in more detail in Chapter 4. The use of isoquant lines is not limited to representing elevation. Anything that varies more or less continuously from place to place can be mapped in this way. Two of the many examples are temperature where the isoquant lines are called *isotherms* and barometric pressure where they are called *isobars.*

Manual Preparation of Maps. Maps may be prepared in a variety of ways ranging from hand drafting with very simple tools to completely computerized plotting. Whatever the method of preparation, the goal is to make the map clear, concise, and accurate.

If hand methods are to be used, it will be necessary to do the drafting in at least two stages. The first draft, which may be the field map, should be drawn with a relatively hard lead pencil so that the lines do not smear. The pencil must be kept sharp to improve the accuracy of plotting. If lines are not drawn too heavily alterations will be easier to make. If the rough map is drawn directly in the field, a heavy paper should be used that will withstand the rigors of weather and brush. It is often helpful if the rough map sheet has a preprinted grid system on it. Convenient forms for preparing field maps are available from forestry and engineering suppliers.

Even in the field, a french curve or flexible curve, straight edge, and protractor should be used to draft the lines on the map. The intended location of a wiggly line is somewhat uncertain, and this reduces the inherent accuracy of the map. It is important to exercise a little foresight in choosing the size of the map sheet and the point at which to begin the map on the sheet. It is somewhat embarrassing to find halfway through the job that the remaining area to be mapped does not fit on the sheet.

If the rough map is prepared from airphotos, the usual approach is to mark on clear acetate sheets taped over the photos themselves. In this case one must choose the marker carefully since lead pencils will not write on acetate, and china markers make lines that are too wide. The lettering on the rough map is not particularly critical, but it still must be done neatly.

The rough map is seldom adequate for presentation, and a final version must be prepared. The cost of preparing the final map should be commensurate with its intended use. It also makes a difference whether or not multiple copies are to be produced by automatic duplicating processes.

If a single copy is needed for office use and possibly a few simple photocopies or blueprints, it is satisfactory to trace the rough map on one of the several dimensionally stable drafting materials in waterproof ink. Even on a map intended for such limited use, neatness is important since a messy map is not only ugly but also easy to misinterpret. A straight edge and similar aids to drafting should be used whenever possible. The symbols and lettering should be in proportion to the size of the feature they explain, and all lettering should be readable without rotating the map. Careful hand lettering is sufficient for a map that is not meant for public distribution. The title, legend, north arrow, and scale bar should be placed with an eye toward balancing the appearance of the map. The map should be finished by being framed in neatly ruled lines commensurate with its size.

When the map is to be reproduced for subsequent distribution, however, considerable care must be taken in its preparation. Hand lettering is not adequate for this purpose, unless one happens to be an exponent of hand lettering as an art form like some professional cartographers. A Leroy lettering set or similar template method can be used, or commerical transfer letters may offer a convenient alternative. Similar transfer and adhesive materials are also available to aid other aspects of cartographic drafting. These materials include shadings of different densities and patterns as well as many types and thicknesses of lines. Technical fountain pens designed for use with India, acetate, and other drawing inks are the standard instruments for drafting on most materials. Points for these pens are available in a whole series of line thicknesses. A word of caution, however, is that these pens must be kept clean and handled carefully in order to avoid a sticky mess.

A map that is to be photographically reproduced should be drawn at a scale somewhat larger than that planned for the final product. The photographic reduction will smooth out minor irregularities in the original drafting. One should not attempt to show excessive detail on a single map. If detail is so abundant that it cannot be absorbed easily by the reader, the possibility of using an acetate or Mylar overlay system should be considered. Color is desirable for overlay systems, but color is also expensive for mass reproduction. If numerous black and white patterns are used on overlays, there is a danger that patterns will wash out or become solid when the overlays are superimposed. A little experimentation beforehand will help to avoid this pitfall.

Finally the alternative of contracting map preparation to professional cartographic companies should not be overlooked. Such organizations have an impressive array of sophisticated equipment as well as a staff of skilled cartographers. Maps of this caliber are often drafted by etching an emulsion coated on a stable transparent base. Federal agencies and other large organizations often have special cartographic offices that prepare maps destined for public distribution or sale.

Figure 2.2 is a hand lettered reconnaissance map showing the shoreline configuration of the fictitious Mirage Lakes introduced earlier in this section. Although highly simplified by the use of hypothetical data, Figure 2.2 contains all the basic elements that should be present in a map.

Computerized Map Preparation. The use of computers for preparing maps is still in its infancy, but is rapidly coming of age. The first requirement is to get the spatial information to be mapped into computer readable form. If the data are in the form of field survey notes, this is a simple matter of coding positions, classifications, and measurements onto one of the standard computer input media such as punched cards, punched paper tape, or magnetic tape. The output of some remote sensors is already coded in numeric form on magnetic tape and thus computer compatible.

In other cases the data may be taken from existing maps or aerial photos. In these latter cases information is extracted from the graphic source by *digitizing,* which involves the use of automatic or semiautomatic devices to convert the positions of map points into X and Y coordinates and classifications associated with these points into numeric codes. Linear features are handled by digitizing more or less closely space points along the path. Spacing of the digitized points determines the smallest deviation in the path that can be shown. A computerization of area data presents special problems relative to digitizing boundaries between categories and making sure the boundary lines meet so that there are no holes. Use of a square grid to create mapping *cells* often provides the most expedient way of avoiding such problems.

The type of output device available largely determines the appearance of the computer map product. The least expensive and most readily available type of output device is a printer with capabilities limited to placing letters, numbers, and punctuation marks at fixed positions. The print positions can be equated to the positions of area cells on the map, and printed characters are used to indicate the classification of features that occupy the cell. If the set of available characters becomes limiting, additional codes or shadings can be created by printing several different characters on top of each other (overprinting). The printing can also be done in several passes with different colored ribbons. Continuous lines cannot be produced with this type of output device. Figure 2.3 is the result of using a typewriter to simulate this type of computer map for the Mirage Lakes.

Incremental plotters are a more versatile type of output device for mapping purposes and are available in many computer centers. They consist of a computer driven pen that moves in small straight steps (increments) on the order of 0.01 in. or less. These steps are so small that the lines appear to be smooth curves. There may be multiple pens loaded with different colors of ink. These plotters can be used to produce maps with a fairly conventional appearance.

Cathode ray tubes (like television screens) are also becoming more common

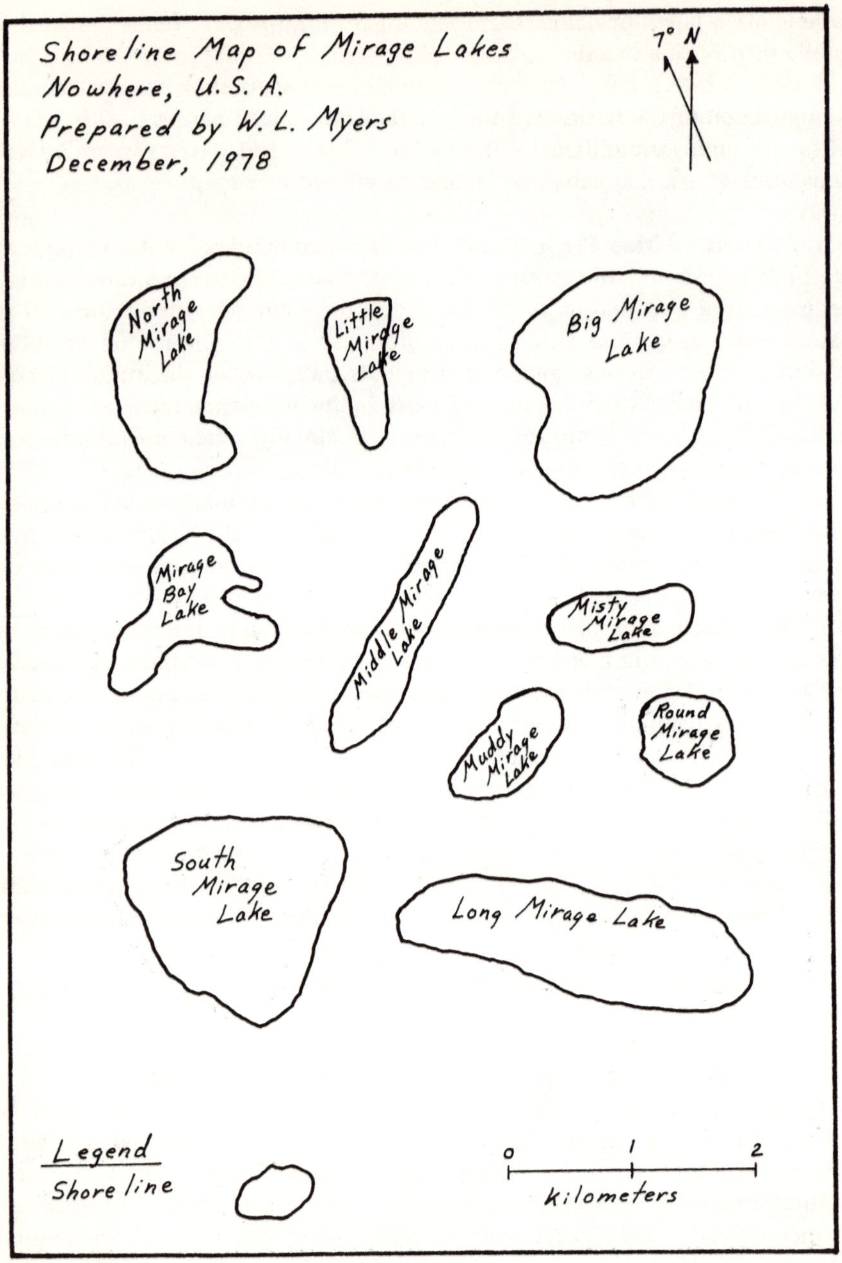

Figure 2.2 Map showing shoreline configuration of hypothetical Mirage Lakes.

```
++++++++++++++++++++++++++++++++++++++++++++++++++++++++++++++++
+                                                              +
+    MAP OF MIRAGE LAKES                          N            +
+    NOWHERE, U.S.A.                              +            +
+    PREPARED BY W. L. MYERS                      | DECL 7W    +
+    DECEMBER 1978                                |            +
+                                                              +
+                                                              +
+      N. MIRAGE L                                             +
+              ...    LITTLE              BIG MIRAGE L         +
+           ......    MIRAGE L             ..........          +
+       ........        .....           ...............        +
+       ........        .....         ................         +
+       .......         ....          ................         +
+       .......          ....           .............          +
+       .......           ...            ............          +
+        ........          .            ...........            +
+         ......                        ..........             +
+                         MIDDLE  .      .....                 +
+       MIRAGE BAY L     MIRAGE ....                           +
+           ....            L  ....                            +
+         ........            ....        MISTY MIRAGE L       +
+         ......             ....         ..........           +
+       ............        ....         ...........           +
+       ....               ....                                +
+      ...               .....       MUDDY      ROUND          +
+                       .....        .....      ......         +
+                       ....        ......      .......        +
+                                  .......      .......        +
+                                 ......                       +
+         SOUTH MIRAGE L          MIRAGE L       MIRAGE L      +
+         ................                                     +
+        .................                                     +
+        .................      ............. LONG MIRAGE L    +
+          ...............    ........................         +
+          .............         ........................      +
+            ...........                ................       +
+               ......                      ............       +
+                                                              +
+                                                              +
+    LEGEND                                                    +
+                ...                  +--1 KM--+               +
+      LAKE     .....                                          +
+                ...                                           +
+                                                              +
++++++++++++++++++++++++++++++++++++++++++++++++++++++++++++++++
```

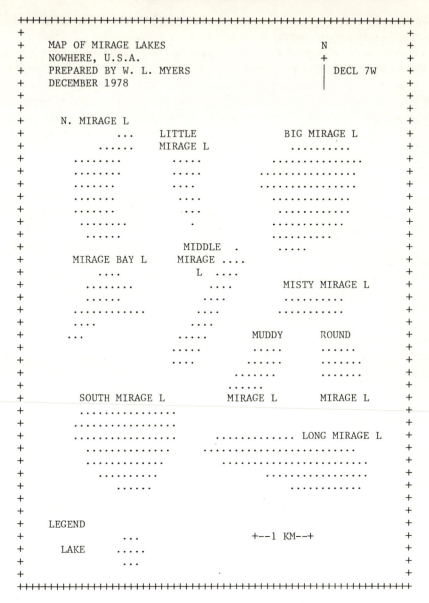

Figure 2.3 Simulated line printer map of hypothetical Mirage Lakes.

as computer display devices. In their simplest form they function much like the printers described above, except that the characters appear on the screen. More sophisticated cathode ray terminals designed especially for computer graphics have a grid of points so fine that the lines appear to be continuous. This latter type gives fairly good maplike displays. Another step up in sophistication adds color to the display screen. The display screen can be photographed, or hardcopy attachments are available to produce photocopies automatically on sensitized paper.

A good point of departure for exploring the world of computer mapping is the SYMAP system of programs developed at the Laboratory for Computer Graphics and Spatial Analysis at Harvard University for making printer maps. This system is well documented, available on many large computer systems, does not require the user to have an extensive background in computer science.

2.6.2 Tabulations

Maps provide a very effective way of showing the locations and spatial relationships of environmental features. Maps are not so effective, however, when it comes to a critical examination of measurable features. A case in point is the Mirage Lakes of Figure 2.2. Two of the measurable features of these lakes are surface area and length of shoreline. The map contains all the information that exists on area and shoreline length; yet which lake is largest in terms of either area or shoreline is not readily apparent by looking at the map. Furthermore this is a situation in which the features are easy to map. Maps are even more awkward when one deals with quantities like growth rates or disease damage which are harder to map.

Details can usually be extracted more readily from tabulations than from maps. For example, Table 2.4 is a tabular listing of surface area and shoreline length by name of lake for the Mirage Lakes. Metric units are used in this table, with the surface area in hectares (10,000 sq m or 2.471 acres/ha) and shoreline length in kilometers (1000 m or 0.6214 mi/km). The area was measured by the grid method described in Section 3.7.2, and the shoreline length was measured by the string method described in Section 3.5.7. This

Table 2.4 Surface areas and shoreline lengths for Mirage Lakes from Figure 2.2

Name of lake	Area (ha)	Shoreline length (km)
South Mirage Lake	210	5.7
Long Mirage Lake	202	6.8
Muddy Mirage Lake	48	2.9
Round Mirage Lake	45	2.6
Mirage Bay Lake	84	5.2
Middle Mirage Lake	77	5.0
Misty Mirage Lake	47	3.1
North Mirage Lake	123	5.3
Little Mirage Lake	42	3.0
Big Mirage Lake	240	5.9

tabular presentation is easier to interpret than the map, since a quick glance through Table 2.4 reveals that Big Mirage Lake has the largest surface area and Long Mirage Lake has the longest shoreline.

Still it would be even easier to study the question in terms of any variable quantity if the lakes were arranged in the order of size according to that variable. Such ordering calls for separate tables for the two variables since the sequence is different for the area than for the shoreline length. Along with an ordered list it is also helpful in studying size questions to give the percentage of items that are the same size or smaller. This gives an easy grasp of the item's relative position in the *distribution* of sizes. Table 2.5 shows the lakes ordered by increasing area along with "same size or smaller" percentages. Table 2.6 contains similar information for shoreline length.

Tables 2.5 and 2.6 contain complete listings of the data for the respective variables of surface area and shoreline length. Interpretation is quite easy in cases like these where the list contains relatively few items. The presence of a great many items in the list, however, would tend to make interpretation less easy. In such cases it may be desirable to do a little condensing by regrouping adjacent data values in the ordered list into *classes*. Judgment must be exercised in choosing the class size since excessively broad classes may conceal important information, whereas very narrow classes do little to ease the problem of interpretation. The results of such a regrouping are usually presented in a *frequency table*. Although regrouping is not necessary with the Mirage Lake data, we will go through the exercise anyhow to illustrate the process.

The regrouping starts with an ordered list like those in Tables 2.5 and 2.6. The presence of distribution percentages is helpful in choosing an appropriate

Table 2.5 Distribution of Mirage Lakes in order of increasing surface area, with percentages of lakes that are the same size or smaller than any given lake

Name of lake	Area (ha)	% of lakes same size or smaller
Little Mirage Lake	42	10
Round Mirage Lake	45	20
Misty Mirage Lake	47	30
Muddy Mirage Lake	48	40
Middle Mirage Lake	77	50
Mirage Bay Lake	84	60
North Mirage Lake	123	70
Long Mirage Lake	202	80
South Mirage Lake	210	90
Big Mirage Lake	240	100

Table 2.6 Distribution of Mirage Lakes in order of increasing shoreline length, with percentages of lakes that are the same size or smaller than any given lake (with respect to shoreline)

Name of lake	Shoreline length (km)	% of lakes same size or smaller
Round Mirage Lake	2.6	10
Muddy Mirage Lake	2.9	20
Little Mirage Lake	3.0	30
Misty Mirage Lake	3.1	40
Middle Mirage Lake	5.0	50
Mirage Bay Lake	5.2	60
North Mirage Lake	5.3	70
South Mirage Lake	5.7	80
Big Mirage Lake	5.9	90
Long Mirage Lake	6.8	100

class size. When a class width has been chosen, either the class limits or the class mark (midpoint of the class) or sometimes both are placed in the first column of the frequency table. This is shown for the Mirage Lakes shoreline data in Table 2.7 using 1 km classes. We suggest that borderline cases be included in the lower class, but there is nothing sacred about this. Where possible it is best to show class limits structured so that borderline cases do not occur. If this is not convenient, then the treatment of borderline cases can be explained in a footnote or in column headings. The second column of the frequency table shows the number of data items falling in the class (frequency). The third column shows the cumulative number of data items (cumulative frequency) included in that class and previous classes (classes with lower class

Table 2.7 Frequency table showing Mirage Lakes shoreline data (from Table 2.6) regrouped into 1 km classes

Class limits	Frequency	Cumulative frequency	% cumulative frequency
2.1–3.0	3	3	30
3.1–4.0	1	4	40
4.1–5.0	1	5	50
5.1–6.0	4	9	90
6.1–7.0	1	10	100

marks). A fourth column may also be included showing cumulative frequencies in percentage form.

Likewise data on two variable quantities can be cross-tabulated by groups to bring out relationships between the two variables. Cross-tabulations of this type are called *contingency tables.* A cross-tabulation of area and shoreline length for the Mirage Lakes will serve to illustrate the format of contingency tables.

Suppose that we use two categories of size for both surface area and shoreline length. Lakes with surface areas of 100 ha or less will be considered "small" lakes with respect to surface area, and those with surface areas greater than 100 ha will be considered "large" lakes with respect to surface area. The "small" category for shoreline length will contain lakes with shorelines of 4 km or less, and the "large" category will contain lakes with shorelines greater than 4 km. This gives four possible cells in the cross-tabulation: lakes with small surface area and small shoreline, lakes with small surface area and large shoreline, lakes with large surface area and small shoreline, and lakes with large surface area and large shoreline. One would expect lakes with small surfaces to have small shorelines, and those with large surfaces to have large shorelines. Table 2.8 is a contingency table showing the number of lakes (frequencies) that fall in each cell. There are no lakes with large surface area and small shoreline, and only two with small surface area and large shoreline. Therefore the contingency table confirms the general nature of the relationship that would be expected on an intuitive basis.

Contingency tables are usually easier to interpret if percentages are shown along with frequencies. Percentages can be calculated either across the rows or down the columns. In this case there is no particular reason to prefer one way of percentaging to the other. Dependng on the nature of the relationship under study, however, one way of calculating percentages may be preferable to the other. Table 2.9a shows percentages calculated by rows, and Table 2.9b shows percentages calculated by columns. One need only see which way the percentages total to 100 in order to determine how the percentages were calculated.

Table 2.8 Contingency table showing cross-tabulation of surface area and shoreline length for Mirage Lakes. Numbers in body of table show how many lakes (frequency) fall in each cell

	Small shoreline ($\leq$4 km)	Large shoreline ($>$4 km)	Marginal total
Small area ($\leq$100 ha)	4	2	6
Large area ($>$100 ha)	0	4	4
Marginal total	4	6	10

Table 2.9a Contingency table showing cross-tabulation of surface area and shoreline length for Mirage Lakes, with percentages calculated across rows

		Small shoreline (≤4 km)	Large shoreline (>4 km)	Marginal total
Small area (≤100 ha)	#	4	2	6
	%	66.7	33.3	100.0
Large area (>100 ha)	#	0	4	4
	%	0.0	100.0	100.0
Marginal total	#	4	6	10

Table 2.9b Contingency table showing cross-tabulation of surface area and shoreline length for Mirage Lakes, with percentages calculated down columns

	Small shoreline (≤4 km)		Large shoreline (>4 km)		Marginal total
	#	%	#	%	#
Small area (≤100 ha)	4	100.0	2	33.3	6
Large area (>100 ha)	0	0.0	4	66.7	4
Marginal total	4	100.0	6	100.0	10

2.6.3 Distribution Graphs

For those who prefer to visualize things instead of examining numerical listings, graphical equivalents of Tables 2.5–2.7 can be prepared quite readily. Graphical analogs of Table 2.5 and 2.6 can be constructed by plotting the "same size or smaller" percentages on graph paper against the size of lake. This type of display, called a *cumulative frequency distribution,* is illustrated for the Mirage Lakes data in Figures 2.4 and 2.5. A cumulative frequency distribution presents the data in a form that makes size factors easy to analyze visually. The information on lake names could also be included by placing the names in position along the horizontal (size) axis.

The percentage of lakes falling in any given size range is easily obtained by projecting the limits of the range up vertically from the horizontal axis to the line and then across to the vertical axis where the relevant percentages can be read and subtracted, as shown for the range of shoreline lengths from 4 to 6 km in Figure 2.5. Note that it might also be necessary to move up a step or down a step to allow for lakes falling right at the ends of the range, depending on whether the limits of the range are to be excluded or included.

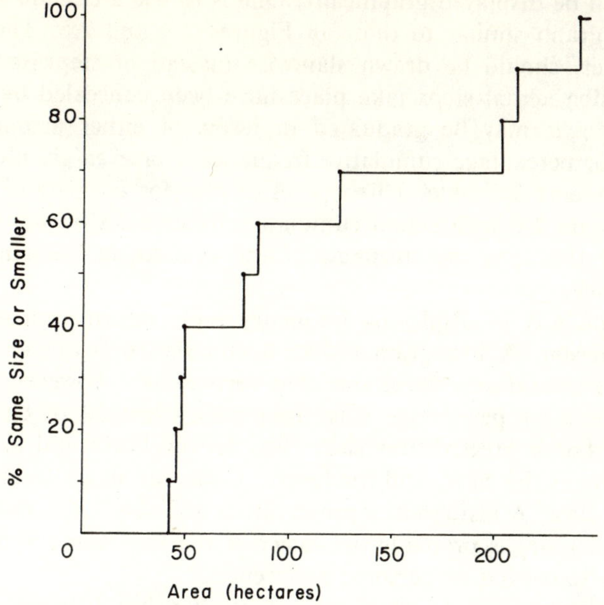

Figure 2.4 Cumulative frequency distribution of areas for Mirage Lakes (see Table 2.5).

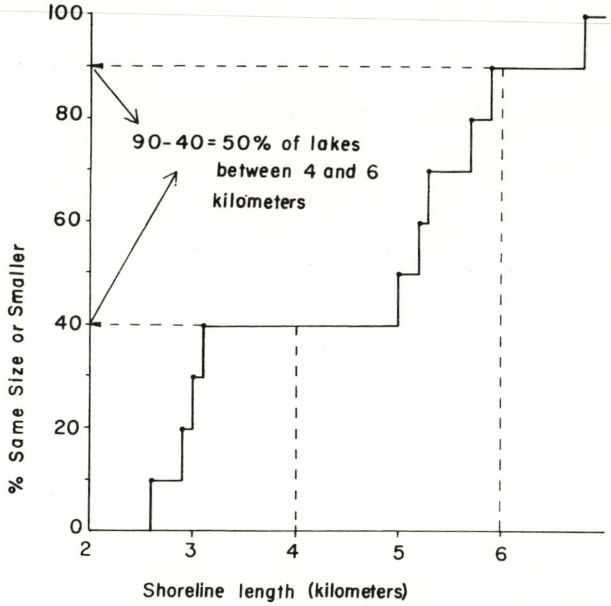

Figure 2.5 Cumulative frequency distribution of shoreline lengths for Mirage Lakes (see Table 2.6).

There are two convenient ways that frequency table information like that in Table 2.7 can be displayed graphically. One is to use a cumulative frequency distribution graph similar to those in Figures 2.4 and 2.5. The connecting lines, however, should be drawn slantwise instead of stepwise because the points at which actual steps take place have been concealed by regrouping. The vertical axis may be graduated in terms of either actual cumulative frequencies or percentage cumulative frequencies, or even given dual graduations if this seems desirable. This type of display for the data of Table 2.7 is shown in Figure 2.6 with actual cumulative frequencies on the vertical axis. In a plot of this type the steepness of the connecting lines increases with class frequency.

The second way of displaying frequency table information graphically is with a *histogram.* A histogram is like a cumulative frequency distribution where the horizontal axis shows size. The vertical axis, however, is graduated in either actual or percentage class frequencies instead of cumulative frequencies. A bar is erected over each class on the horizontal axis. The base of the bar covers the class, and the height of the bar shows frequency of data items in the class. A histogram representation of Table 2.7 is shown in Figure 2.7. The histogram is probably the easier of the two forms to interpret, but this depends somewhat on personal preferences.

The usual approach to presenting a relationship between two variable quantities graphically is to prepare an X–Y plot with one variable on each axis. This gives a visual impression of the way the paired data scatter about a line. Fitting an approximate line through the scatter of data points by eye with a flexible curve may help make it easier to get an idea of the nature of the relationship, but one should not place too much confidence in such a roughly drawn curve. A *scatter diagram* of this sort for surface area and shoreline length of the Mirage Lakes is shown in Figure 2.8.

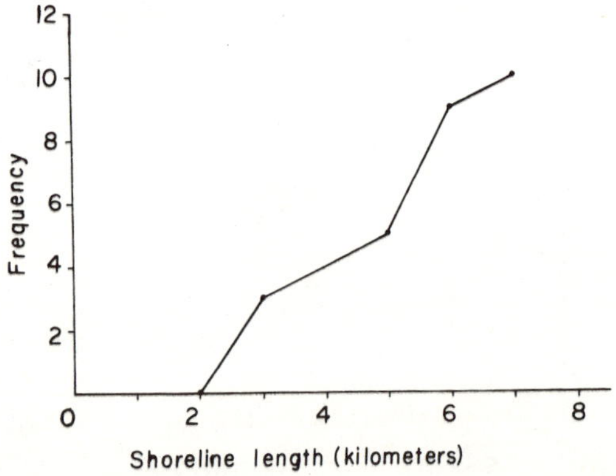

Figure 2.6 Cumulative frequency distribution of grouped shoreline data for Mirage Lakes (see Table 2.7).

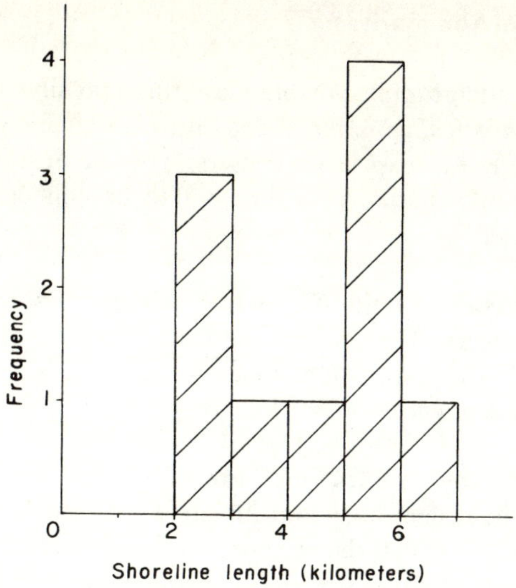

Figure 2.7 Histogram of grouped shoreline data for Mirage Lakes (see Table 2.7).

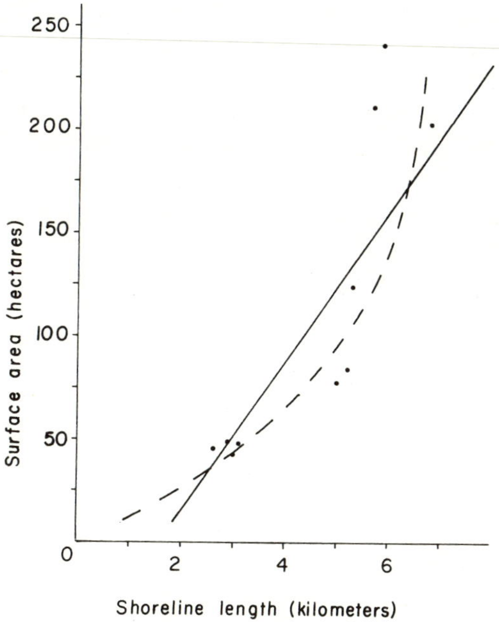

Figure 2.8 *X–Y* plot or scatter diagram of surface area versus shoreline length for Mirage Lakes, including freehand curves.

2.6.4 Statistical Summaries

Maps, tabulations, and graphical displays are three possible ways of presenting data to the user. *Statistical summaries* constitute a fourth type of data product which may be preferred by experienced users. The use of statistical summaries allows most of the information in a data set to be condensed into relatively few numerical figures.

Statistical Expressions of the Typical or Average Value. The user's first question usually concerns the (numerical) value that is typical or average for the data set. Statistical expressions that speak to this question are called *measures of central tendency.*

Median. Perhaps the simplest measure of central tendency is the value of the variable in the ordered list that has an equal number of data items on either side of it. This is called the *median* value. If there is an odd number of items in the list, the median is the middle value in the ordered list. If there is an even number of items in the list, the median is the average of the two middle values in the ordered list. Since there are 10 lakes in the Mirage Lakes example and this is an even number, the median for either area or length of shoreline is the average of the fifth and sixth values in the respective ordered lists. For the lake area data from Table 2.5, the two middle values in the ordered list are 77 and 84 with the average (sum divided by two) being 80.5. For the shoreline data from Table 2.6 the two middle values in the ordered list are 5.0 and 5.2 with the average of the two being 5.1.

The median can also be found from a graph of the cumulative frequency distribution as shown for the shoreline data in Figure 2.9. Project horizontally across from the 50% point on the vertical axis to the plotted line. If you hit the plotted line on a riser (odd number of data items), drop vertically down and read the median at this point on the horizontal axis. If you hit the plotted line on a level with the top of a stair step (even number of data cases), continue on across to the middle of the step and then drop down to read the median.

It often helps to get a better feel for statistical expressions if they can be given a physical interpretation. To this end suppose that we have a thin rod that is rigid but virtually weightless. The first thing we will do is graduate the rod with the same scale used for the horizontal axis of the cumulative frequency distribution; that is, in terms of hectares or kilometers or whatever the units of measure happen to be for the variable under consideration. Next we find a set of cubical (or cylindrical) beads drilled so that they can be strung on the rod in the manner of a shish kebab. One bead represents each data item or measured unit, and the beads are positioned along the rod according to the measured value of the data items on the scale. If there happened to be two or more identical items, the beads could be glued or tied together so that they hang in a column from the rod. A model of this kind for the shore-

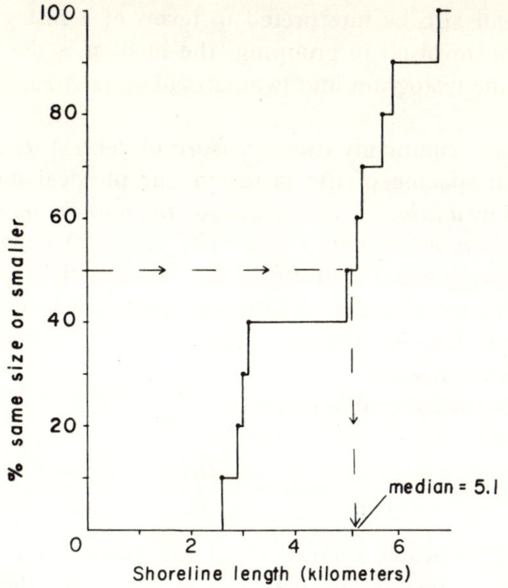

Figure 2.9 Graphical procedure for locating median from plotted cumulative frequency distribution.

line data of the ten Mirage Lakes is shown in Figure 2.10. The median is the value or position on the rod that has equal weights of beads on either side, assuming that the beads are all of the same weight. If the number of data items (beads) is even, this point will fall midway between two beads. Actually, any point between the two middle beads would satisfy the condition of equal bead weights on either side, but convention says that the halfway point between the two beads should be used. If the number of data items (beads) is odd, the median point will fall right on the center of the middle bead.

An important characteristic of the median is that spacing of beads lying outside the middle pair makes absolutely no difference. The outer four beads on either side in Figure 2.10 could be moved tight against the center pair and still leave the same weight of beads on either side. Thus the median is not affected by extremely large or small values in the data set.

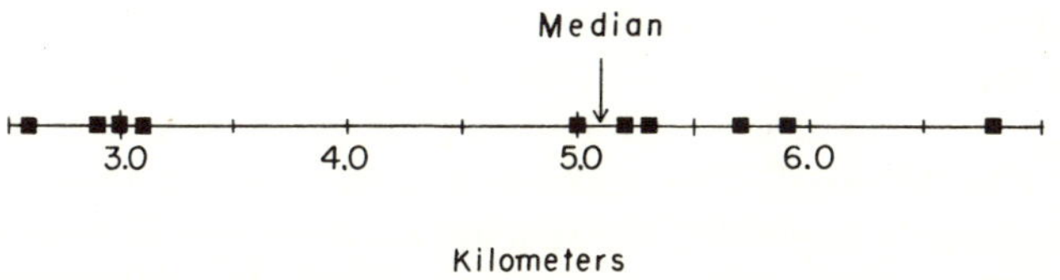

Figure 2.10 Physical interpretation of median for shoreline data of Mirage Lakes (see Table 2.6) as point on rod and bead model having equal weight of beads on either side.

The median can also be interpreted in terms of a histogram. Aside from the approximation involved in grouping, the median is the foot of a vertical line that divides the histogram into two parts of equal *area*.

Mean. The most commonly used measure of central tendency is one that *is* affected by the spacing of the beads in our physical model. This is the average obtained by *adding up the values for the data items and then dividing this sum be the number of data items.* The technical name for this type of average is *arithmetic mean,* but the word "arithmetic" is usually omitted. The reason that we mention the "arithmetic" part is that more exotic kinds of "means" do exist. Unless otherwise state, however, the word "mean" will imply the arithmetic mean.

The formula for calculating a mean is

$$\text{mean} = \frac{\Sigma X}{N}$$

In this formula Σ indicates summation of the items in the list, X stands for any individual data item in the list, and N stands for the number of data items in the list. The mean value for the area in the set of Mirage Lakes is

$$\text{mean} = \frac{210 + 202 + \cdots + 240}{10} = \frac{1118}{10} = 111.8 \text{ ha}$$

which is larger than the median (80.5 ha). The mean value for the shoreline length is 4.55 km, which is smaller than the median (5.1 km).

The physical interpretation of the mean in terms of a rod and bead model is that the mean would mark the balance point of the shish kebab if we could actually find a weightless rod. A diagram of this is shown in Figure 2.11 for the shoreline example. Even though the rod may not be weightless, a little experimenting with a light rod and a couple of heavier weights will demonstrate that the balance point does change as the weights are moved along the rod. Moving a weight in one direction will move the balance point in the same direction.

The conditions under which the mean and median will be the same can also be demonstrated with such a device. Take a light rod and balance it at the center, perhaps by tying it on a string. Now add equal numbers of identical weights on both sides. The equal numbers of identical weights make the original balance point correspond to a median. The rod will remain in balance as long as the spacing of the weights on the left side is a mirror image of the spacing on the right side. As long as this symmetry exists, the mean and median will be identical. The mean and median being identical, however, does not necessarily imply that the spacing is symmetric. A little more experimentation will show that a long movement of a single weight can be balanced by shorter movements of several weights on the other side. In terms of a histogram, the mean is the balance point along the horizontal axis.

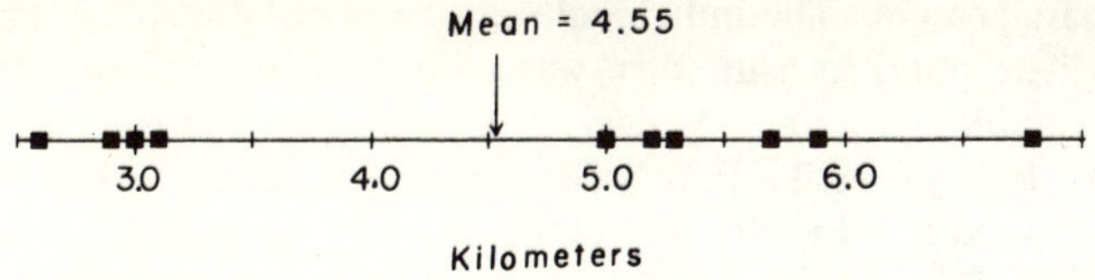

Figure 2.11 Physical interpretation of mean for shoreline data of Mirage Lakes (see Table 2.6) as balance point of beads on a weightless rod.

Midrange. Two other measures of central tendency will also be mentioned even though they have less utility than either the mean or the median. One of these is the *midrange*, which is the average or halfway point between the highest and lowest values in the data set. Thus the midrange for the lake area in the example data is 141 ha, and the midrange for the shoreline length is 4.7 km. In terms of a rod and bead model, the midrange is the point that lies halfway between the two beads that are at opposite ends. As long as these two beads lie the same distance from the median, the median and the midrange are the same. The difficulty with the midrange as a measure of central tendency is that it considers only the two extreme values and ignores everything else.

Mode. The other measure of central tendency that one may encounter with some regularity is the *mode*. The mode is the *value that occurs most frequently*, and the number of times that it occurs is the *modal frequency*. A mode cannot even be determined unless some value occurs more than once, which makes it inappropriate for the example of the Mirage Lakes unless we use the regrouped data from the frequency table (Table 2.7). The *modal class* (class that has the most items) in Table 2.7 is that covering the interval from 5 to 6 km. It is usually satisfactory to use the midpoint of the modal class as an approximation to the mode with grouped data, although this can be adjusted according to the relative frequencies in the neighboring classes. It is possible to have multiple modes if several different values occur with equal frequency.

Statistical Expressions of Variability. Another characteristic of any data set that cannot be ignored is the amount of *variability* that it contains. Two data sets can have the same mean, median, and midrange, yet be vastly different with regard to spread or variability. Anyone who wants to minimize the importance of variability should consider the choice between two cans of mushrooms, both of average quality, when the container is taken as a whole. If one is composed of uniformly mediocre specimens from a culinary standpoint while the other contains a mixture of delectable and poisonous specimens, there will be little argument about the importance of variability.

Although not quite so extreme, the situation is similar with respect to ecosystem management. The individual units in populations that exhibit little variability will respond in a uniform way to treatment or disturbance, which makes that population relatively easy to manage. In contrast, a population that exhibits wide diversity may well contain individuals that respond in diametrically opposite ways to the same kind of treatment. If undetected through too much emphasis on the average, the presence of these unusual units can spell disaster for the manager almost as effectively as the poisonous mushrooms. If one has any thoughts of sampling a population, the greater difficulty of getting a good cross section in a diverse population should be apparent. In statistical jargon, expressions that reflect variability in a population are called *measures of dispersion.*

Range. The first thought that usually comes to mind is to use the difference between the greatest and the least values in the data set to gauge variability. This difference is called the *range*. The range in size of the Mirage Lakes when measured in terms of surface area is 198 ha. Likewise the range in shoreline lengths is 4.2 km. The range reflects the extremes but conveys little information about the pattern of variability in the data set as a whole. One aberrant individual is sufficient to alter the range drastically, even though the population may contain many thousands of individuals. It is important to be aware of the extremes, but it is unwise to ignore the rest so completely.

Quartiles, Interquartile Ranges, and Percentiles. One way of taking into account more of the items is to use something less than the complete range as a measure of variability. For instance, one might choose the range that encompasses the central half of the items. This is the *interquartile range.* The name arises from the necessity of first finding *quartiles,* which are the three points that divide the ordered list into four parts such that one quarter of the data items lie in each part. Since the median breaks the ordered list in half, the median is also one of the quartiles. The first quartile breaks the lower half in half again to form quarters; the median is the second quartile separating the two halves; and the third quartile breaks the upper half into two quarters. The interquartile range is calculated by subtracting the first quartile point from the third quartile point.

The interquartile range, then, covers the half of the items lying in the second and third (or middle) quarters. For the example of the lake data there are five data cases on either side of the median. The middle one of the five on each side marks the quartile point. Therefore the interquartile range for the shoreline data is $5.7 - 3.0 = 2.7$ km. This is shown in the form of a rod and bead model in Figure 2.12. Note that one quarter of the total weight of the beads falls in each quartile segment.

It is also worth noting at this point that the idea behind quartiles can be extended readily. The points that divide the ordered list into five parts with one fifth of the data items in each part are called *quintiles*; the points that make a ten part division are called *deciles*; and the points that make a one

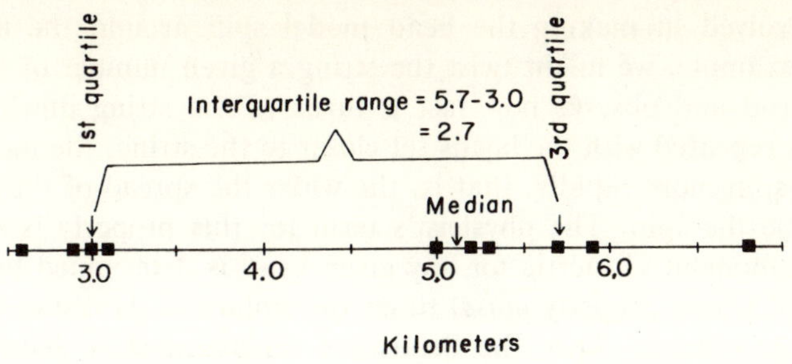

Figure 2.12 Rod and bead model of shoreline lengths for Mirage Lakes (see Table 2.6) showing interquartile range.

hundred part division are called *percentiles.* Thus the median is the second quartile, the fifth decile, and the 50th percentile. Likewise the first quartile is the same as the 25th percentile, and the third quartile is the same as the 75th percentile.

Variance, Standard Deviation, and Coefficient of Variation. Although the interquartile range gives a better picture of overall variability than the total range, it still leaves a lot to be desired. We can mix around the data cases between the first and third quartiles in any way we like without affecting the interquartile range at all. Alterations above the third quartile or below the first quartile will also go undetected. What we really need is an expression of variability that is sensitive to a change in any individual item. Such a measure is available, and its statistical name is *variance.*

The variance can also be interpreted in terms of a rod and bead model, but it reflects a less familiar physical property than the other expressions discussed so far. Imagine the bead model suspended from a string tied at its balance point (mean) as in Figure 2.13. The variance is a measure of the

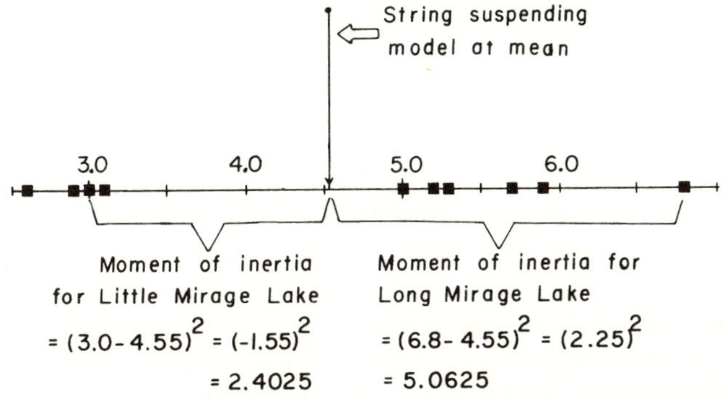

Figure 2.13 Rod and bead model showing moments of inertia used in determining variance for shoreline data of Mirage Lakes (see Table 2.6).

difficulty involved in making the bead model spin around the axis of the string. For example, we might twist the string a given number of turns, then release the rod and observe how fast it turns as the string unwinds. If the procedure is repeated with the beads set closer to the string, the model will be observed to spin more rapidly, that is, the wider the spread of the beads, the slower will be the spin. The physicist's term for this property is *moment of inertia.* The moment of inertia for any given bead is determined by multiplying its weight (more properly *mass*) times the square of its distance from the balance point (mean). Since the beads are all assumed to have the same weight, we can take the weight of each bead as being 1. With this simplification variance is the average moment of inertia for the set of beads that represent the data items (see Figure 2.13) in the population.

To calculate the variance of a population, first determine the distance between each data item and the mean by subtraction. Next square each of these differences. Then average the squares. Stated as a formula this is

$$\text{variance} = \frac{\Sigma(X - \text{mean})^2}{N}$$

where Σ indicates summation of the squared differences, X stands for any individual data item in the list, mean is the arithmetic mean for the population listed, and N is the number of data items in the population or list.

Thus the variance of the surface area for the population of the Mirage Lakes is

$$\text{variance} = \frac{(210 - 111.8)^2 + (202 - 111.8)^2 + \cdots + (240 - 111.8)^2}{10}$$

$$= \frac{98.2^2 + 90.2^2 + \cdots + 128.2^2}{10}$$

$$= \frac{9643.24 + 8136.04 + \cdots + 16{,}435.24}{10}$$

$$= \frac{53{,}927.6}{10} = 5392.76 \text{ sq ha}$$

and the variance of the shoreline lengths is 2.0425 sq km.

Notice that the squaring process involved in finding the variance also makes the units of measure attached to the variance come out squared. When one squares a unit such as hectares which is already square in the sense that it measures an area, the result is a little hard to visualize. This is one reason that most practitioners prefer to use the square root of the variance, called the *standard deviation*, as the statistical expression for variability. Standard deviations have the same units of measure as the original data items. Thus the standard deviation for the lake surface area is 73.44 ha, and the standard deviation for the shoreline length is 1.43 km.

Another thing to notice is that changing the units of measure also changes both the variance and the standard deviation. For instance, the standard deviation of the shoreline length as measured in miles is 0.89 mi. For some purposes it is more convenient to have an expression for variability which is independent of the units of measure. This can be obtained by expressing the standard deviation as a percentage of the mean so that the units cancel out. This percentage is called the *coefficient of variation*. For the shoreline data, then, the coefficient of variation is 31.43%. For completeness it should be noted that the coefficient of variation will not be independent of the units of measure if a shift in the zero point takes place, as it occurs between Farenheit and centigrade (or Celsius) temperature scales.

Degrees of Freedom. The method of arriving at the variance as described above requires a little modification when the data set is only a sample instead of a complete census. Instead of simply averaging the moments of inertia for the data items, the sum of the moments (squared differences from the mean) is divided by *one less than the number of data items.* The number of data items less 1 is called the *degrees of freedom* for estimating the variance. The use of the degrees of freedom provides a correction for the bias which counteracts a tendency to underestimate the variance from samples.

Statistical Expressions for Relationships Between Variables. The use of scatter diagrams to show relationships between variables was introduced in Section 2.6.3, and the relationship between surface area and shoreline length for the Mirage Lakes was plotted in Figure 2.8. The trend of this relationship is upward, but the line is definitely not straight. Unfortunately, however, the simpler statistical approaches to describing relationships are based on the assumption of a straight line type of relationship. A preliminary step, therefore, is to cast about for some logical way of transforming the curved trend into a straight line trend. The emphasis in the foregoing statement is on the word *logical*, since little is to be gained by finagling with the data in ways that cannot be explained in terms of factors underlying the relationship.

The geometric relationship between surface area and shoreline length in either circular or square water bodies provides a logical basis for transforming the example data into a more linear form. In both circular and square water bodies the surface area is proportional to the square of the shoreline length. A reasonable approach then is to try plotting the surface area against the square of the shoreline length in a scatter diagram. This plot is shown in Figure 2.14. The plot of the surface area against the squared shoreline length does appear to be more linear than the one with actual shoreline lengths in Figure 2.8. However, there are still two lakes noticeably above the line and two below which tend to make the line appear curved. Before we try further transformations, an examination of these unusual cases is in order.

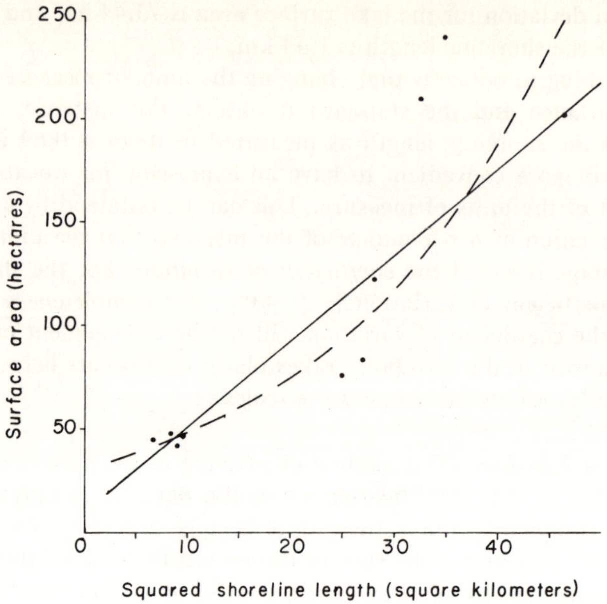

Figure 2.14 *X–Y* plot or scatter diagram of surface area verses square of shoreline length for Mirage Lakes, including freehand curves.

The two points above the general trend in Figure 2.14 represent South Mirage Lake and Big Mirage Lake. These two lakes are more nearly circular than most others, which gives them unusually large surface areas relative to their shoreline lengths. In hydrological terms, these lakes have a low *shoreline development index.* The shoreline development index is the ratio of the actual shoreline length to the shoreline length of a circular lake having the same area. Round Mirage Lake also has a low shoreline development index, but its small size tends to make this less obvious. The two points below the trend line in Figure 2.14 represent Mirage Bay Lake and Middle Mirage Lake. Both of these lakes have much larger shoreline development indexes than the other lakes. Thus we can safely conclude that the lakes which make the trend in Figure 2.14 appear somewhat nonlinear are actually atypical and should not cause concern about linearity. A different pattern of shoreline development indexes could have given the impression of downward curvature instead of upward curvature.

Correlation Coefficient. When satisfied that the relationship is roughly linear, one can proceed to use a statistical expression called the *correlation coefficient* to describe the strength of the relationship. The correlation coefficient can range between $+1$ and -1. A perfect upward sloping relationship with all data points falling exactly on the trend line has a correlation coefficient of $+1$. A perfect downward sloping relationship with all data points falling exactly on the trend line has a correlation coefficient of −1. A correlation coefficient of 0 indicates no apparent relationship between the variables. Correlation coefficients between 0 and +1 indicate that the two variables increase together (upward sloping), but the relationship is less than perfect in the sense that points do not all fall exactly on the trend line. Likewise, correlation coefficients between 0 and −1 indicate that one variable decreases as the other increases (downward sloping), but the relationship is less than perfect. The closer the correlation coefficient gets to either +1 or −1, the stronger the relationship with points clustering more closely about a straight line.

Interpretation of the correlation coefficients does require a little caution. The existence of a strong correlation does not necessarily imply a *cause and effect* relationship between the two variables. It is quite possible that both variables may be controlled by a third factor or a complex of factors which creates the correlation as a secondary effect.

Covariance. Three components go into the calculation of the correlation coefficient. Two of these are the standard deviations for the respective variables, which have already been discussed. The third component is the *covariance* between the two variables. The covariance is the average product of the differences between the variables and their means for the respective data items. For each data item find the difference between the value of the first variable and its mean, then the difference between the second variable and its mean. Next find the product of these two differences for each data item. Then add up these products. Finally divide this sum of products by the number of data points to complete the calculation of the covariance if the data set is a complete census, or by the degrees of freedom if the data set is a sample.

The product of the differences (often called cross product) for Mirage Bay Lake is shown in terms of a rod and bead model in Figure 2.15. Note that the signs of the differences are included in calculating cross products. If both variables tend to be on the same side of their respective means, the cross products will be large and positive, thus averaging out to give a large positive covariance. If the variables tend to be on opposite sides of their respective means, the cross products will be numerically large but negative, thus averaging out to give a large negative covariance. If there is no consistent pattern in the way the two variables deviate from their means, then positive cross products will approximately equal negative cross products, giving a covariance close to zero. For the Mirage Lakes example the covariance between the surface area and the square of the shoreline length is 827.78 (sq km)(ha).

When the covariance has been calculated, the correlation coefficient is obtained by dividing the covariance by the product of the standard deviations for the two variables. The standard deviation for the square of the shoreline length is 12.91 sq km, and the standard deviation for the surface area is 73.44 ha. Therefore the correlation coefficient between surface area and squared

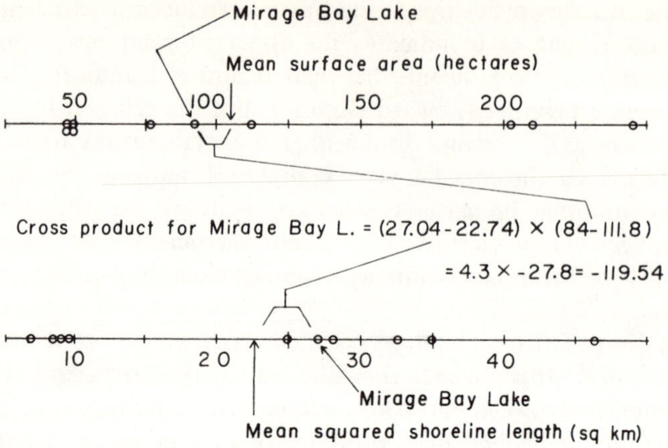

Figure 2.15 Rod and bead model showing product of differences from means for Mirage Bay Lake. Upper rod shows surface area (ha). Lower rod shows square of shoreline length (sq km).

shoreline length is 0.87, which indicates a fairly strong relationship. Note also that the units cancel out, making the correlation coefficient a dimensionless number. Thus the same correlation coefficient will be obtained regardless of the units of measure.

Neither simple correlation coefficients nor scatter diagrams are appropriate for working with relationships involving more than two variables. *Nomograms* (also called *alignment charts*) provide a graphical approach for relationships involving more than two variables. *Multiple regression* and *multiple correlation* provide statistical approaches to relationships involving more than two variables. These topics are covered in the Appendix.

Confidence Intervals. *Parameter* is the technical term for a statistical expression that summarizes a particular aspect of the information about a *population*, such as a measure of central tendency. Technically the term *statistic* is reserved for the companion expression determined from a *sample.* In general every population parameter has a companion sample statistic which serves as an estimator of that parameter. The important difference between a parameter and a statistic is that the parameter is a characteristic of the population which will be the same from one complete census to another as long as the censuses are taken in the same way and the population does not change, but chance variation in the composition of samples will make the computed value of a statistic vary from one sample to the next.

Because of this variation between samples, very few samples will yield estimates that are exactly the same as the population parameter. Thus reporting a single number as a sample based estimated of a population parameter tends to give an unwarranted appearance of accuracy. It is much more informative to provide the user with a low estimate and a high estimate which should bracket the true value of the population parameter, unless a rather unusual sample has been obtained. Furthermore, "should bracket" and "rather unusual" are just weasel words unless they are quantified.

These words can be quantified by stating the percentage of similar samples in which high and low estimates computed by a given procedure will bracket the true value. The high and low estimates form the ends of a *confidence interval,* and the percentage of samples in which the interval is expected to contain the true value is called the *confidence percentage.* This is the preferred form for reporting statistical summaries from sample data, and methods of determining confidence limits (ends of the interval) are presented in the Appendix. The user must understand, however, that there is no quarantee that the interval contains the true value for any given sample. If one uses a 95% confidence percentage, they can be just as certain that the true value lies outside the limits in 5% of the possible samples.

2.7 MAINTENANCE OF THE SYSTEM

The old saying "a stitch in time saves nine" may be trite, but it fits the situation regarding survey information systems quite nicely. No such system can continue to serve the needs of its users very long unless it is conscientiously maintained; and preventive maintenance is far less expensive than the crisis oriented alternative. In closing this chapter we review three important aspects of system maintenance; documentation, equipment, and aging of information. Changes in users' needs for information are also considered.

2.7.1 Documentation

The importance of having good documentation and keeping it current is one aspect of system maintenance that bears strong emphasis. Technical descriptions of system structure, catalogs of data bank contents, handbooks of data collection and coding procedures, users' manuals, and logbooks of system usage are key items of documentation. It is virtually impossible for one person to remember all the technical details regarding the structure and function of even the simpler systems. Furthermore it is inevitable that the personnel charged with operating the system will change over time as present employees move on to other jobs or retire. Though it is to be avoided whenever possible, replacements may not arrive until after their predecessors have departed. Having people who do not know exactly what they are doing tinker with the internal workings of a system is a sure route to chaos, so the importance of complete technical descriptions should be apparent.

Much the same sort of thing applies to catalogs of contents in data banks. Without records of the source of information, extent of coverage, details of how the information was obtained, and specifics of data formats, the information in the system becomes virtually useless. Handbooks of data collection and coding procedures also help to answer questions of this type as well as to reduce the cost and effort involved in updating old information or adding new information to the system.

Clearly written and up-to-date user's manuals can save countless hours of consulting, not to mention dissatisfaction on the part of the users who misunderstand or misinterpret information they obtain from the system.

Likewise logbooks of system usage are essential for justifying expenses to sponsors of the system and/or billing users. These logbooks assume special significance when users are allowed to remove maps, photographs, equipment, or other materials from the central site. It is frightfully easy to lose track of such materials under these circumstances, and locating them for other users can be a major source of headaches.

With a fully computerized system it is tempting to make the documentation part of the data bank itself so that minor alterations can be made easily without retyping major portions. There is nothing wrong with this approach, but backup copies must be made after each alteration as a precaution against loss of information through system malfunctions. This applies equally well to the entire contents of the data bank. Backup files in the form of magnetic tapes or removable disk packs provide the only real insurance against having relatively minor system malfunctions turn into major catastrophes because of lost information.

2.7.2 Equipment

The problem of equipment wearing out or becoming obsolete is one that must be faced in any system, and analytical techniques tend to become obsolete in much the same way as equipment since research is continually producing improved methods. Wear and tear on equipment can be minimized by proper use and good care, and the best means to this end is well trained personnel.

Users or sponsors will settle for second best only when there is a commensurate cost savings involved. Newer equipment and methods, however, tend to be more economical in addition to having greater accuracy and processing capacity. The net result is that users will soon abandon a system that does not stay abreast of the state of the art.

It frequently happens that changes are dictated by external forces, with computer systems being a prime example. Some survey systems are large enough to justify a private computer, but most are not. When an external computer service center makes changes in its processing system, there is no choice but to adapt to the changes. In some cases this adaptation may even require major revamping of the survey information system, particularly in the storage and processing components. The best that can be done in these circumstances is to anticipate such changes and not bind the system too closely to special features available on only one piece of computational equipment. Users should be alerted to impending changes since a more or less extended period is usually necessary to work the bugs out of the alteration.

2.7.3 Aging of Information and Changing User Needs

Aging of information provides the most obvious motive for updating, and one that is of great concern to users. If one is willing to measure change on geologic time scales, there is no such thing as a static component of the ecosystem. The human planning horizon usually is not quite that broad, however, and the relevant question is how rapidly the combination of human activities and natural processes induces changes of interest to users. This depends on both the type of information and the nature of the user, which makes it a little difficult to generalize. As with so many other aspects of survey information systems, users constitute the best barometer of need for updating. This again underscores the importance of close liaison with the user community.

For environmental features like meteorology, hydrology, and vegetation which are subject to rapid change and require detailed measurements, the best approach may be to establish a network of permanent sampling locations which are instrumented for continuous recording or revisited at regular intervals. Remeasurement at permanent sample locations gives excellent data on change, but there is also a possibility that the permanent sample locations may become less representative of the area as a whole with the passage of time. There are several variations on the permanent network theme which help to detect such drift. One is to change the location of part of the sampling stations at regular intervals. This is called *sampling with partial replacement.* Another is to supplement the permanent network with an additional set of temporary sampling locations. Either of these two approaches can also help to overcome high costs of maintaining permanent stations or to obtain additional data in areas known to be undergoing unusually rapid change. Still another way to reduce the costs of updating is to do partial resurveys in order to develop correction factors for the more complete data from an earlier survey.

The decay of information present in the system is only half the updating problem. It must also be anticipated that users will develop needs for information that does not currently reside in the system. In this respect a survey information system is like the national economy—lack of expansion is a sign of stagnation. New surveys are expensive, however, and prudence dictates a close check with other ongoing survey systems to explore the possibility of sharing costs in obtaining the additional data.

2.8 SUGGESTIONS FOR INDIVIDUAL STUDY

We hope that the reader will attempt to reinforce the text material wherever possible by applying the techniques to simple situations which hold personal interest. To encourage such individualized study, in Part I we offer a few suggestions for projects based on materials at hand.

First, select some activity that you plan for the near future and include an objectives-needs analysis in the planning for the activity. The process can be adapted readily to include needs for things other than information. A needs-methods analysis can then be used to decide on the best way of satisfying the needs.

A second suggestion is to select some personal collection of information or objects such as books, maps, or photographs and design a system for handling both the current collection and additions to the collection. Try to relate your system to the diagram of an information system in Figure 2.1.

Third, select some population of things readily available to you on which at least two measurements can be made for each item in the population. Work up tabular, graphical, and statistical descriptions of the population as described in this chapter. It will be helpful to refer to the Appendix during the course of this project.

There are many other possibilities for similar projects which will occur to the alert reader. Performing such projects will prove immensely helpful in gaining a feel for the material in this book and learning how to adapt the techniques to special purposes.

2.9 BIBLIOGRAPHY

American Society of Photogrammetry. *Manual of Remote Sensing,* Vols. 1;2. Falls Church, Va.: American Society of Photogrammetry, 1975.

Boyle, A., H. Calkins, C. Fry, D. Marble, R. Shelton, and R. Tomlinson. *Information Systems for State Land Use Planning.* A Report of the International Geographical Union Commission on Geographical Data Sensing and Processing prepared for Argonne National Laboratories. March 1975 draft.

Cardenas, A. *Data Base Management Systems.* Boston: Allyn and Bacon, 1979.

Cougar, J., and F. McFadden. *Introduction to Computer Based Information Systems.* New York: Wiley, 1975.

Davis, J., and M. McCullagh. *Display and Analysis of Spatial Data.* New York: Wiley-Interscience, 1975.

Dougenik, J., and D. Sheehan. *SYMAP User's Reference Manual.* Cambridge, Mass.: Laboratory for Computer Graphics and Spatial Analysis, Graduate School of Design, Harvard University, 1975.

Freese, F. *Elementary Forest Sampling.* Washington, D.C.: U.S. Department of Agriculture Handbook 232, 1962.

Greenhood, D. *Mapping.* Chicago: University of Chicago Press, 1964.

Haseman, W., and A. Whinston. *Introduction to Data Management.* Homewood, Ill.: Irwin, 1977.

Heaslip, G. *Environmental Data Handling.* New York: Wiley-Interscience, 1975.

I.N.S.-E.E. *Data Banks for Development. Proceedings of the International Expert Meeting, Saint-Maximin, France, May 24–28, 1971.* 10 rue Léon Paulet, 13 Marseille 8, France: Observatoire Economique Méditerranéen, 1971.

Kroenke, D. *DATABASE: A Professional's Primer.* Chicago: Science Research Associates, 1978.

Lund, H. G. "So We Know What We Have—But Where Is It?" In T. Cunia, Ed. *Proceedings of Symposium on Monitoring Forest Environment Through Successive Sampling.* Syracuse, N.Y.: State University of New York, International Union of Forest Research Organizations, and Society of American Foresters, 1974.

Meadow, C. *The Analysis of Information Systems.* New York: Wiley, 1967.

MITRE Corporation. "Information Sources." In *Resource and Land Information Program: Complementary Federal Data Programs and User Charge Policies.* Washington, D.C.: Report to the U.S. Geological Survey, MTR-6539, 1973. Sec IV, pp. 45-63.

Mittman, B., and L. Borman. *Personalized Data Base Systems.* Los Angeles: Melville, 1975.

Monkhouse, F., and H. Wilkinson. *Maps and Diagrams.* London: Methuen, 1971.

Peucker, T. *Computer Cartography.* Washington, D.C.: Association of American Geographers, Resource Paper 17, 1972.

Robinson, A., and R. Sale. *Elements of Cartography,* 3rd ed. New York: Wiley, 1969.

Shelton, R. "Data Sources and Characteristics." In *Information/Data Handling: A Guidebook for Development of State Programs.* Washington, D.C.: U.S. Department of the Interior, Office of Land Use and Water Planning; and U.S. Geological Survey, Resource and Land Investigation Program, July 1975. Draft copy, pp. 37-55.

Shelton, R., and E. Hardy. "Design Concepts for Land Use and Natural Resource Inventories and Information Systems." In *Proceedings of the Ninth International Symposium on Remote Sensing Information and Analysis,* Vol. 1. Ann Arbor, Mich.: Willow Run Laboratories, Environmental Research Institute of Michigan, 1974. pp. 517-535.

Slonim, M. *Sampling.* New York: Simon & Shuster, 1960.

Spiegel, M. *Theory and Problems of Statistics.* New York: McGraw-Hill, Schaum's Outline Series, 1961.

Stone, R., and K. Ware, Eds. *Proceedings of a Workshop on Computer and Information Systems in Resource Management Decisions.* Washington, D.C.: U.S. Department of Agriculture, Forest Service, Cooperative State Research Service, 1971.

Strandberg, C. *Aerial Discovery Manual.* New York: Wiley, 1967.

Tanur, J., F. Mosteller, W. Kruskal, et al., Eds. *Statistics: A Guide to the Unknown.* San Francisco: Holden-Day, 1972.

Tomlinson, R., Ed. *Geographic Data Handling, Symposium Edition. Proceedings of the UNESCO/IGU Second Symposium on Geographical Information Systems.* Ottawa, Ont.: International Geographical Union, Commission on Data Sensing and Processing, 1972.

U.S. Army. *Map Reading.* Washington, D.C.: U.S. Department of Defense, FM 21-26, 1961.

Wessel, A. *Computer-Aided Information Retrieval.* Los Angeles: Melville, 1975.

CHAPTER 3–BASIC MEASUREMENTS

This chapter covers three aspects common to most surveys. The first is the nature of numerical data. The second concerns techniques for locating points in the field and describing their geographic positions. The third is that of identifying features and measuring their physical dimensions.

Most results of survey measurements either are or can be expressed in numerical form. However, the information content of numerical data varies considerably, and one must be careful not to read too much into the simple fact that a number has been recorded. A brief look at the nature of numerical data serves as a prelude to the more pragmatic aspects of location and measurement.

The need to locate features or data collection points in the field is common to surveys of all ecosystem components. Rather obviously, one must first find a feature before it can be measured. Beyond this simple fact, the importance of recording location as an item of data was discussed in Section 2.4.3. Techniques of locating points in the field and describing the geographic position comprise a major portion of this chapter. One will likely find this material rather dry if one approaches it in a passive fashion as something to be read and memorized. However, such most certainly need not be the case. The challenge involved in actually getting out into the field and skillfully navigating unfamiliar and especially rugged terrain has sparked the rapidly growing sport of *orienteering*. There is a great deal of satisfaction in the ability to know your map position at all times without having to rely on manmade landmarks. The possession of this ability can transform wildlands from inhospitable places where one might easily become lost into scenic and fascinating realms of nature that beckon to be explored first hand. By all means, then, take these techniques along with a compass and map (or make your own map as you go) into the great outdoors and learn them by exploring and mapping an area that holds personal interest—whether it be your backyard, nearby hiking trails, a wilderness area, or an underground world of caves.

A companion problem to that of orienting and locating a field position is the need to measure the physical dimensions of features. This constitutes the other broad subject area of the chapter, but the two topics of orientation/location and dimensional measurement are interwoven to show that they are really just different applications of a common set of geometric principles. Furthermore we prefer to carry the several methods of achieving the same end, such as photo measurements and ground measurements, along together as much as possible to bring out their parallelism. Hopefully this will make it easier for the reader to see the comparative advantages and disadvantages of the alternative approaches to a survey problem. In so doing we also hope to emphasize a fact that is often hard for the neophyte to appreciate, namely, that there is usually no one way of doing a particular survey job which is best under all circumstances. The conditions surrounding the application must be evaluated to determine which way is most appropriate.

In the presentation of this chapter we assume that the reader has access to or will purchase the following equipment; rulers, protractors, tape measures, hand held compass for orienteering, sighting level, clinometer, and pocket stereoscope. Rulers, protractors, and tape measures are so readily available in bookstores and hardware stores that these items should present no problems. An inexpensive orienteering compass suitable for learning purposes can be obtained from any Boy Scout or Girl Scout supplier. Simple hand held sighting levels can be obtained from such ready sources as building suppliers or a Sears catalog. Clinometers are a little more expensive and difficult to obtain. They can be purchased from forestry and engineering suppliers, but the techniques we will describe can be learned almost as readily by attaching a weighted string to a protractor. There is no such homemade substitute for a pocket stereoscope. If a pocket stereoscope is not already available, one should seek out a catalog for forestry suppliers, engineering suppliers, or airphoto equipment suppliers and put in an order. An inexpensive plastic pocket stereoscope will serve most needs of the beginner.

Procedures for using the above named equipment are described in step-by-step laboratory manual fashion in this chapter since this equipment is basic to virtually all surveys. The reader may feel that we should have included photographs of the equipment. However, there are so many minor variations in design that the photos would be unlikely to match the specific piece acquired by the reader. Therefore such photos would be of little instructional value and have not been included. More sophisticated, expensive, or less easily available equipment is mostly described in rather general terms without step-by-step instructions for use.

3.1 NATURE OF NUMERICAL DATA

Numbers can be used in several ways. They can be used as substitutes for qualitative names; as positions on a rating scale which could just as easily be set up in terms of letters of the alphabet; or to record more quantitative determinations such as length and weight. Even length and weight can be measured accurately or inaccurately. It is patently absurd to record the average of a series of single digit numbers to eight decimal places, even though a computer may print it out that way.

3.1.1 Numbers as Labels

Numbers carry the least amount of information when they are used as substitutes for names. Numbers are used in this way mainly as a convenience in data processing and storage. Numbers are often shorter to record than names and are easier to handle than alphabetic data with several of the more common "languages" for writing computer programs. The number may serve to identify an individual unit, as for example the use of the social security number to index a person's record. Or the number may serve to identify a category of individuals such as male or female. The technical term for numbers used to replace names is *nominal data.*

When numbers are used as substitutes for labels, the kinds of data processing operations that can logically be performed on the data are rather limited. For instance, male might be coded as 1 and female as 2. As long as there is a mixture of males and females, the mean of the sex codes would lie somewhere between 1 and 2. However, such an average is rather meaningless since it would seem to indicate that the typical individual is neither male nor female but something in between, whatever that might be. The only data processing operations that make sense with nominal data are those based on counting. Data processing operations of the counting type are considered briefly in the next section.

3.1.2 Frequencies, Fractions, and Percentages

The basic data processing operations for nominal data are *sorting* on the basis of like codes and *counting* to determine the number of individuals (or frequency) that fall in a coded category. Any other numerical expression of nominal data is secondarily derived from the frequencies. *Fractions* or decimal fractions are obtained by dividing the frequency for a given category by the total frequency. *Percentages* (number of individuals of a given kind per hundred) are obtained by multiplying the fraction of individuals in a given category by 100. Likewise the number per thousand or per million can be computed by multiplying the fraction by 1000 or 1,000,000, respectively. The chemical composition of water and air is often expressed as parts per million. Percentages are often displayed graphically in the form of a "pie chart" in which a circle represents 100% (the whole pie) and the sizes of the "pieces" correspond to the percentages.

The only other statistical summarization (see Section 2.6.4) that can be used with nominal data is the mode, since it is determined solely on the basis of the category with the greatest frequency.

3.1.3 Ratings, Rankings, and Indexes

The next step up in terms of information content is data of the rating, ranking, or ordering type. The important characteristic of rating scales is that they are ordered in the same sense that the alphabet is ordered. We know that B follows A and C follows B, but there is no quantitative basis for comparing the difference between A and B with the difference between B and C. If there is such a quantitative basis for the comparison of steps on the scale, then there is no consistent pattern in the size of successive steps along the scale. In other words each point on the scale is defined in terms of a different arbitrary standard.

Geology furnishes a prime example of such a rating scale in Moh's scale of mineral hardness. A mineral with a given hardness on this scale will scratch a mineral with a lower hardness rating but will not scratch one with a higher hardness rating. There are ten points on Moh's scale ranging from talc as the standard for a rating of 1 to diamond as the standard for a rating of 10. The steps along this scale, however, are not of equal size and do not form a mathematical sequence of change in step size. A second example of a rating scale is the Beaufort scale of wind speed as judged by its effect on smoke, trees, boat sails, and so on. Still a third example is the scale of star brightness as expressed in magnitudes. Many other examples can be found in any field of study.

Rating scales often represent early attempts at quantification. Even though subsequent research produces new scales with consistent steps defined in terms of a single standard, the older rating scales often linger either because of familiarity or because they do not require sophisticated instrumentation. The Beaufort wind scale is an example of the case where rating has been superceded by measurement with anemometers. When one does not have an anemometer, however, the Beaufort scale may still come in handy. The intelligence quotient (IQ) is an example of the case where a rating scale has not yet been superceded by a more quantitative system of measurement. The technical name for data based on a rating scale is *ordinal data*.

Because of the increased information content, ordinal data support a greater variety of data processing operations than nominal data. Since the number of data items lying between any two points on the rating scale can be counted, all the count based operations applicable to nominal data are also appropriate for ordinal data. Thus one can compute frequencies, fractions, percentages, and modes. Since one can further determine whether any given data item is "greater than" or "less than" another item, any operation based on ordering can also be applied to ordinal data. In particular percentiles can be used to compare relative frequencies of data items lying in major sections of the scale. Furthermore statistical summaries that are derived from percentiles such as the median and ranges can also be used. An interpretation of minor differences between medians and ranges of data sets is complicated, however, by the unequal scale steps.

Although it is not uncommon to find means and standard deviations reported for ordinal data, the nature of a rating scale makes these statistics inappropriate. The computations underlying means and standard deviations (refer to the rod and bead models of Section 2.6.4) assume a mathematical regularity of the graduations along the scale. This regularity need not exist on a rating scale, so means, variances, standard deviations, coefficients of variation, and correlation coefficients may be rather meaningless and should be viewed very skeptically.

3.1.4 Measurements and Measurement Scales

The next step up in terms of information content brings us to the type of data that have full measurement strength, and therefore support the greatest variety of data processing operations. The intervals or steps along a scale of this type are all defined in terms of a single standard, so that one can meaningfully compare the difference between any pair of data items with the difference between any other pair. Length, weight, volume, and temperature are familiar examples of this type of data. Even within this type of data, however, there are some distinctions which must be made.

One distinction concerns the nature of the zero point on the scale. Variables such as length, weight, and volume have a true zero point. With a true zero point an item measured as zero in terms of one unit will also be zero in any other unit. Thus a weight of zero is still zero whether measured in pounds, ounces, kilograms, or whatever. On the other hand variables such as temperature are treated as having an arbitrary zero point. The zero point of the centigrade or Celsius scale is set at the freezing point of water, whereas that of the Fahrenheit scale is somewhat below the freezing point of water. Actually, there is a true zero of temperature called absolute zero (−273°C) at which molecular motion ceases, but we will disregard it for purposes of the present discussion. In technical terms scales with a true zero point are called *ratio* scales, and those with an arbitrary zero point are called *interval* scales.

All the data processing operations that have been or will be considered in this book can be applied to data from ratio scales. Strictly speaking, however, the coefficient of variation should not be used with data from interval scales. This is because scales with different zero points will give different coefficients of variation for the same set of data. Part of the utility of the coefficient of variation lies in its independence of units of measure. This independence does not hold when the zero point changes between scales. One can demonstrate this by calculating the coefficient of variation for a set of temperature readings, first with the readings expressed in degrees Fahrenheit (°F) and again with the readings expressed in degrees Celsius (°C). As long as the zero point does not shift, as when length is expressed either in meters or feet, the coefficient of variation will not change with a change in units of measure. With variables measured on an interval scale, then, the utility of the coefficient of variation is limited to making comparisons between data sets in which the same unit of measure has been used.

The difference between ratio and interval scales actually goes beyond the effect on the coefficient of variation. As long as scale intervals are uniform and there is no zero shift between systems, a pair of readings will stand in the same ratio to each other, no matter what scale is used. This is not true when a zero shift takes place.

Another distinction to be made is that steps along a scale do not have to be uniform, although they must be defined in terms of a common standard and must form a mathematical sequence. Logarithmic scales are a fairly common example of scales that are not uniform in terms of basic units. Thus the size of soil particles may be expressed as the logarithm of the diameter in millimeters, or on a phi scale which involves a geometric series. Likewise the acidity of a solution is usually measured in terms of pH, which is the negative log of the hydrogen ion concentration in dilute solutions. When a logarithmic scale (base 10) is used, each successive step constitutes a tenfold change from the previous step.

A final point with respect to scales of measurement concerns the use of metric or SI (système international) units versus units of the English system. The metric system has been in common use for scientific studies in the United States for some time. The beginnings of metric instruction in elementary schools will lend impetus to the trend toward use of the international system. The question is no longer whether the metric system will be adopted, but how quickly. However, the land survey records are all in the English system. This fact alone will mean that anyone involved in surveys of ecosystems in the United States must remain conversant with the English system for some time to come. Therefore the two systems will be carried in parallel through this book. Some of the examples will be given in the English system, others in the metric system. Units of measure for specific purposes (length, area, and so on) will be treated in the sections that deal with the respective types of measurement.

3.1.5 Accuracy, Recording, Rounding, and Significant Figures

The information content of measurements also varies according to the standards of accuracy used in taking the measurements. Measurements should be recorded so as to indicate their accuracy clearly. This is the matter of *significant figures*. The rule is that the last digit should be the only one in doubt. Suppose, for instance, that measurements are being taken to the nearest tenth of an inch. Then a recorded measurement of 2.5 in. actually falls somewhere between 2.45 and 2.55 in., since it would have been recorded as 2.4 in. if it fell below this range and as 2.6 in. if it fell above this range. Leading zeros that are used only to locate the decimal point are not significant digits. For instance, only the italicized digits are significant in the following three numbers: 0.0*25*, 0.00*25*, and 0.000*25*. Trailing zeros after a decimal point are significant as in 0.250 where all three digits after the decimal are significant. When a number does not have a fractional part, scientific notation can be used to avoid showing trailing zeros which are not significant. Thus when a measurement of 125 m taken to the nearest meter is converted to millimeters, the result should be written as 1.25×10^5 mm.

Strictly speaking the results of all computations should be rounded back to reflect the significant figures involved. When a list of measurements is added, make sure that the positions of the decimal points are aligned vertically, then mentally drop a line down from the last significant digit of each number in the list. The leftmost line marks the position of the last significant digit in the sum.

For example, in adding

$$\begin{array}{r} 125 \\ +\ \ 3.4 \\ \hline 128.4 \end{array}$$

only the first three digits of the sum are significant. Accordingly, the sum should be rounded to three figures and reported as 128.

The usual rules of rounding are as follows. If the part to be dropped is greater than one-half unit in the preceding digit, add 1 to the preceding digit before dropping the nonsignificant figures. If the part to be dropped is less than one-half unit in the preceding digit, simply drop the nonsignificant figures without altering the preceding digit. If the part to be dropped is exactly one-half unit in the preceding digit, add 1 to the preceding digit only if it is odd.

When two measurements are multiplied, the product only has as many significant digits as the factor with the fewest number of significant digits. In multiplying $125 \times 0.35 = 43.75$, for example, only the first two digits of the product are significant. Accordingly the product should be rounded up and reported as 44.

When one is doing a series of computations on a calculator, however, rounding intermediate results can be very inconvenient because most calculators require that the number be deleted from the register and reentered without the nonsignificant figures. Likewise the rules for rounding are almost impossible to apply in computer programming. Furthermore complicated computations can be very sensitive to rounding. A workable compromise is to round the final answer, but not round intermediate results.

3.2 IDENTIFICATION OF FEATURES

The processes of identification and classification generate nominal data as a direct result, which makes them a form of measurement. They also constitute a necessary preliminary step in collecting more quantitative types of data. Since the identification/classification phase of ecosystem surveys can be both time consuming and an appreciable source of errors, it deserves a little closer scrutiny before we become too engrossed with instrumentation and analytical techniques. We shall look at the problem first in the context of field surveys and then in terms of remote sensing.

3.2.1 Identifying Field Specimens

In most ecosystem surveys it would be economically prohibitive to staff field crews entirely from the ranks of highly trained experts. The usual approach is to rely heavily on college students seeking experience and income during the summer recess. Crew leaders are drawn either from the technical staff or from the more experienced students. The remainder of the crew often consists of students with little or no prior experience in field work. These inexperienced students are given a concentrated training session before being sent into the field. Time limitations, however, make it virtually impossible to bring trainees to the point where they can identify or classify the full range of field specimens on a sight recognition basis. This creates the need for supplementary guides to identification and classification which can be taken into the field and used without the guidance of an expert.

For much the same reasons it would be completely infeasible to attempt even a general coverage of the taxonomy and the classification of ecosystem components in a book such as this. We shall treat identification and classification as being simply another type of determination which must be made. In the next few paragraphs we shall discuss briefly the more commonly used aids for guiding the processes of identification and classification. Further details on the identification and classification of particular ecosystem components can be obtained by consulting the references listed at the ends of the appropriate chapters. These reference lists cover a wide range of technical detail, including materials written primarily for the layman as well as those requiring advanced knowledge of the ecological discipline under consideration.

Regardless of the particular ecosystem component involved, experience has shown that *keys* usually provide the most efficient format for guides to identification and classification. Keys are usually constructed in one of two forms. Elimination or nested keys are the most effective for systematizing the process of identification or classification. Such a key reduces the identification process to a series of well-defined steps. At each step the user is forced to choose one of a limited number of alternatives which reduces the possibilities. The dichotomous key is the most popular form of elimination key. In a dichotomous key the user is offered only two choices at each step.

The utility of an elimination key for inexperienced persons is determined primarily by two factors. One is the extent to which technical terminology is used in phrasing the choices. If the prospective user is likely to be unfamiliar with technical terminology, it must be carefully avoided. The second factor is the degree of subjectivity involved in the alternative choices. Inexperienced users lack mental standards for comparison, so subjective terms like "strongly rounded" are likely to lead them astray. Diagrams and a glossary of terms are both helpful additions to an elimination key. Dichotomous keys are used in Section 4.2.2 for presenting information relative to soil surveys.

The second type of key is the selection or matching key. Keys of this type provide the user with several series of diagrams or photographs accompanied by verbal descriptions from which he or she selects the one that best fits the specimen in question. Actually the difference between an elimination key and a selection key is not really so clearcut. Most selection keys use a few steps of the elimination type to reduce the list for selection to a manageable size. The selection approach is often faster for the user who can narrow the possibilities down to a few without the aid of the key, since it is easier to turn directly to a subsection of the key. The selection approach is also somewhat more natural than the elimination type since it makes more use of specimens in the whole as opposed to homologous parts of diverse specimens.

Regardless of how well the field keys are constructed, there will always be some cases in which uncertainty remains. This is particularly true of things like rocks, minerals, or minute organisms in which laboratory apparatus such as a microscope may be required to make a final determination. Therefore provisions must be made for collection, labeling, and preservation of samples to be returned to the laboratory. Periodic collection of samples is also a good way to check on the accuracy of field determinations so that the need for further training of personnel or improvement of keys can be detected.

3.2.2 Identifying Features on Airphotos

The problems that arise in identifying and classifying features from airphotos have many parallels with the field situation, but there are also some important differences.

The most obvious difference is that the functions of the laboratory and field are essentially reversed. Identification and classification are carried as far as possible in the laboratory on the basis of the photographs. When uncertainties arise which cannot be resolved from the photos, field checks become necessary.

Keys play the same role in photointerpretation that they do in the field. However, the number of characteristics on which keys can be based is greatly reduced. All types of features must be analyzed from the same basic set of characteristics that comprise the photo image. As set forth in several standard texts on airphoto interpretation, the salient characteristics of photo images are shape, size, pattern, texture, tone or color, shadows, and location/association.

The *shapes* of individual objects are probably the most important clues to their identification. Highly regular shapes are characteristic of manmade (cultural) features like buildings, roads, fields, and such. Natural features are usually somewhat irregular in shape since they tend to follow curving boundaries of soil types, moisture patterns, topography, and similar characteristics of the physical environment. However, shape is still an important aid to identification. The crowns of different tree species, for example, often have distinctive shapes when the trees are growing far enough apart to show as individuals on the photo. The shapes of watercourses reflect their gradients and the physical characteristics of the streambeds. Likewise the shapes of other landforms are highly indicative of geologic origins. It must be remembered, however, that objects often appear quite differently from the air than they do on the ground.

The *size* of a feature provides a second major clue to its identity. Size serves to differentiate trails from roads, streams from rivers, ditches from ravines, trees from shrubs, cows from sheep, and so on. The combination of shape and size narrows the possible identity of a feature considerably.

Geometric *patterns* are often characteristic. For instance, orchards or other plantations of trees usually have a regular spacing, whereas natural stands are spaced more or less at random or in clusters. The pattern of contour farming in rolling terrain is unmistakable. The drainage pattern of an area reflects the slope of the terrain and the erodibility of the soil. Many other geological formations such as strata of sedimentary rocks and dune areas also have characteristic patterns.

When clustered objects are too small or too closely spaced to show as individuals, then shape, size, and pattern merge into *texture*. If the cover is composed of large and loosely spaced objects, such as the crown canopy formed by mature trees in a forest, the image will have a rough or coarse texture which might be described by a term such as "cobbled." With small objects which are closely spaced, such as agricultural crops or young trees, the texture will be relatively smooth.

In black and white photos *tone* refers to how light or dark an image appears. It is the black and white analog of color. The tone of an image will depend on the amount of solar radiation reflected by the object. Light substances like sand are reflective and will usually appear in light tones. Organic soils, on the other hand, absorb radiation and so appear in dark tones. With areas such as calm water, which have a mirrorlike surface, the angle of the sun is important in image tone. If the sun's angle is such that the light is reflected directly into the camera lens, the water will be very light in tone. If the camera is not in the path of the reflected rays, the water surface will appear very dark. Image tone also depends on the type of film and filter used. Coniferous forests are relatively hard to differentiate from broadleaf (hardwoods) on the basis of tone in black and white panchromatic photos; however, conifers are much darker than hardwoods in black and white infrared photos.

When *color* films are used, color replaces tone as a clue to image identity. Color depends on the wavelength as well as on the amount of the light reflected. Like tone, the color of an object's image also depends on the film–filter combination used. Some natural color film–filter combinations will highlight certain shades and suppress others. False-color infrared film even shows features in a completely different color scheme than they appear to the eye.

Shadows may also be useful aids to interpretation. For one thing, an object must stand above the surrounding terrain to cast a shadow. Therefore shadows give an indication of relief. Distinctive profiles which would not normally be visible from above are often revealed quite clearly in the shadow of an object. On the negative side, dense shadows obscure the details of the terrain which they cover.

Many mistaken identifications are caused by the failure to observe characteristic *locations* and *associations* among features of a landscape. Swamps are rare on hilltops. Most types of industry are characterized by distinctive layouts and combinations of equipment. The absence of a critical component must cast doubt on the tentative identification. Flora, fauna, and physical features also occur in distinctive associations. A knowledge of the normal composition and topographic placement of natural ecosystems occurring in an area will allow an interpreter to write surprisingly detailed and accurate descriptions from an examination of an airphoto. Likewise it is often possible to learn a great deal about the earth's surface from known associations with natural coverings. Soil types, for instance, are reflected in the vegetative cover; the same applies to geological formations. Information on the nature of lake and stream beds can also be inferred from the conformation of the banks and marginal vegetation.

3.3 ANATOMY OF AIRPHOTOS

As stated at the beginning of this chapter, airphoto survey techniques and field survey techniques will be presented in parallel. To make this possible, the reader must have at least a passing acquaintance with the geometry and terminology of airphotos. The purpose of this section is to provide such a background.

3.3.1 Airphoto Geometry and Terminology

Photos taken with precision aerial cameras are usually printed in a 9 in. square format. This is the actual width of the film. Four so-called *fiducial* or *collimation* marks are located at the edge of this type of photograph. These fiducial marks are used to locate the center of the photograph, which is called the *principal point.*

The principal point is important in working with photographs, and the first step is usually to locate it carefully. The principal point is located at the intersection of the lines connecting opposing fiducial marks, as illustrated in Figure 3.1. Lay a straight-edge across between one pair of fiducial marks and make a slit through the emulsion (but not through the paper or film base) near the center with a single edge razor blade. Move the straight-edge to the other

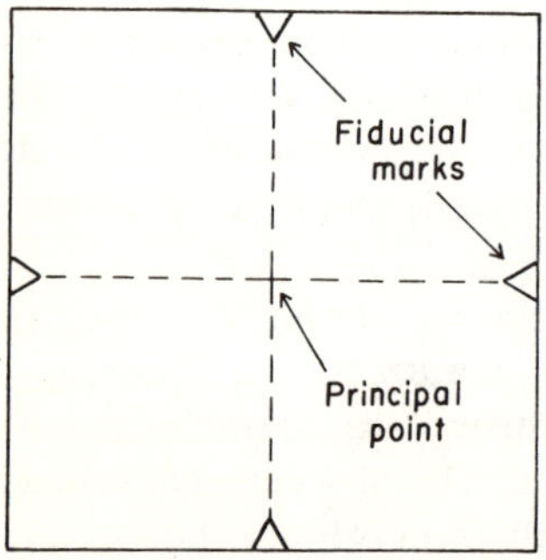

Figure 3.1 Location of principal point from fiducial marks.

pair of fiducial marks and repeat the procedure. The intersection of the two incisions marks the principal point. The principal point may also be circled with a bow pen (pen mounted on one arm of a drafting compass) in colored ink for easy location in later use of the photo.

Loosely speaking, the principal point is where a "bore sight" through the camera lens hits the terrain shown in the picture. More specifically it is where the perpendicular line through the optical center of the camera lens intersects the film plane. It is the optical center of the photograph, and any distortion due to lens defects gets worse as you move away from the principal point. Therefore the best image quality is usually found near the center of the photo.

Some aerographic cameras have the fiducial marks in the corners instead of along the sides. Smaller cameras do not provide fiducial marks. In the absence of fiducial marks a reasonable approximation to the principal point can be obtained by using the corners of the photo as fiducial marks. Corners cannot be used, however, if the photo has been cropped either mechanically or by enlargement.

In a vertical photo (camera pointed straight down), the principal point also marks the base of a plumb line dropped directly down from the center of the camera lens. The foot of this plumb line (or the equivalent image point on the photo) is called the *nadir*. In a vertical photo the principal point is also the nadir.

A photo that is not vertical is called *oblique.* In an oblique photo the camera is tilted away from the vertical. If the camera is tilted so far that the horizon (skyline) shows in the photo, the photo is called a *high oblique.* If the tilt is not sufficient to show the horizon, the photo is called a *low oblique.* The line that passes through both the principal point and the nadir in an oblique photograph is called the *principal line* of the photo.

The coincidence of principal point and nadir makes vertical photos much easier to work with than obliques. For practical purposes a photo is usually considered to be vertical if the camera was not tilted more than 3° away from the vertical when the picture was taken. The amount of tilt in an airphoto taken with a precision camera can be determined by reference to the image of a level bubble which appears along the margin of the film. Other useful items of marginal information usually include a clock face, altimeter dial, date, mission code, roll number, and frame number within the roll.

3.3.2 Scale

Very little information of a numerical nature can be determined from an airphoto until the scale has been determined. As defined in Section 2.6.1, the scale of a photograph or map is the relationship between the actual length of an object on the ground and the length of its image on the photo or equivalent representation on a map. Recall that the scale is commonly expressed as the fractional relationship between the image length as numerator and the object length as denominator, that is,

$$\text{scale fraction} = \frac{\text{length of image}}{\text{length of object}}$$

where both the numerator and the denominator are in the same units so that the units cancel. The scale fraction is expressed in simplified form so that the numerator is unity.

It should be reasonably apparent from the definition that the scale of a particular area on a photograph may be determined by measuring the actual length of an object on the ground and also the length of its corresponding image on the photo. For example, the actual length of a footbridge might be 50.0 ft. If the image of this bridge in a photograph is 0.06 in. long, the scale is

$$\text{scale fraction} = \frac{\text{length of image}}{\text{length of object}} = \frac{0.06\text{ in.}}{50\text{ ft}}$$

$$= \frac{0.06\text{ in.} \times 1\text{ ft}/12\text{ in.}}{50\text{ ft}} = \frac{0.005\text{ ft}}{50\text{ ft}}$$

$$= \frac{0.005}{50} = \frac{(0.005/0.005)}{(50/0.005)}$$

$$= \frac{1}{10{,}000}$$

By following through this simple example, we find that the first step is to identify an object in the photograph, and then measure both the length of its image and its actual length on the ground. If a suitable map is available, the actual length can sometimes be scaled from the map in order to avoid a trip into the field. The next step is to convert both measurements to the same units if this is not already the case. It often helps avoid mistakes to write out the conversions completely as in the example and then cancel units. The final step is to simplify the fraction so that the numerator is 1. This is accomplished by dividing both the numerator and the denominator by the numerator.

A vertical photograph of level terrain has a uniform scale, provided that there are no distortions due to lens defects. This situation is depicted graphically in Figure 3.2 using three objects AA′, BB′, and CC′ of equal length. The uniformity of scale for vertical photos of level terrain can be proved mathematically by a simple analysis of similar triangles in Figure 3.2.

The scale of a vertical photo over level terrain can also be calculated from the focal length of the camera lens and the flying height of the aircraft without mak-

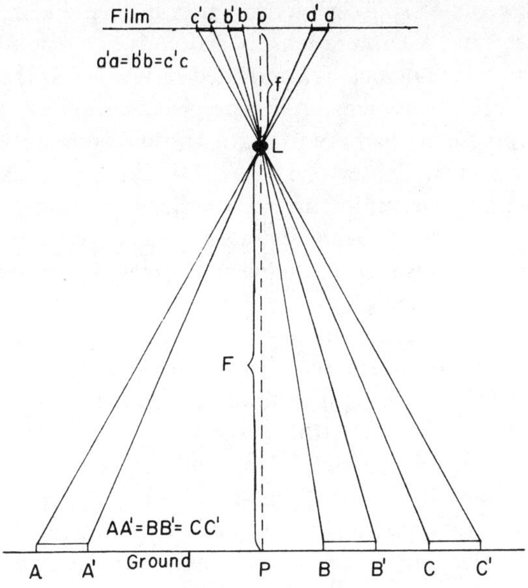

Figure 3.2 Diagrammatic representation of uniform scale property of vertical airphotos over level terrain. Equal ground distances AA ′, BB ′, and CC ′ give equal length images a ′a, b ′c, and c ′c. Solid arrows represent light rays from ends of objects. Dotted line connects principal point p of photo with terrain point P pictured at principal point. *F* is flying height above ground; and *f* is focal length of lens L.

ing any ground or photo measurements. The relationship (again based on similar triangles) is

$$\text{scale fraction} = \frac{\text{focal length}}{\text{flying height}}$$

where focal length and flying height must be expressed in the same units. For example, a vertical photo taken from a flying height of 10,000 ft with a 6 in. focal length lens has the following scale:

$$\text{scale fraction} = \frac{\text{focal length}}{\text{flying height}} = \frac{6\text{ in.}}{10{,}000\text{ ft}}$$

$$= \frac{0.5\text{ ft}}{10{,}000\text{ ft}} = \frac{1}{20{,}000}$$

3.3.3 Distortions

In order to make effective use of airphotos in survey work, one must be aware of distortions and their severity. For present purposes a distortion will be defined as anything that causes the scale to vary within a photograph.

There are three major sources of distortion in airphotos. The first source, consisting of defects in the camera lens, has already been mentioned. The effect of distortions due to lens defects is controlled in two ways, the most obvious of which is to use a high quality lens. When precision aerographic cameras (sometimes called mapping cameras) are used, distortion from lens defects is seldom a problem. Defects in the lenses of less expensive cameras can be a problem, and are likely to limit measurements to a rather low order of accuracy. Since distortions due to lens defects become more serious as one moves away from the principal point, a second way of reducing such errors is to make measurements on the central portion of the photo.

The second source of distortion comes from tilting the camera away from the vertical. The effect of tilt on the scale within the photo is illustrated diagrammatically in Figure 3.3. This figure shows that scale increases in the direction from the principal point toward the nadir, and it decreases in the opposite direction. The scale is uniform only along lines perpendicular to the principal line. The variation in scale due to tilt makes oblique photographs very inconvenient for making survey measurements. Therefore survey measurements are usually made on vertical airphotos. As mentioned earlier, airphotos in which tilt does not exceed 3° from the vertical can be treated as vertical for most purposes. Those who must work with oblique photos should refer to the *Manual of Remote Sensing*, vol. 2 (Reeves, Anson, and Landen, Eds., 1975) for use of perspective grids.

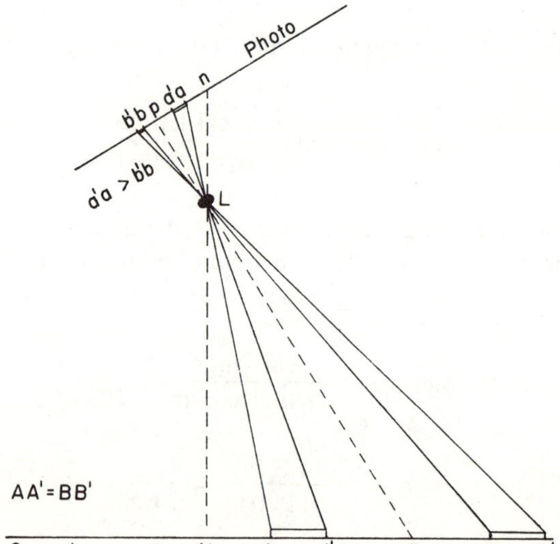

Figure 3.3 Diagrammatic representation of varying scale due to tilt. AA ′ and BB ′ are of equal length on the ground, but their photo images a ′a and b ′b are not of equal length. N is nadir point on terrain, and n is nadir point on photo. P is principal point on terrain, and p is principal point on photo. L is camera lens.

The third major source of distortion in airphotos arises from relief in the terrain. The tops of hills are closer to the camera and therefore appear at a larger scale than valley bottoms. In areas with little relief this effect can be ignored for most practical purposes. In strongly rolling or mountainous areas, however, the variation in scale due to relief cannot be ignored. The scale should be determined separately for each area of the photo that differs appreciably in elevation. The determination of these separate scales from image and object measurements in the respective areas can be accomplished as discussed in Section 3.3.2.

If elevation data are available for the area photographed, the relationship of scale to focal length and flying height offers an alternative method for correcting the scale. Recall that

$$\text{scale fraction} = \frac{\text{focal length}}{\text{flying height}}$$

We shall assume for the sake of illustration that the flying height above the principal point is known for use in the above equation. A change in elevation of the ground is effectively a change in flying height. Therefore the scale of an area that differs in elevation from the principal point will be

$$\text{scale fraction} = \frac{\text{focal length}}{(\text{flying height above pp} \pm \text{elevation difference})}$$

where pp stands for principal point. Since the focal length of the lens does not change, the percent change in the denominator of the scale fraction is equal to the change in elevation expressed as a percentage of flying height.

For example, suppose that an airphoto includes both upland and lowland which differ in elevation by 500 ft. The flying height of the aircraft was 10,000 ft over the principal point lying in the lowland. The scale fraction in the lowland is determined to be 1/20,000. Then the difference in elevation between upland and lowland expressed as a percent of flying height is

$$\text{elevation difference as \% of flying height} = \frac{500}{10{,}000} \times 100 = 5\%$$

The scale fraction for the upland is therefore:

$$\text{scale fraction for upland} = \frac{1}{20{,}000 \times 0.95} = \frac{1}{19{,}000}$$

Conversely, if the scale had been determined as 1/19,000 for the upland, the scale for the lowland would be

$$\text{scale fraction for lowland} = \frac{1}{19{,}000/0.95} = \frac{1}{20{,}000}$$

Still another approach to the correction is to note that the actual (not percentage) change in the denominator of the scale fraction is proportional to the difference in elevation. If the scale fractions and elevation difference are known for two points in the photo, one can determine the change in scale denominator per 100 ft of elevation difference. In the preceding example the change in scale denominator per 100 ft of elevation difference is

$$\text{scale denominator change per 100 ft} = \frac{20{,}000 - 19{,}000}{5}$$

$$= \frac{1000}{5} = 200 \text{ units/100 ft}$$

where the 5 in the denominator arises from the 500 ft difference in elevation. This change per 100 ft can then be used to find the scale fraction at other elevations.

3.3.4 Enlargement

If a photograph has been enlarged in the process of printing, the relationship between scale, focal length, and flying height given in Section 3.3.2 no longer holds. Enlargement is usually stated in terms of diameters using a "×" notation. Thus an enlargement of 2× means that the original diameter of a circular image has been doubled. Likewise a 2× enlargement doubles the length of a linear object's image. Figure 3.4 shows that the effect of enlargement is similar to an increase in focal length of the lens.

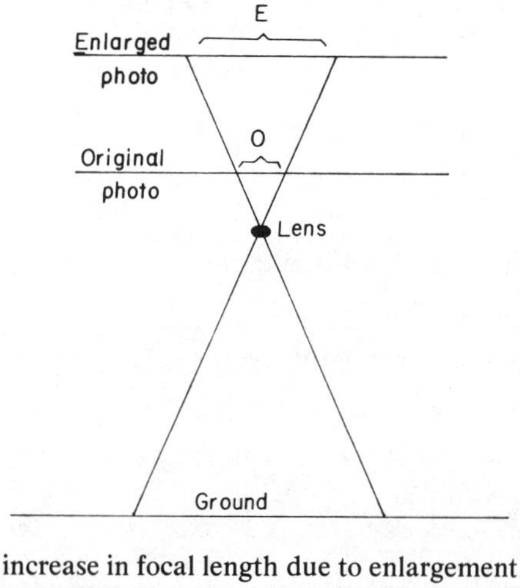

Figure 3.4 Apparent increase in focal length due to enlargement. *O* is original length of image; *E* is length of enlarged image when film plane is moved upward.

Enlargement by a factor of $n\times$ has the effect of dividing the denominator of the scale fraction by n. Thus

$$\frac{1}{(\text{original denominator})/n} = \frac{\text{apparent focal length}}{\text{flying height}}$$

Since the original relationship was

$$\frac{1}{\text{original denominator}} = \frac{\text{focal length}}{\text{flying height}}$$

a little algebra shows that

$$\text{apparent focal length} = n \times \text{original focal length}$$

When working with enlargements, then, the original focal length must be multiplied by the enlargement in diameters before it is used in computations.

3.3.5 Stereo

As discussed in Section 3.3.3, distortions due to relief create some extra work in determining the scale. Relief distortion is not all bad, however, because it also makes three-dimensional or *stereo* viewing possible. Although there are some other visual cues, depth perception arises primarily from the fact that one's eyes view a scene from slightly different angles. The brain "fuses" these two different images into a single three-dimensional picture. The same effect can be achieved by taking two photographs from different angles, and then arranging some optical trickery so that each eye sees a different photo. The brain then thinks that each eye is where the respective photo was taken and proceeds to combine the two photos into a three-dimensional image. Furthermore the amount of depth perception is determined in part by the distance between the eyes. When photos are used to produce a stereo image, the eyes are effectively as far apart as the camera stations from which the two photos were taken. Consequently the sense of depth can be greatly exaggerated, causing hills to look like mountains.

A little better feeling for the nature of relief distortion can be gained from Figure 3.5. This figure shows that a vertical object will appear to lie on its side in the photo, with the top of the object farther from the nadir of the photo than the base. Also, the length of the image becomes progressively smaller for objects closer to the nadir. For an object situated right at the nadir the images of top and base would be superimposed. The photo resulting from the situation depicted in Figure 3.5 might appear as shown in Figure 3.6. Note that the left side of the photo as shown in Figure 3.5 becomes the right side in Figure 3.6 when the photo is turned face up.

Vertical airphotos are usually timed so that each photo overlaps the area

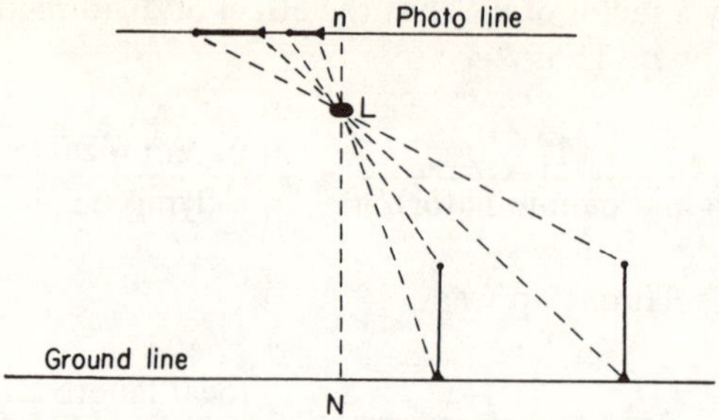

Figure 3.5 Relationship between vertical objects and their images on an airphoto. N is nadir point on terrain, n is nadir point on photo, and L is camera lens.

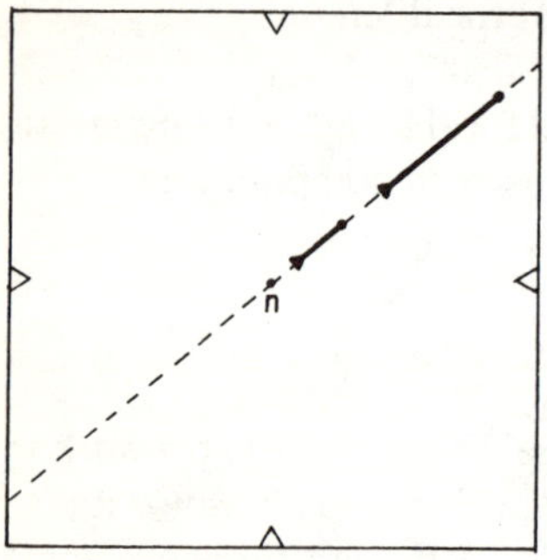

Figure 3.6 Diagrammatic representation of photo resulting from situation depicted in Figure 3.5. Dotted line corresponds to ground and photo lines of Figure 3.5, and n is nadir of photo.

shown in the previous photo by about 60%. This overlap ensures that every terrain point will appear on more than one photo for stereo viewing. Any two such overlapping photos are called a *stereopair*. The two members of a stereopair must be oriented correctly with respect to each other in order to obtain a proper stereoscopic image. This positioning of the stereopair for viewing is our next concern.

Due to the 60% overlap between successive photos, each photo should show the images of the principal point (nadir) locations for the adjacent photos along the flight line. This is illustrated in Figure 3.7, which is a diagrammatic representation of three adjacent photos along a flight line. The overlap causes the second photo to take in the principal point locations of all three. The images on one photo of the principal point locations for adjacent photos are called *conjugate principal points* (see also Figure 3.8).

The procedure for locating the principal points has already been discussed in Section 3.3.1. When the principal point has been located and marked, the next step ordinarily is to find and mark the conjugate principal points on the photo. For each conjugate this is done by looking at the adjacent photo to see what object is imaged at its principal point. Then look for this same image on the photo to be marked and prick its location with a pin. The conjugate principal points

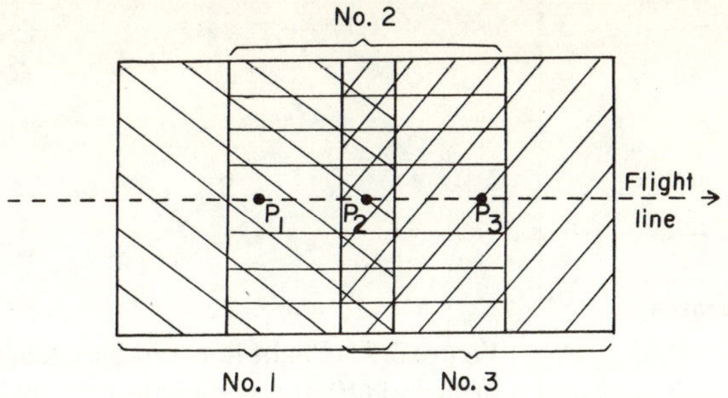

Figure 3.7 Diagrammatic representation of three overlapping photos along a flight line. The first photo is crosshatched downward to the right, the second photo is hatched horizontally, and the third is hatched upward to the right. P_1, P_2, and P_3 are the principal points of the respective photos. Notice that the area of the second photo takes in the principal points of all three.

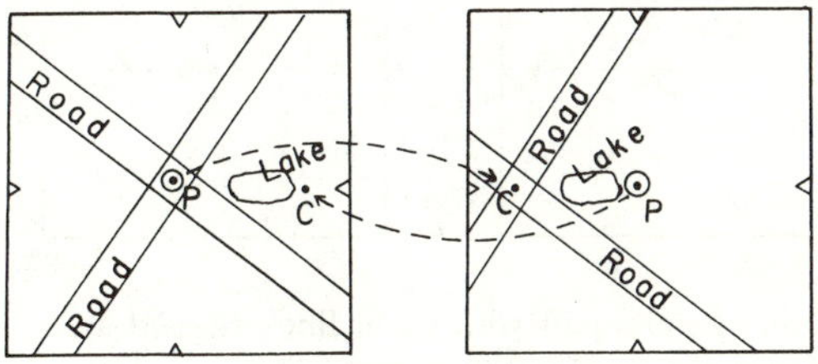

Figure 3.8 Diagram of relationship between principal points and conjugate principal points. The dotted arrows connect corresponding images on the two photos. P denotes principal point, and C denotes conjugate principal point.

are usually circled in a different color of ink than is used for the principal points.

A line connecting the principal point with a conjugate principal point on the same photo shows the location of the flight line on the photograph (see Figure 3.9). The distance along the flight line between the principal point and a conjugate principal point is called the *base length* of the photo. Since conjugate principal points lie to both the left and the right along the flight line, two base lengths can be measured (see Figure 3.9). The two base lengths may not be quite the same because of irregularities in the terrain and flight pattern.

When the principal and conjugate principal points have been marked, the next step is to place the two photos of the stereopair side by side on a table with the area of overlap toward the inside. Tape down the outside edge of one of the photos with masking tape, and then use a straight-edge to line up the flight lines as shown in Figure 3.10.

The normal tendency of the eyes is to converge so that they both focus on the same object. In order to "see stereo" from photographs it is necessary to make

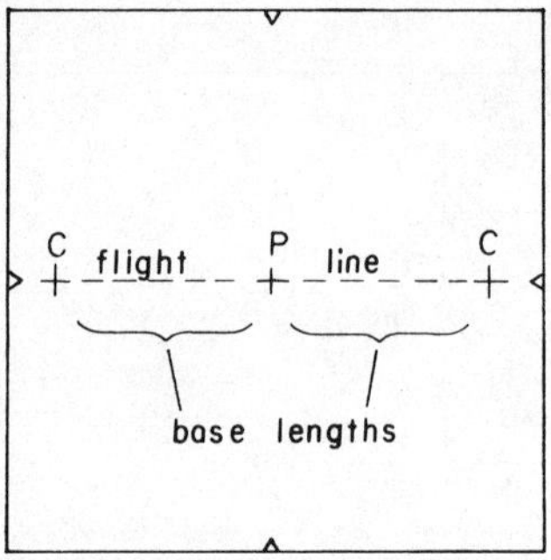

Figure 3.9 Flight line and base lengths in relation to principal (P) and conjugate principal (C) points.

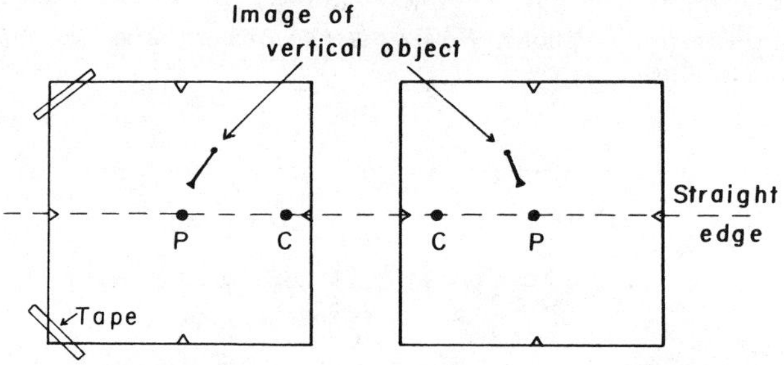

Figure 3.10 Diagram of stereopair with flight lines aligned along a straight edge. The straight edge is represented by the dotted line. The image of a vertical object is also shown schematically on both members of the stereopair.

the eyes diverge so that each eye sees only one member of the stereopair. This divergence is best achieved by means of a *stereoscope*. Many people are able to "see stereo" without the aid of a stereoscope by consciously letting their eyes drift apart. However, this is likely to lead to eyestrain and is not recommended as routine practice. Stereoscopes are designed to diverge the eyes optically and should not cause eyestrain when properly adjusted.

Separation of the lines of sight in a stereoscope may be accomplished by lenses, mirrors, prisms, or a combination of the three. Consequently stereoscopes come in a wide variety of shapes and sizes. Lenses have the additional advantage of allowing magnification. The simple and inexpensive pocket stereoscope is just a pair of lenses to separate the lines of sight. The lenses are held the proper distance above the photos by a pair of folding legs, and the distance between the lenses can be adjusted to fit the separation of the individual's eyes. Pocket stereoscopes, however, permit stereo viewing of only a portion of the overlap area at a time. Larger mirror stereoscopes have the advantage of allowing the viewer to see the entire overlap area in stereo without repositioning the photos or the stereoscope.

Different types of stereoscopes require different spacing of the photographs. This is the reason that only one of the photos is taped down in Figure 3.10. To determine the proper spacing for a particular stereoscope, first adjust the eyepieces to match the separation between the pupils of your eyes (eye base). The eye base varies between people, but is commonly around 2.2 in. When the eyepieces have been adjusted, lay a ruler under the scope along the line defined by the centers of the eyepieces. While looking through the scope, close the right eye and read the graduation on the ruler which appears to be directly below the left eye. Then close the left eye and read the ruler directly below the right eye. The difference between the two readings gives the approximate spacing needed between corresponding images of the two photos, as for instance the principal point on one photo and the corresponding conjugate principal point on the other photo. An alternative method of finding the approximate spacing is to place a sheet of paper under the stereoscope, and make a pencil mark in the center of each field of view by closing the eyes alternately as with the ruler. The distance between the dots as measured with a ruler is the approximate spacing.

After the approximate spacing has been determined, place the stereopair as shown in Figure 3.10. Slide the free photo toward or away from the other to achieve the desired spacing between image points. Take care, however, to keep the flight lines aligned with the straight-edge. With a pocket stereoscope the spacing will be about 2.2 in., and the photos will physically overlap.

Now place the stereoscope over the stereopair. The stereoscope should be oriented so that a line connecting the centers of the lenses is parallel to the flight line. Canting the stereoscope with respect to the flight line will give a distorted image and will make it difficult to "see stereo." Pick out a prominent feature which appears on both photos. Close one eye and center this object in the field of view for the eye that is open. Close this eye, open the other, and make sure that the same object is also near the center of the field of view for the other eye. With the image thus centered in both fields of view, look through the stereoscope with both eyes open.

When one first looks through the stereoscope with both eyes open, two images may be seen near the center of the field of view instead of one. However, the two images should not appear to be very far apart. Adjust the position of the stereoscope or, if necessary, the distance between the photos to bring the two images together. When the two images finally come together, they should fuse into a single stereoscopic image. It is helpful to start with a stereopair that has considerable relief so that there is no doubt as to whether you are "seeing stereo." When a person has once succeeded in seeing stereo from photographs, stereo vision will be much easier to achieve thereafter.

As mentioned above, a pocket stereoscope requires that the two photos be so close together that they physically overlap to some extent. The fact that one photo masks part of the other means that only a portion of the potential stereo area can be seen at one time. The remainder of the stereo area can be seen, however, by bending the inside edge of the top photo upward in order to expose the lower photo to view.

If a stereopair is to be used repeatedly for demonstration or teaching purposes, the photos can be fixed permanently in stereo position with the nonstereo areas cut away. This kind of permanently mounted stereopair is called a *stereogram*. The permanent mounting does restrict the type of stereoscope used. A stereogram prepared for viewing with a pocket stereoscope cannot be viewed with a mirror stereoscope, and vice versa.

Refer back for a moment to Figure 3.10 and note the two images of the vertical object shown in the diagram. Notice that the tops of these images are closer together as measured across the stereopair than are the bottoms. It is this difference in spacing between the tops and bottoms of the images that gives the stereo impression. The closer the corresponding image points lie (as measured across the stereopair parallel to the flight line), the greater the elevation that they will appear to have in the stereo image. In fact, this difference is greatly exaggerated in Figure 3.10 to the extent that a person's eyes would not "believe" that the object could be so tall. Therefore one should not expect to actually see stereo in Figure 3.10.

Two other things bear mention at this point. One is the dependence of stereo vision on a balanced eyesight. One eye being stronger than the other is likely to cause difficulty in stereo perception. A person who has lost the sight in one eye will not be able to see stereo at all. The other thing is the effect of reversing the two photos of a stereopair left for right. This produces a *pseudoscopic* image that inverts the topography, that is, hills become valleys and valleys become hills. Experienced photointerpreters sometimes use a pseudoscopic image as an aid in detecting minor depressions in the terrain.

The background relative to airphotos which has been covered in this section (3.3) should be sufficient to allow the presentation of photo measurements in parallel with ground measurements through the remainder of this chapter.

3.4 DIRECTIONS AND ANGLES

The need to determine directions and angles is common to most types of survey work which involve data collection either in the field or from remote sensors. Therefore it is not overstating the case to say that such measurements constitute a fundamental component of survey methodology. The most frequently used methods of determining directions and angles are covered in this section.

3.4.1 Terminology and Units of Measure for Angles and Directions

Hopefully the reader will be sufficiently familiar with the nature of angles and directions to make basic definitions unnecessary. Therefore we will proceed directly with units of measure.

Angular Measure

Degrees. We can approach the measurement of angles by degrees if we imagine a square corner divided into 90 equal parts by radiating lines. Each of the resulting pieces is 1 *degree*. Since four square corners can be fitted together at the center of a circle, there are $4 \times 90 = 360$ degrees in a full circle. The degree measure is usually indicated by floating a little bubble at the upper right-hand corner of the number, as in 360°. Degrees are further subdivided into minutes and seconds. A minute is 1/60 of a degree, and a second is 1/60 of a minute. An angle of 10 degrees, 20 minutes, and 30 seconds is usually written as 10°20′30″. The current trend, however, is to dispense with minutes and seconds and express fractional degrees directly in decimal form.

Radians. The more mathematically inclined lean toward a different unit of measure called the *radian*. There are 2π radians in a full circle, where π (pi) = 3.14159 correct to five decimal places. Consequently 1 radian is equal to 57.296 degrees correct to five figures, and 1 degree is equal to 0.01745 radian correct to five decimal places. Physically a radian is the angle covered by an arc that has the same length as the radius of the circle.

Mils. The U.S. military has still a different system for measuring angles, in which the unit of measurement is called a *mil*. As usually applied, a mil is 1/6400 of a full circle, or exactly 0.05625 degree. The term "mil," however, is also used at times as an abbreviation for milliradians which is 0.001 radian. Therefore one must be sure which way the term "mil" is being used in any specific case.

Grads. Yet another system of angular measure is sometimes used in European countries. In this system a square corner (or right angle) is divided into 100 equal parts called *grads*. A full circle, therefore, contains 400 grads. The grad is further broken down into 100 minutes, and the minute into 100 seconds. Since the latter system has not been used in North America and is not part of the proposed conversions to the metric system, it will not be considered beyond this point.

Expressing Directions. A direction is an angle measured from a reference line. There are several common systems of expressing directions and some variations on the basic systems.

Azimuths. The azimuth system is perhaps the simplest and easiest to use in many applications. As used in natural resource surveys, an azimuth is the angle in degrees measured clockwise from north. Thus azimuths can range from 0 to 360°. East is 90° azimuth, south is 180° azimuth, west is 270° azimuth, and north is either 0 or 360° azimuth. Astronomers and some others, however, use a variation of the azimuth system in which the reference direction is south instead of north.

Bearings. Another system of expressing directions which is often used in land surveying and to some extent in navigation is that of *bearings*. In the bearing system a direction is identified as being in one of four *quadrants*. Directions between north and east comprise the northeast quadrant, those between east and south comprise the southeast quadrant, those between south and west comprise the southwest quadrant, and those between west and north comprise the northwest quadrant. The quadrant is identified by a leading and a trailing letter as follows:

Quadrant	Leading letter	Trailing letter
northeast	N	E
southeast	S	E
southwest	S	W
northwest	N	W

Depending on the quadrant, directional angles are measured from either north or south toward either east or west as follows:

Quadrant	Measurement of directional angles
northeast	clockwise from north toward east
southeast	counterclockwise from south toward east
southwest	clockwise from south toward west
northwest	counterclockwise from north toward west

The directional angle in degrees is placed between the letters that identify the quadrant. Thus an azimuth of 160° lies 20° to the east of south, and this same direction in bearing notation is S20°E. In the bearing system, cardinal directions are usually written as "due N," "due E," "due S," and "due W."

Military Directions. The military directional system works like the azimuth system, except that mils are used instead of degrees. Thus, due east (or 90° azimuth) is 1600 mils in the military system.

Backazimuth and Backsight. It is often convenient to have a short way of saying that one direction is opposite another direction. The term used to indicate opposite direction in the azimuth system is *backazimuth.* Thus if you face in one direction then turn around and face the other way, you end up looking along the backazimuth from your original direction. Backazimuths of directions between 0 and 180° are obtained by adding 180° to the azimuth. Between 180 and 360° one must subtract 180° to obtain the backazimuth. Terms corresponding to backazimuth in the bearing system are *backsight* and *backbearing*. To obtain backsight bearings, simply replace N by S or vice versa and E by W or vice versa. For instance, N45°W is the backsight of S45°E.

Declinations. A slight complication arises from the fact that there is more than one "north." First, there is the point defined by the earth's rotation called the north geographic pole. A line pointing toward the north geographic pole is called *true north.* Second, there is the direction that a compass needle points. This direction is called *magnetic north*. In most localities magnetic north does not coincide with true north. The difference (angle) between true north and magnetic north at a particular locality is called the *magnetic declination*. Magnetic declination arises from two sources. One is that the north magnetic pole is not situated at the north geographic pole. The second source of magnetic declination consists of local or regional deposits of magnetic materials.

A line runs along the east side of Lake Michigan on which a compass needle does point true north. This line is called the *agonic line*. The agonic line is somewhat irregular because of regional differences in the distribution of magnetic materials. When one is west of the agonic line, a compass needle points east of true north, and the locality is said to have an *east declination*. When one is east of the agonic line, a compass needle points west of true north, and the locality is said to have a *west declination*. The magnetic declination for a particular locality can be determined from published charts called isogonic charts and is also shown on many maps.

To convert magnetic azimuths to true azimuths one must add an east declination to the magnetic azimuth and subtract a west declination. To convert true azimuths to magnetic azimuths, the corrections are reversed. It should also be noted that magnetic declinations change slowly with time.

There is still a third type of "north" which is an artificial creation of map makers rather than a physical thing. The task of mapping and computerizing data from maps is often simplified by defining a conceptual square grid on the earth's surface and then locating points according to the coordinate system formed by the grid lines. One axis of such a grid is usually designated as *grid north*. Since grid north differs from both true and magnetic north, the result is a *grid declination* with respect to the true north. The amount and direction of the grid declination depends on the placement of the grid and the location of the point with respect to the origin of the grid. A map that shows grid coordinates should also give the grid declination. Given the grid declination, then, one can convert grid directions to true or magnetic directions in a manner parallel to that already outlined for magnetic declinations.

Having considered the various ways of expressing directions, our next concern is with the use of compasses for determining directions in the field.

3.4.2 Use of Compasses for Determining Directions in the Field

All compasses designed for use in field surveys have four basic parts: (1) a magnetic indicator on a low friction pivot, (2) a graduated circle, (3) a sighting line, and (4) a housing or case. These basic components can be assembled in quite a variety of ways, however, and a brief discussion of the more common variations is in order before getting into the details of usage.

Types of Compasses. It should be apparent from the foregoing discussion that one way compasses may differ is in graduation of the circle. Most commercially available models are offered in a choice of either azimuth or bearing (quadrant) graduations. A few compasses even have dual graduations in both azimuths and bearings. Compasses that originate from the military are usually graduated in the mil system or carry dual graduations in mils and degrees.

The steadiness with which a compass can be held in use is a major factor in determining the accuracy of the readings. Other things being equal, the more firmly the compass is supported, the better will be the accuracy. The most accurate compasses, therefore, are designed to be mounted on a tripod. Some tripod compasses are even equipped with a telescope for sighting. Somewhat faster to use but a little less steady is the Jacobs staff type of support. This consists of a wooden staff with a metal tip which is jabbed into the ground. The top of the staff is fitted with a mount for the compass. The suitability of such a staff mounting depends on how firmly the staff can be planted by jabbing it into the ground. Where the soil is deep and soft, the staff is quite satisfactory. Where the soil is hard or rocky, however, the staff will be less suitable.

A caution in the use of a staff mounted compass is to remove the compass from the staff when moving from one sighting station to the next. Sensitive components of the compass are almost sure to be damaged if the staff is jabbed into the ground while the compass is attached. Likewise it is good policy to demount tripod compasses when moving from one station to the next.

On virtually all staff compasses and on most tripod compasses the sighting line is defined by a pair of folding sighting vanes. Compatible mountings are often provided so that the compass may be mounted on either a tripod or a staff.

When field work must be performed rapidly and the standards of accuracy are relatively low, hand held compasses are usually employed. The older types of hand held compasses were mostly designed to be held centered in front of the body at waist level. Such compasses are often called "box" compasses because the housing is typically boxlike and the construction is simple. Since the eyes are several feet above a compass held at waist level, the sighting line tends to be rather poorly defined and a fair amount of practice in holding the head straight is needed to achieve consistency. To overcome this difficulty, many of the newer compasses are designed to be held at eye level. This requires some sort of adaptation so that the face of the compass is visible when held before the eyes. This may be accomplished in several ways, giving rise to a variety of appearances in eye level compasses.

The adaptation for eye level viewing usually used in military compasses and some civilian models is to place a wide angle lens in the rear sight. This makes the front part of the compass face visible when held at eye level. Compasses of this type are called *lensatic* compasses.

Perhaps the most common adaptation for eye level viewing in civilian compasses is to place a mirror on the inside of the compass lid. This permits the user to view the face of the compass in the mirror when it is held at eye level. Compasses of this type are called *prismatic* compasses.

Still another approach to implementing eye level viewing is to replace the compass needle with a magnetic disk. The edge of the disk is bent down and the graduations are placed on this overhanging edge. This disk is enclosed completely except for windows to admit light and a lens through which the user sees a magnified view of the edge of the disk with one eye. The other eye is held open, but does not look into the compass. The result is a kind of optical illusion in which the compass graduations appear to be superimposed on the terrain.

A factor that is much more important with hand held compasses than with staff or tripod compasses is the tendency of the needle to swing on its pivot. After a compass is mounted on a tripod or staff and left untouched for a moment, the needle settles quickly and remains steady. The unsteadiness of the hand, however, makes it very important to have a way of damping the swing of the needle without reducing its sensitivity. The presence of such a damping system is one way of distinguishing the better hand held compasses from the cheaper models.

Damping may be accomplished in two ways. The most common way is to fill the compass housing with a clear, noncorrosive liquid. The liquid filled case reduces the needle swing to practically nothing. The other system makes use of magnetic induction forces which act to counter swing or random motion of the needle but disappear when the needle is at rest. Any good compass that is not liquid filled should be equipped with a needle lock to prevent damage to the pivot when the compass is being transported. With some practice one can use the needle lock as a brake for damping the swing of the needle. On cheap compasses one must either tip the compass gently so that friction of the needle's tip on the base of the case dampens the swing (which is difficult to do without introducing error), or simply interpolate halfway between the extremes of the needle's side to side swinging motion.

Reading the Compass. The way one actually goes about finding a direction with a compass depends on two things. One is the construction of the compass. The other is whether a person is attempting (1) to follow a predetermined direction or (2) to read the direction of some terrain feature relative to the person's present position.

Let us first consider the use of staff compasses, tripod compasses, and box compasses. These types of compasses are all graduated in the same relatively simple manner. The graduations are engraved on a circle along the inside edge of the case, and the magnetic element is a simple needle with north and south ends. Since the needle always points magnetic north-south, one must rotate the case with its attached circle around the compass needle in order to read or follow directions. The sighting line is marked on the case or defined by vanes attached to the case and, therefore, turns with the graduated circle. These compasses are constructed so that the north end of the needle points to the place on the graduated circle where the reading should be taken. The direction so indicated is the direction in which the sighting line points.

The first thing one should do with such a compass is make sure that one knows which is the north end of the needle and which end of the sighting line should be directed away from the user. To read the direction of an object, then, one simply points the sighting line in the desired direction and reads the graduated circle at the point indicated by the north end of the needle.

To follow a predetermined direction, one rotates the compass until the north end of the needle points at the desired directional value on the graduated circle. The sighting line then shows the proper direction of travel. Notice also that a compass of this type has east and west reversed on the graduated circle. This is necessary to make the system work as just described, and it should not cause confusion as long as one works through the sighting line.

Keep in mind, however, that the directions indicated are relative to the magnetic north. Many compasses have an adjustment that can be used to correct automatically for the declination by a slight rotation of the graduated circle relative to the sighting line. When the adjustment is properly set, the readings will automatically be relative to the true north or grid north instead of the magnetic north. To make the setting that shifts the directions from magnetic north to true north, first set the adjustment at zero and then position the compass so that it shows true north. With the compass in this position, manipulate the adjustment so that the compass shows the declination instead of the true north.

The second category of compasses with respect to the procedure for taking readings is exemplified by the military lensatic compass. In compasses of this type the magnetic element is a flat disk instead of a needle. The graduations are engraved on the outer edge of this disk. The sighting line is attached to the case. The case also bears an index mark situated along the sighting line.

In order to read the direction of a terrain feature with this type of compass, one first directs the sighting line toward the feature and then reads the graduation on the magnetic disk which falls opposite the index mark. To follow a predetermined direction, one looks along the sighting line and turns until the desired graduation lies opposite the index mark. The sighting line is then oriented along the proper direction of travel.

The adjustment for declination in this type of compass is usually accomplished by moving the index mark relative to the sighting line. The adjustment is at zero when the index mark lies right on the sighting line. The declination setting can be accomplished in the same manner as for the previous type of compass.

The familiar automobile compass is only a slight modification of this design in which the magnetic disk is replaced by a small sphere, the index mark takes the form of a vertical line, and the front of the auto defines the sighting line. This basic design has the advantage of allowing east and west to appear in their natural positions on the graduated circle instead of being reversed. This avoids one possible source of confusion for beginners.

The third and last type of compass which will be considered has become popular in recent years for hand held use. This design is somewhat more complicated than the other two, but has the advantage of not requiring the compass to be held steady while the numerical value of the direction is read. The magnetic element of such a compass is a simple needle. The case has two parts, one is fixed in position and the other can be moved (rotated) around the needle pivot by being twisted with the fingers. The stationary part of the case bears the sighting line and an index mark. The movable part of the case bears the graduated circle and an engraved alignment arrow that moves with the circle.

To follow a predetermined direction with this type of compass, first twist the movable part by grasping the graduated circle until the desired directional reading on the circle lies opposite the index mark. Then hold the compass in sighting position and turn your body until the north end of the needle coincides with the head of the alignment arrow. The sighting line is now oriented in the proper direction of travel.

To read the direction of a terrain feature, hold the compass in sighting position with the sighting line directed toward the terrain feature. While holding the compass in this position, twist the movable part so that the head of the alignment arrow coincides with the north end of the needle. Now relax from the sighting position and read the direction on the graduated circle opposite the index mark.

Declination adjustments with this type of compass usually involve the movement of the graduated circle relative to the alignment arrow. This design, like the previous one, does not involve a reversal of east and west on the graduated circle. The design is easily adapted for night use by placing luminous spots on the north end of the needle and the head of the alignment arrow. Since readings or settings do not have to be done in sighting position, a flashlight can be used for these operations.

Other designs can also be adapted for night use by providing fully luminous dials, using radioactive "lamps," or counting "clicks" as a luminous pointer is moved between the sighting line and the north point of the magnetic element. These other types of adaptations, however, tend to be more awkward.

Cautions in the Use of a Compass. There are several precautions that must be observed in using any compass if gross errors are to be avoided. The need to know which end of a compass needle is north and which end of a sighting line to direct away from the observer has already been mentioned.

It is essential that the compass be approximately level when held in sighting position. If the compass is not level, the needle or magnetic disk may bind against the case and give an incorrect reading. Tipping has an additional effect in compasses with the sights elevated above the case. When a compass of this type is tipped, the sighting line is slanted away from its correct position.

Most compasses that are not liquid filled will have a lock to hold the magnetic needle or disk away from its pivot when the compass is being transported. This prevents damage to the point of the pivot, and the lock should be engaged unless a sight is actually being taken. On many compasses the lock is engaged automatically when the lid of the compass is closed. A good compass will last almost indefinitely if it is well cared for. On the other hand, rough treatment is likely to cause damage even if the needle is locked.

The user of the compass must always be on guard against local sources of magnetic attraction which might deflect the needle from the magnetic north. Vehicles, railroads, and steel structural components of buildings are obvious sources of trouble. However, smaller articles carried on the person can also cause difficulty. A mechanical pencil, for example, can deflect the needle if held in the same hand as the compass. Articles like cameras, hand axes, or a geologist's hammer should be carried behind the body in a belt case or knapsack rather than on the front or side. Likewise shovels, picks, soil augers, and such should be placed well away from the compass. Even steel components in the frames of eyeglasses may be suspect. When in doubt, smaller objects can be tested for attraction by being held close to the compass and then gradually moved away to see if the reading changes.

When features of the terrain are suspected to be sources of magnetic attraction, always check the readings by taking a backsight from the next station. If the two readings do not agree, local attractions are probably present at one or both stations. When local attractions are prevalent, it may be necessary to turn angles as discussed in the next section instead of working with simple directions.

3.4.3 Instruments for Measuring Horizontal Angles in the Field

Directions as discussed in the previous section are angles measured with respect to some standard reference line such as true north, magnetic north, or grid north. In many survey situations, however, the primary need is to know the angle between two intersecting lines on the terrain when the observer's position is at their intersection. "North" is usually irrelevant to this problem, so it is not essential to use an instrument that has a compass needle.

We shall limit our attention in this section to *horizontal* angles. In effect, this means that the graduated scale of the instrument must be held level while the angle is being measured. A nontechnical way of visualizing the situation is to imagine that the terrain features which define the angle are raised or lowered to the level of the instrument for purposes of determining the angle. A more technical way of describing a horizontal angle is to say that it is the angle between two vertical planes, each of which contains the observer's position and one of the two terrain features that define the angle. Vertical angles will be considered in later sections of this chapter.

Determining Angles by Compass. Although a compass needle is not necessary for measuring horizontal angles, nevertheless a compass can serve the purpose. Since we have already dealt with most aspects of compass usage in the previous section, we may as well finish the job before moving on to other kinds of instruments for measuring angles.

The procedure is simple. First measure the directions of the two lines that form the angle, then subtract the smaller direction from the larger one. The difference is the angle. The subtraction is relatively straightforward when the directions are expressed as azimuths. The only complication with azimuths arises when the difference between the two azimuths is greater than 180°. In this case the difference must be subtracted from 360° to obtain the angle. For example, if an angle has sides with azimuths of 30 and 320°, first subtract to get $320° - 30° = 290°$. Since this difference is greater than 180°, subtract from 360° to obtain $360° - 290° = 70°$. There will usually be less chance for error if the bearings are converted to azimuths before the subtractions are done.

There is no need to worry about declinations or local attractions when a compass is used to measure angles. Since the directions of the two sides are read from the same point, the effect of declination or attraction will be the same for both sides of the angle and will disappear in the process of subtraction.

The main limitation in using compasses to measure angles is the relatively limited accuracy that can be achieved. Most compasses designed exclusively for hand held use are graduated in 2° steps and can be read by interpolation to the nearest degree. Compasses designed to be mounted on a tripod or staff are usually graduated in 1° steps and can be read by interpolation to the nearest half degree. For accuracy greater than this, one must go to an instrument designed to measure angles by "turning," which we shall consider next. Instruments of this type are usually equipped with telescopic sights.

Instruments for Measuring Angles by "Turning." The essential features of the mechanism in this type of instrument are (1) a graduated circle attached to the baseplate of the instrument, and (2) a (telescopic) sighting device and attached index mark which rotate over the graduated circle. When the baseplate has been leveled, a reading is taken along each side of the angle. These readings do not represent azimuths because the baseplate is not oriented with north. Nevertheless the angle is computed by the difference between readings in exactly the same manner as for azimuth readings from a compass.

In most instruments of this type the sighting tube can be tilted up and down to permit sighting on objects above or below the level of the instrument. The fact that the graduated circle remains level, however, gives readings in the horizontal plane. Many instruments of this type also have a compass needle for reading directions, but the compass needle does not enter directly into the measurement of angles.

The need for actually reading the graduated circle when the sight is taken on the first side of the angle can be eliminated by designing the instrument so that the graduated circle can be rotated over the baseplate. This makes it possible to obtain a zero reading for the first side of the angle. The reading for the second side then gives the size of the angle directly without subtraction. The movement of the graduated circle over the baseplate is called "lower motion," and the movement of the telescope (and index mark) over the graduated circle is called "upper motion." Clamps are provided so that the graduated circle can be fixed relative to either the baseplate or the telescope. In practice most instruments are constructed with both upper and lower motions. The better instruments are also equipped with so-called "slow motion" or "tangent" screws which allow for fine adjustments in the alignment using either upper or lower motion. The main clamp for the respective motion must be tightened before a slow motion screw becomes operative.

Verniers. Instruments for measuring angles as described above will usually be equipped with a *vernier* for reading the graduated circle. This is a device that makes interpolation between graduations a mechanical process rather than guesswork. Since verniers occur on various types of instruments, we provide a brief introduction to their use.

The vernier itself is a short scale adjacent to the index mark which rides alongside the main scale. In fact, the zero of the vernier scale coincides with the index mark. The graduations on the vernier are spaced closer together than those on the main scale, and it is this difference in spacing that makes it possible to use the vernier for interpolation.

To use a vernier, first find the two graduations on the main scale between which the zero of the vernier lies. Read the value for the lesser one. Now look in the direction of increasing main scale graduations for the vernier line that most nearly matches a mark on the main scale. Add the vernier reading for this line to the value previously read from the main scale.

If there is any doubt as to the units in which the vernier reads, divide the increase represented by the smallest division on the main scale by the number of divisions on the vernier scale upward from zero. This will give the unit for the smallest vernier division. Suppose, for example, that the smallest division on the main scale represents 30 minutes, and the vernier has 15 divisions not counting zero. Then the vernier unit is $30/15 = 2$ minutes.

In this case the vernier would be constructed so that each vernier division is $14/15$ of the length of the smallest main scale division (a difference of $1/15$). When the zero of the vernier is exactly lined up with a main scale mark, the fifth mark on the vernier (for example) will lie $1/15 \times 5 = 1/3$ of a main scale division short of the next mark on the main scale above it. When the zero of the vernier lies $1/3$ of the way between two main scale divisions, the fifth vernier mark will be moved up $1/3$ unit and will lie directly opposite a main scale mark. Thus $5 \times 2 = 10$ minutes should be added. The 10 minute vernier reading is $1/3$ of the 30 minute main scale graduation.

Improving Accuracy by Repetition. There are several ways that the person using an instrument may allow small errors to creep into the measurement of an angle, such as failure to align the scope exactly in sighting or slight errors in reading. One way to reduce these types of operator errors is to average the results of several measurements. If the instrument has both upper and lower motions, several measurements can be averaged without the necessity of recording each individual measurement separately. The scale is set to zero only for the first measurement. For the second and subsequent measurements the scale is allowed to accumulate rather than having to be reset to zero. If the angle is measured three times, for instance, the reading is taken at the end and divided by 3 to obtain the average.

The one caution in using this procedure is to note that 360° must be added to the final reading each time the accumulation moves the scale past zero. For instance, measuring an angle of 130° three times will give a final reading of 30° rather than 390°, since the scale moves past zero on the third repetition. Therefore it would be necessary to add 360° to the final reading before dividing by 3.

Transits and Double Centering. The repetition in measuring angles as described above helps to average out certain errors on the part of the person using the instrument, but it does nothing to reduce errors internal to the instrument. In the class of instruments called *transits*, this problem is resolved by constructing the instrument so that the sighting scope can be flipped over to point in the opposite direction. When the telescope is reversed in this manner, most errors internal to the instrument are also reversed. Therefore such internal errors will be self-compensating if half the repetitions are done with the telescope in the normal position and the other half with the telescope in the reversed position. Balancing errors in this manner is called *double centering.* Double centering is advisable whenever the standards of accuracy are high.

In years past most transits were pretty similar. More recently the variety has increased considerably, both in outward appearance of the housing and in available accuracy as reflected in the quality of construction and the fineness of graduation. In view of this variety there seems to be little point in describing a specific model. For specifics it is best to have hands-on training with the model in question.

Since the purchase prices vary widely according to accuracy, one must exercise discretion when acquiring an instrument that is suitable for the needs of the job. The instrument should be well built and ruggedly constructed, but there is little point in paying a great deal extra for more accuracy than is really needed to accomplish the specific survey task. It may well be possible to equip several crews with adequate instruments for the same expenditure involved in buying a single instrument of much higher accuracy.

Good transits, for instance, have a maximum accuracy on the order of 1 minute of angle, but a highly precise class of instruments known as *theodolites* is available with the accuracy exceeding 1 second of angle. Environmental surveys would normally not justify the expense of theodolites.

Range Poles. Many of the stations along survey lines do not have natural features that provide definitive sighting points. This makes it necessary to hold a pole over the point as a sighting target. Poles used for this purpose are called *range poles.* Commercial range poles are wooden or metal rods painted in contrasting colors. If the need arises, homemade range poles can usually be fashioned from materials at hand. The only caution in using range poles is that they must be held vertically. Other than this, their use should be quite straightforward.

Determining Angles Graphically with Plane Table and Alidade. The *plane table* and *alidade* offer quite a different approach to determining angles than the other methods which have been considered thus far. The basic idea is to draw the angle to scale on paper. The plane table is essentially a drawing board mounted on a tripod. The head of the tripod is built to allow two kinds of motions. One type of motion makes it possible to level the board without moving the legs of the tripod. The other type of motion allows the board to spin after leveling so that a given edge can be oriented in any desired direction. Clamps are provided for both motions. The map sheet or drawing paper is attached to the top of the plane table. The alidade is essentially a ruler with a sighting device attached. In the simplest type of alidades a pair of folding sighting vanes forms the sighting device. In more sophisticated alidades the sighting device is a telescope.

In order to draw an angle on the (map) sheet, one begins by setting up the tripod and leveling the plane table. Level bubbles may be built into the plane table or into the alidade. If level bubbles are not built into either board or alidade, it is a simple matter to set a surface level on the board and use it for the leveling operation.

A north arrow is drawn on the sheet, and the next step is to orient that north arrow with the north on the terrain. This can be done with a compass built into the board, built into the alidade, or simply set on the map sheet. When the map sheet is properly oriented with the terrain, the clamp is set to maintain that orientation.

The map position of the plane table relative to the terrain is marked on the sheet by pricking with a pin. The ruling edge of the alidade is pivoted around the pin mark until the sights are aligned along one side of the angle, and a line is drawn along the ruling edge from the pin mark. The alidade is then pivoted about the pin mark until the other side of the angle appears in the sights, and a second line is drawn. These lines meet at the pin mark and represent the two sides of the angle correctly oriented with respect to each other and also with respect to map north. If it is desired to express the angle in numerical terms, this can be done by using a protractor to make the measurement. More commonly, however, the angle is simply left in graphical form as part of the control network for a field map that is being developed on the plane table.

As with compasses and transits, the appearance and cost of the plane table and alidade vary considerably according to the accuracy that they are meant to produce. A small plane table mounted on a light tripod is often called a *traverse board.* The alidade used with a traverse board is usually just a ruler with folding vanes attached to the ends. The accuracy to be expected from a traverse board is similar to that for a hand held compass. As a matter of fact, the simple traverse board is becoming rather scarce, and it may be difficult to find a supplier. With a few simple tools and a little ingenuity, however, one can usually fashion a workable traverse board and a simple alidade for oneself.

The primary utility of this rather primitive equipment is in preparing rough field maps and introducing students to mapping. Considerably better accuracy can be achieved with a heavier board mounted on a sturdy tripod fitted with a special "Johnson head." When a plane table of this caliber is used with a telescopic alidade and low shrinkage paper, the accuracy attainable approaches that of the simpler transits. This type of equipment is used quite frequently in geologic mapping and is available from several major suppliers of surveying equipment. However, one can expect the cost of a telescopic alidade to be similar to that of a good transit.

The use of the plane table and alidade in mapping is explained more fully in Section 3.8.

Determining Angles with a Sextant. All the methods previously mentioned for measuring angles in the field require that separate sights be made along the two sides of the angle and that the instrument be held stable in a level position while the readings are taken. This poses certain problems for reading angles from boats during hydrographic surveys. The motion of the boat makes plane tables, transits, and similar instruments virtually useless for this type of work. A rather ancient instrument called the *sextant* is useful in these circumstances.

A skeleton diagram of a sextant is shown in Figure 3.11. The frame of the sextant consists of two arms and a graduated arc. A telescope is attached to one arm, and the horizon glass is attached to the other. The index arm is pivoted at the juncture of the two arms of the frame. An index mirror is attached to the upper end of the index arm. The horizon glass is half-silvered so that it splits the line of sight through the telescope. The object or target marking one side of the angle is seen through the horizon glass. When the index arm is positioned properly, the image of the object or target marking the other side of the angle will be reflected from the index mirror onto the horizon glass and then into the telescope. One simply directs the telescope at the fainter object (the one that is

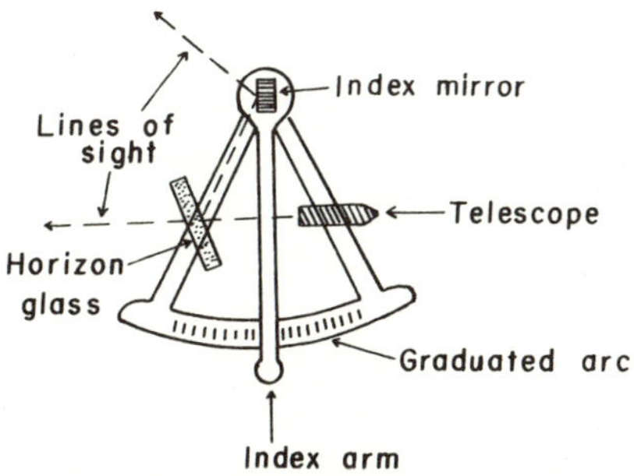

Figure 3.11 Skeleton diagram of a sextant.

hardest to see) and then moves the index arm until the image of the second object seems superimposed on that of the first. The angle between the two objects is then read on the graduated arc.

The unique feature of the sextant is that the angle is measured in a single operation. Other instruments require that the sighting line be "turned" from one side of the angle to the other for two separate sightings.

The objects which form the angle must not be too close to the sextant, or the accuracy will be affected. There may also be an "index error" due to slight maladjustments of the sextant. The index error (if any) should be determined before an angle is measured. To check for index error, select an object at a considerable distance and direct the telescope toward it. Move the index arm until a single image is obtained. The end of the arm should now read zero on the graduated arc; and any departure from zero is the index error. Subsequent readings should be adjusted accordingly.

3.4.4 Angles and Directions from Maps and Vertical Airphotos

Determining angles from maps and vertical airphotos is a relatively simple matter. The basic procedure is to draw the sides of the angle on the map or photo or on an overlay, then to measure the angle with a protractor. The assumption underlying this procedure is that the scale is uniform, but minor variations in scale due to relief in airphotos will not affect the results greatly. Angles measured from the principal point as a vertex are not affected by relief at all, and relief effects will be small for angles in which the vertex lies near the center of the photo.

If the direction of one line on the map or photo is known, the directions of other lines can be determined by measuring angles. For purposes of illustration suppose that the direction of the road in Figure 3.12 is known to be N10°E and that the photointerpreter's tentative identification of a feature lying some distance off the road needs to be field checked. We want to know what bearing a field party should follow to go across country from the cabin on the road to reach the feature.

The first step is to draw the desired line of travel on the photo or an overlay. This is represented by the dashed line in Figure 3.12. Since the desired line of travel and the road intersect, the next step is to measure the angle between them with a protractor. Assume that the angle is measured to be 45°. The desired line of travel is then 45° to the left of N10°E, or N35°W.

If the known line and the desired line had not intersected on the photo, an intermediate line could have been drawn to connect them. The direction of the intermediate line would be determined by the procedure described, and then the intermediate line would replace the known line.

If a map is being used to determine directions, the north arrow will provide the necessary directional reference. There are several ways to establish the direction of a baseline when airphotos are being used. Obviously one way is to make a trip into the field with a compass. Very often a reference direction can

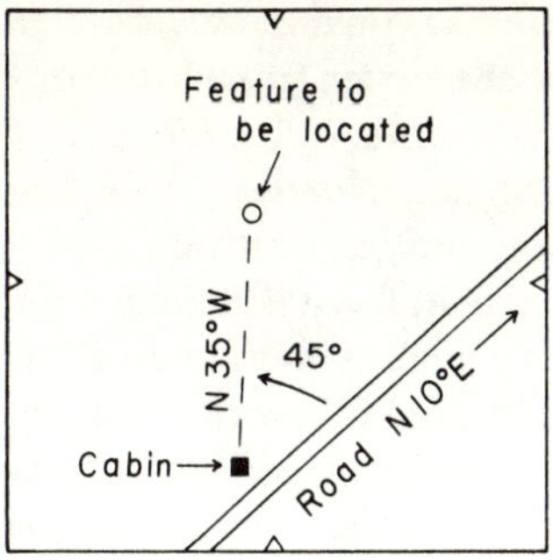

Figure 3.12 Determining directions from an air-photo or map.

be determined from an existing map without making a trip into the field. Still another way is to work from the time that the photos were taken, the locality of the photo, and the direction in which the shadows are cast. Although the latter procedure can get a little complicated, it may be worthwhile if the area cannot be visited and maps are not available.

3.5 DISTANCE

Determination of distance is a necessary component of virtually all survey operations. Quite a variety of methods for measuring and expressing distances are available. The choice among alternative ways will depend on accuracy requirements coupled with time, personnel, and equipment available.

It is important to note that the standard procedure in most surveys is to measure distances along the horizontal rather than along the sloping surface of the ground. Ways of obtaining the horizontal distance as opposed to the slope distance will be considered at several points in the remaining part of this chapter.

3.5.1 Units of Measure for Distance

The most familiar units of measure for distance in the English system are *inches* (in.), *feet* (ft), *yards* (yd), and *miles* (mi); there are 12 in./ft, 3 ft/yd, and 5280 ft or 1760 yd/mi. There are several other English system units that are less familiar to the layman, but are nevertheless commonly used in land surveys. One of these units is the *chain.* A chain is 66 ft long and is subdivided into 100 *links* of 7.92 in. each. A mile, therefore, contains 80 chains. The *rod* (also sometimes called *pole* or *perch*) is 16 ft in length. Thus there are 4 rods/chain. It should also be noted that the *nautical mile* differs from the more familiar *statute mile.* The international standard nautical mile used for air and sea navigation as well as hydrographic surveys is 6076.097 ft in length as compared to 5280 ft for the statute mile.

Because of its decimal nature, the metric or SI system is considerably simpler than the English system. The basic unit of length is the *meter* (m) which contains 39.37 in. Longer distances are measured in 1000 m units called *kilometers* (km). Smaller lengths are measured in $1/10$ m units called *decimeters* (dm), $1/100$ m units called *centimeters* (cm), or $1/1000$ m units called *millimeters* (mm).

Surprisingly enough the chain makes a fairly convenient mental standard for linking the metric and English systems. A chain is just slightly more than 20 m long, with the actual conversion being 20.11684 m/chain. Likewise there are approximately 50 chains (actually 49.71)/km. The figure of 20 m/chain is usable for converting relatively long distances from miles to meters (or kilometers) if one adds an additional correction of 10 m/mi. Use of this conversion gives 161,000 m (or 161 km)/100 mi, which is only about 65 m too large.

3.5.2 Pacing

Pacing is undoubtedly the oldest way of measuring distances. Since it is pretty hard to walk off and forget your feet, one need never be without a yardstick if one learns to utilize the regularity of the natural stride in measuring distances. A *pace* is generally considered to be 2 steps. Accordingly a pace is counted every time the right foot hits the ground. Some people may prefer to use the left foot for counting paces, and this is perfectly alright too. The length of the natural stride differs from one person to the next, and experience has shown that more consistent results are obtained by using this natural stride than by attempting to develop a pace of convenient length such as 6 ft. Therefore the first step in measuring distances by pacing is to calibrate the length of your natural pace. Mark out a known distance on some relatively level piece of ground and pace this distance several times to develop conversion factors from paces to meters or chains and vice versa. You may want to make a small pacing table to help in doing the conversions.

The natural tendency is for the pace to lengthen on open, firm ground and to shorten on soft ground with dense vegetation. Likewise, both uphill and downhill slopes tend to shorten the effective horizontal pace, but uphill slopes tend to shorten it much more than downhill slopes. Be aware of these tendencies at first, and you will gradually find your pace becoming quite stable under most conditions. In very hilly terrain it may be necessary to take more direct corrective action. One approach is to drop a pace count every so often according to the steepness of the slope and whether it slants up or down. A second approach is to cut a stick the same length as your pace. Use this stick to measure off horizontal paces as you scramble up or down steep hills.

As with anything else, practice makes perfect. An experienced person should be off by no more than 1 chain in a mile, or 1 part in 80, provided that he or she has kept count of the paces correctly. If losing count of paces seems to be a problem, a hand operated tally meter can be used as an aid. Instruments called *pedometers* are also available which will count the number of steps taken by the person carrying it. Pacing is widely used in reconnaissance surveys (those where accuracy requirements are low) because it is quick and convenient for a person working alone. However, the low order of accuracy rules it out for more critical survey operations.

3.5.3 Taping

Almost everyone is pretty familiar with the general nature of measuring tapes since simple versions may be found in any well equipped household. Like most other survey equipment, the quality and construction of tapes vary according to the accuracy required. Fabric tapes are light and inexpensive, but give a rather low order of accuracy because they do stretch to some degree. Metal or fiberglass strands may be incorporated in fabric tapes to give better stability. Metal tapes are more serviceable for most applications since they withstand heavy use, and their stretch under tension is almost imperceptible. For extreme accuracy one must also take into account the expansion and contraction of metal tapes with changing temperature. When such extreme accuracy is needed, tapes made of special alloys with a low coefficient of thermal expansion should be used.

There is a similar variety in provisions for storing and carrying tapes when not in use. Short tapes which will be used primarily for measurements that do not exceed the length of the tape usually have an enclosed self-storing reel from which the tape cannot be detached easily. These tapes are often called "box" tapes. When the tape is to be used for measuring distances in excess of its length, having the case attached to the tape becomes quite cumbersome. In this situation the tape is usually stored on an open reel from which it can be removed easily. An alternative to the use of reels for metal tapes is to form the tape into a double looped coil for carrying and storage. The use of such a double looped coil is called "throwing" the tape. Throwing a tape is something of an art which requires a little practice if proficiency is to be attained. The process is difficult to describe in words and is best learned by a demonstration from someone who has acquired the knack. Regardless of the construction and mode of storage, however, any tape must be used and cared for carefully. Kinks in the tape are to be strictly avoided, and the tape must be cleaned and dried after use.

As stated earlier, the object in using a tape is normally to measure horizontal distance rather than distance along the sloping surface of the ground. If the slope is too steep to allow holding the entire length of the tape horizontally, it is necessary to take the measurement in several parts each of which is made with whatever fraction of the tape can be held horizontally. Using less than the total length of the tape in this manner is called "breaking chain," but the term obviously should not be taken to mean that the tape is physically broken.

Two persons are required for the taping operation, which is also frequently called "chaining." The head chainman advances along the line to be measured holding the zero end of the tape and carrying a set of wire markers called chaining pins. When the tape has been run out the required distance, the rear chainman alerts the head chainman by calling "chain." The tape is then tightened and aligned carefully, during which it is held up off the ground and horizontal. When the tape is in correct position, the rear chainman calls out "stick" to indicate that the head chainman should mark the position of the end of the tape with a chaining pin. After placing the chaining pin in position, the head chainman responds with "stuck" to signify that the rear chainman should remove the last chaining pin and prepare to advance.

If the full tape is being used each time, the tally of tape lengths is equal to the number of pins held by the rear chainman. Since there are 11 pins in a set and one pin must be left in the ground as a marker, a new tally is started after 10 tape lengths. If regular chaining pins are not available, they can be fashioned easily from heavy wire. For less accurate work it is common practice to dispense with chaining pins entirely and mark the end of the tape by scuffing the ground.

The point bears repeating that the tape should not be allowed to rest on the ground while a measurement is being taken, because one cannot apply enough force to tighten it properly in this position. When accuracy requirements are high, tape clamps should be used to provide a better grip on the tape which in turn allows use of greater tension in straightening the tape. Likewise plumb bobs can be used for greater accuracy in projecting the tape graduations to the ground for positioning chaining pins. A plumb bob is simply a pointed weight suspended from a string.

3.5.4 Measuring Wheels

Measuring wheels provide a very rapid and inexpensive approach to measuring distances under certain circumstances. The most common type of measuring wheel is basically a bicycle wheel with a handle attached for pushing it along the ground. A tally meter is attached to the frame, and this tally meter is advanced at each revolution of the wheel by a small finger projecting from the hub. Some measuring wheels also have clamps for attaching them to the bumper of an automobile.

Measuring wheels perform quite well on smooth, level surfaces. The margin of error in such ideal circumstances is one revolution (circumference) of the wheel. Over relatively long distances this margin of error is fairly small. Unfortunately there are several factors which limit the use of measuring wheels. The nature of the measuring wheel is to measure the length of the line of travel along the surface, whether or not the line is straight or the surface is level. Therefore the wheel cannot be allowed to stray from the line as it is being pushed, and the measured length will be along the horizontal only if the surface is level. Furthermore roughness of the surface may cause the wheel to bounce and obstructions along the line cause obvious difficulties.

3.5.5 Rangefinders and Stadia

Several types of devices allow one to measure a distance in the field without the necessity of moving directly along the line to be measured. However, all these devices do require an unobstructed line of sight. Most such systems are based on the geometry of triangles.

Short Base Rangefinders. The most common design for rangefinders completely avoids the need to visit the opposite end of the line to be measured. A rangefinder of this type consists of two lines of sight separated by a short distance. The distance between the lines of sight forms the short leg of a right triangle. One of the lines of sight is perpendicular to this leg, forming the right angle. The angle between this short leg and the other line of sight can be changed by moving the dial. This geometry is illustrated in Figure 3.13. When both lines of sight are directed at an object to measure its distance, a right triangle is formed with the object at one vertex. Since the leg within the instrument is very short in comparison to the other two sides, this design is called a "short base" rangefinder. Multiplying the tangent of the variable angle by the length of the short base gives the distance to the object in question. The scale is usually graduated directly in units of distance, making calculations unnecessary.

The major problem with this design is its lack of accuracy for anything other than short distances. The tangent function increases extremely rapidly for angles between 89 and 90°, tending asymptotically to infinity at 90°. Therefore an almost imperceptible change in angle gives a large difference in distance. The result is that accuracy in this type of rangefinder decays rapidly with increasing distance.

The Stadia Principle. If one is willing to hold a graduated rod at the opposite end of the line to be measured, somewhat better results can be obtained by using the stadia principle. The basic idea in a stadia is to project a small fixed angle from one end of the line and then use a graduated rod at the other end to see how far the sides of the angle have spread. This geometry is illustrated in Figure 3.14. Let us use the term "stadia interval" for the length intercepted on the rod and denote the fixed angle by θ. As long as the graduated rod is perpen-

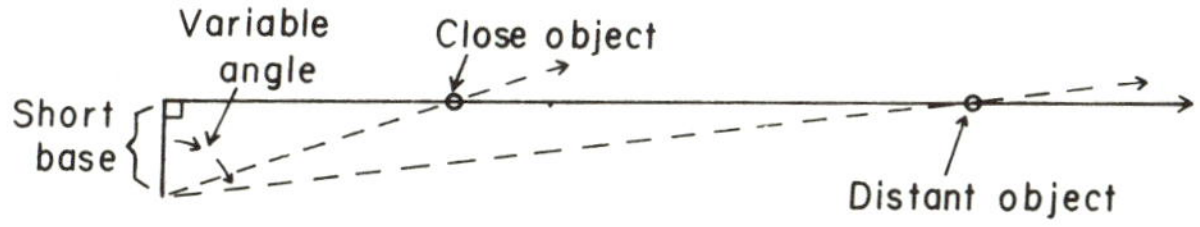

Figure 3.13 Geometry of the short base rangefinder.

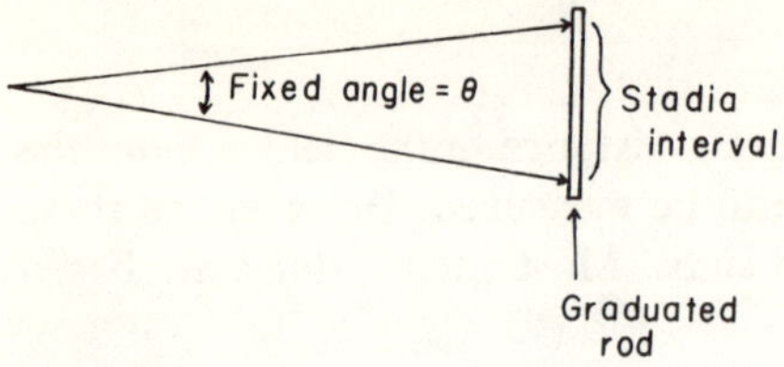

Figure 3.14 Geometry of the stadia system.

dicular to the bisector of the angle θ, the distance from the vertex of the angle to the rod is given by

$$\text{distance} = \text{stadia interval} \times \frac{1}{2} \cot \frac{\theta}{2}$$

The factor $\frac{1}{2} \cot (\theta/2)$ is usually called the "stadia interval factor."

Projection of the angle is usually accomplished by means of a pair of crosshairs in the telescope of a surveying instrument. If the telescope has three horizontal crosshairs, then the upper and lower ones are stadia hairs for reading the length intercepted on a graduated rod (stadia rod). To ensure that the stadia rod is perpendicular to the bisector of the angle the rod is held vertical and the telescope leveled with a level bubble. When a telescope of this type is used, the vertex of the angle does not coincide exactly with the center of the telescope. To correct for this difference, a "stadia constant" is added to the distance as calculated from the formula above. Thus the complete formula for distance by horizontal stadia is

$$\text{distance} = \text{stadia interval} \times \text{stadia interval factor} + \text{stadia constant}$$

Specifications as to the values of the stadia constant and the stadia interval factor should be provided with the instrument.

If the telescope is inclined at an angle ϕ from the horizontal when the stadia interval reading is taken on a vertical rod, correction factors must be applied to both the stadia constant and the stadia interval factor in the above formula to obtain *horizontal* distance. The stadia constant is corrected by multiplying it by $\cos \phi$. The stadia interval factor is corrected by multiplying it by the square of $\cos \phi$.

When the task is to lay out a predetermined distance instead of measuring an unknown distance, one can make a rangefinder for the purpose by using the stadia principle in reverse. Use the stadia formula to determine what length will be intercepted on a rod at the given distance. Mark off this length on a rod, then move the rod away from the instrument (or vice versa) until the angle covers the marked length. This reverse stadia system is most often implemented in small, hand held instruments with a stadia constant of zero. A target of the correct length is placed at one end of the line, and the observer backs away with the instrument until the projected angle covers the target.

Electronic Systems. Both short base rangefinders and stadia systems are derived from the geometry of triangles. Although the stadia has a longer usable range than the short base system, neither of these gives an accuracy comparable to that of careful measurement with a tape. The situation is quite different, however, for a class of systems based on the known wavelength of electromagnetic radiation. A system of this type uses an instrument at one end of the line which generates a signal of known wavelength. The signal is reflected from a target at the other end of the line. The distance can be determined either by timing the signal from emission to return or by measuring the amount by which the generated and returned signals are out of phase. The letters EDM (electronic distance measurement) are commonly used as a label for systems of this type. Radar and sonar also fall within this general category, except that sonar is based on sound waves. EDM systems are capable of giving extremely accurate measurements of long distances. However, such systems are also relatively expensive.

3.5.6 Scaling Distance from Vertical Airphotos and Maps

Given the definitions and examples of Section 3.3.2, the procedure for scaling distances from maps or vertical airphotos of known scale should be rather straightforward. The process consists of three simple steps. First measure the desired distance on the photo or map very carefully. Second multiply the measured photo or map distance by the denominator of the scale fraction to obtain the ground distance. Third perform any necessary conversion of units.

As an example suppose that two road intersections are shown on a vertical airphoto with scale 1/40,000. It is desired to know the distance between these two intersections expressed in kilometers. Suppose further that in carrying out the first step, this distance on the photo is measured to be 12.4 mm. Multiplying by 40,000 in step two gives 496,000 mm as the equivalent ground distance. Dividing by 1,000,000 mm/km gives 0.496 km as the desired distance.

A note of caution is that the measurement on the photo must be done very carefully. At a scale of 1/40,000 each difference of 0.1 mm in the measurement makes a difference of 4 m in the result. Likewise a difference of 0.01 in. makes a difference of 33 ft in the result. Therefore it is important to make photo measurements with a very finely graduated scale and to make the readings with a magnifying glass. An 8× or 10× lens works quite well for this purpose. Likewise one should postpone any rounding of the computations until the end.

3.5.7 Lengths of Curved Lines

There are essentially two approaches to measuring the lengths of curved lines. One of these approaches is to approximate the length of the curved line by a series of short segments (chords). The other is to use a measuring device that will follow the curving course of the line.

When the measurement is being taken directly in the field, any of the methods presented for measuring straight lines can be used for the approximation by chords. The measuring wheel is really the only device that is capable of moving along a curved path in the field.

When measurements are taken on a map or airphoto, there are three strategies for implementing the approximation by chords. One is to mark the turning points along the line; measure the chords between the turning points individually; and add up the lengths of the several chords to obtain the total. The second is to mark the turning points as before; transfer the distances between the turning points to the edge of a piece of paper in cumulative fashion, thus straightening out the connected chords into a single line; then make a single measurement of this cumulated line. The third is to mark the turning points with pins; weave a string along the pins; then mark or clip the string and measure its length. The second strategy is probably the easiest in most circumstances, but the third is also quite workable if pin holes are not objectionable and accuracy is not a major concern.

The equivalent of the measuring wheel for use on maps or photos is a *map measurer* or *opisometer*. This type of device has a small wheel which is guided along the line to be measured. The distance moved along the circumference of the wheel is shown on a dial in inches or millimeters, depending on the graduation of the dial. The distance as read from the dial can then be expanded according to the scale of the map or photo.

A more sophisticated approach for maps and photos is to digitize the line electronically on a very fine grid and then compute its length to an extremely close approximation by numerical methods.

3.5.8 Lengths of Obstructed Lines

When a line is obstructed so that a satisfactory measurement cannot be made directly along its length, one has no alternative but to circumnavigate the obstacle.

Offsets. If the obstacle is relatively small, one can use a simple offset which does not involve much computation. Two types of simple offsets are shown in Figure 3.15.

The one approach to offsetting is to run two sides of an equilateral triangle (all sides of the same length). First turn a 60° angle away from the obstacle and proceed in this direction far enough to clear the obstruction. Then turn a 120° angle toward the obstruction, and go an equal distance in that direction. Finally turn another 60° angle to get back on the original line. The third side of the triangle will be that part of the original line that lies across the obstruction, and its length will be the same as that of either offset line.

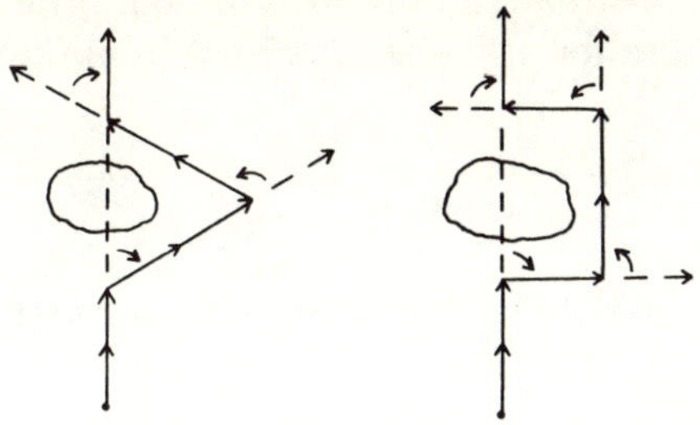

Figure 3.15 Two types of simple offsets for avoiding obstructions on a line. The offset on the left forms an equilateral triangle; the offset on the right forms a rectangle.

The other approach is to use right angle offsets to run three sides of a rectangle. The two sides perpendicular to the original line must be of equal length. The side parallel to the original line of travel is equal to the missing piece of the line.

Traverses. When the nature of the obstruction precludes using a simple offset, then the problem can be resolved by means of a traverse. A traverse is a sequence of straight lines connected end to end, with the length and direction of each line being known. Traverse lines do not cross, and they meet only at a turning point or *station* where one line ends and one other line begins. In short, a traverse is a more or less zigzag path of travel. If the end of the last line in the sequence connects back with the beginning of the first line to "fence in" an area, then the traverse is said to be *closed*. Otherwise the traverse is said to be *open*.

In order to determine the length of an obstructed line, one can run an open traverse which goes around the obstruction and connects the ends of the line. Each line of the traverse is then resolved into north/south and east/west components. These components are used to obtain a right triangle having the obstructed line as hypotenuse. One leg of the triangle is formed by the algebraic sum of the north/south components, and the other leg is formed by the algebraic sum of the east/west components. The Pythagorean theorem can then be applied to find the length of the obstructed line.

The north/south component of a line is called the *latitude* of the line, being positive if the line goes in a northerly direction (northeast and northwest quadrants) and negative if the line goes in a southerly direction (southeast and southwest quadrants). The magnitude of the latitude for a line is calculated from the formula

$$|latitude| = \cos \text{bearing angle} \times \text{length of line}$$

where the vertical bars on either side of latitude indicate *absolute value* (sign ignored). Latitudes in the southeast and southwest quadrants are then given a negative sign.

The east/west component of a line is called the *departure*, being positive if the line goes in an easterly direction (northeast and southeast quadrants) and negative if the line goes in a westerly direction (northwest and southwest quadrants). The magnitude of the departure for a line is calculated from the formula

$$|\text{departure}| = \text{sin bearing angle} \times \text{length of line}$$

A negative sign is then attached to departures in the northwest and southwest quadrants.

As an example, suppose that the ends of an obstructed line are connected by running an open traverse consisting of the following three lines:

Line	Bearing	Length
1	N30°E	50 m
2	due N	80 m
3	N60°W	40 m

The latitudes, departures, and their algebraic sums are as follows:

Line	Latitude	Departure
1	43.3	25.0
2	80.0	0.0
3	20.0	−34.6
Sum	143.3	−9.6

Applying the Pythagorean theorem gives:

$$\text{length} = \sqrt{(143.3)^2 + (-9.6)^2} = 143.6 \text{ m}$$

as the length of the obstructed line.

A traverse also provides the information needed to calculate the direction of the obstructed line. The formula for the bearing angle of the obstructed line is

$$\text{bearing angle} = \arctan\left(\frac{|\text{sum of departures}|}{|\text{sum of latitudes}|}\right)$$

The quadrant in which the bearing lies is determined by reference to the signs of the departure sum and the latitude sum as follows:

both positive = northeast quadrant
both negative = southwest quadrant
negative latitude, positive departure = southeast quadrant
positive latitude, negative departure = northwest quadrant

For example, the direction of the line in the foregoing example is

$$\text{bearing angle} = \arctan(9.6/143.3) = 3.83°$$

in the northwest quadrant, or N3.83°W.

One disadvantage of an open traverse is that there is no check on the accuracy of the work. If one is willing to spend the extra time, a check on the accuracy can be obtained by continuing the traverse so that it closes back on the starting point over a different route. The accuracy can be checked as both the latitudes and the departures in a closed traverse should sum algebraically to zero. If this is not the case, then an *error of closure* exists. The error of closure is the straight line distance between the starting point and the ending point of the traverse as calculated from the directions and distances recorded. The formula for calculation is

$$\text{error of closure} = \sqrt{(\text{latitude sum})^2 + (\text{departure sum})^2}$$

which is the same as that for calculating the length of an obstructed line from an open traverse. The error of closure can also be expressed as a percentage of the total length of the traverse in order to obtain a measure of relative error. If the error of closure is not within acceptable limits for the survey, the traverse needs to be rechecked in the field to determine the source of the inconsistency.

3.6 ELEVATIONS, SLOPES, HEIGHTS, AND DEPTHS

The need to measure some aspect of the third dimension arises in surveys almost as frequently as the need to measure directions, horizontal angles, and horizontal distances. The items most often of interest are elevation above a reference surface, slope of a line or surface, height of a vertical object above the ground, and depth below ground or water surface. The concepts of height and depth are so familiar that they require no preliminary comment. Elevation is the distance between a point and a reference surface (datum) as measured along a line passing through the point and the center of the earth. The datum is most often mean sea level, but other references may be used when interest lies only in relative elevations of features within a restricted area. Since slope and ways of expressing it are less familiar, a more detailed introduction is in order.

3.6.1 Methods of Expressing Slopes

Slope is a measure of steepness, or a rate of rise or fall with a change in the horizontal position. Several alternative types of scales are available for expressing the steepness of a slope.

One way of scaling a slope is to measure the vertical angle (angle in a vertical plane) between the sloping line or surface and the horizontal, with the angle being expressed in degrees. Angles above the horizontal are taken to be positive, and those below the horizontal are taken to be negative. In this system 0° indicates a horizontal line or surface, and 90° indicates a vertical line or surface.

An alternative approach to scaling slopes is to express steepness by the number of units of vertical rise or fall that take place with a given amount of horizontal movement. The several variations on this approach have to do with the amount of horizontal movement that is taken as a base. The *percent slope* scale gives units of rise or fall per hundred units of horizontal distance. Slope as measured in *topographic units* gives units of rise or fall per 66 units of horizontal movement. Likewise scales for other selected horizontal base movements are also used.

Given the slope angle in degrees, the slope reading on any scale of this type can be obtained from the formula

$$\text{slope reading} = \tan \text{slope angle in degrees} \times \text{base distance}$$

Thus a slope angle of 45° when expressed on a percent scale, is

$$\tan 45^\circ \times 100 = 1.0 \times 100 = 100\%$$

Likewise a slope angle of 45° when expressed in topographic units is:

$$\tan 45^\circ \times 66 = 1.0 \times 66 = 66 \text{ topographic slope units}$$

In fact a slope angle of 45° will be equal to the base horizontal distance on any scale of this type. The range of slopes on scales of this type varies from 0 for a horizontal line or surface to infinity for a vertical line or surface.

Readings on a scale having a base of A units can be converted to a scale having a base of B units by the formula:

$$\text{reading with base } B = (\text{reading with base } A) \times \frac{B}{A}$$

For example, a topographic slope reading of 33 is equivalent to a percent slope reading of:

$$33 \times \frac{100}{66} = 50\%$$

3.6.2 Horizontal Distance in Relation to Slope Distance

The methods presented in Section 3.5 are designed primarily for a direct determination of the horizontal distance. There are times, however, when it will be more convenient to make the original distance measurements along the sloping surface of the ground with a subsequent mathematical conversion to the equivalent distance along the horizontal. The slope angle in degrees must also be known in order to calculate the horizontal distance. Instruments for measuring slope angles will be discussed in Section 3.6.5, but the mathematical relationships involved will be considered at this point.

The basic relationship for converting slope distance to horizontal distance is

$$\text{horizontal distance} = \text{slope distance} \times \cos \text{slope angle}$$

As mentioned above, this formula assumes that the slope angle is expressed in degrees. For example, the horizontal component of a 50 m distance measured on a 10° slope is

$$\text{horizontal distance} = 50 \text{ m} \times \cos 10^\circ = 50 \text{ m} \times 0.9848 = 49.24 \text{ m}$$

If the slope angle is measured in some other units such as topographic or percent slope, there are two alternative ways of converting it to a form that is usable in the above formula. One way is to convert the slope reading to degrees by inverting the formula

$$\text{slope reading} = \tan \text{slope angle in degrees} \times \text{base distance}$$

to get

$$\text{slope angle in degrees} = \arctan\left(\frac{\text{slope reading}}{\text{base distance}}\right)$$

where base distance is the horizontal distance on which the slope scale is constructed (for example, 100 units for percent scale). The other way is to make use of the trigonometric identity

$$\sec^2\theta - \tan^2\theta = 1$$

to find the cosine of the slope angle from

$$\cos \text{slope angle} = \frac{1}{\sqrt{1 + (\text{slope reading/base distance})^2}}$$

Although less frequently required, conversions from horizontal distance to slope distance may be accomplished by the formula

$$\text{slope distance} = \sec \text{slope angle} \times \text{horizontal distance}$$

As before, the slope angle must be expressed in degrees for use in this formula. For other slope scales the secant term can be obtained directly from the formula

$$\sec \text{slope angle} = \sqrt{1 + (\text{slope reading/base distance})^2}$$

3.6.3 Elevations by Leveling

The least complicated and most accurate method of determining differences in elevation between points on the ground is to use a *level* and *leveling rod.* Since this method only gives elevation differences, a reference point of known elevation must be located in the vicinity to serve as a starting point. If one only needs to know relative elevations over a local area without expressing them in relation to the mean sea level, any convenient starting point can be assigned an arbitrary base value of elevation.

In brief the procedure is this. The level is positioned at a place from which stations (points) of both known and unknown elevations are visible. The sighting plane of the level must be above both stations, but by not more than the height of the leveling rod. If the difference between stations is greater than the height of the rod, it will be necessary to establish intermediate stations. The rod is held vertically at the station of known elevation and read through the level. The elevation of the sighting plane for the level is calculated by adding this rod reading to the known elevation for the station where the rod is situated. The rod is then moved to the station of unknown elevation, the level is pivoted to face the rod, and the rod is read through the level. The elevation of the unknown station is calculated by subtracting this second reading from the elevation of the sighting plane for the level. This station now becomes one of known elevation, and the process can be repeated for other stations for which elevations are to be determined.

Most applications will call for determining the elevations of a sequence of points along a line of travel. This is called "running a line of levels." The pattern of alternating the movement of level and rod in the process of running a line of levels is illustrated in Figure 3.16. The rod stations correspond with positions for which the elevation is to be determined. The level positions are selected arbitrarily for ease of intervisibility between rod stations and do not have to be located on a straight line between adjacent rod stations. This pattern of alternately calculating the elevation of the sighting plane and the elevation of a rod station is shown in Figure 3.17.

When the acceptable error is on the order of a foot rather than a fraction of an inch, it is satisfactory to use a simple and inexpensive hand level. In its simplest form the hand level is just a sighting tube with a level bubble. The sighting tube contains a mirror or prism so that the bubble is visible through the tube. The use of a hand level is pretty straightforward. In this case the height of

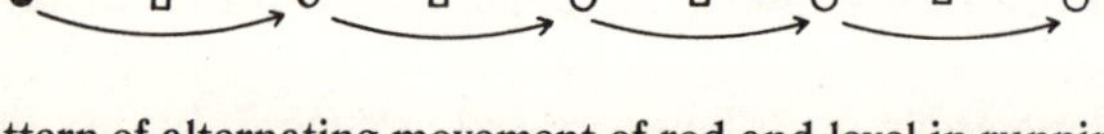

Figure 3.16 Pattern of alternating movement of rod and level in running a line of levels. Solid circle indicates station of known or assumed elevation which provides the starting point. Open circles indicate other rod stations for which elevations are to be determined. Open squares indicate positions occupied by the level.

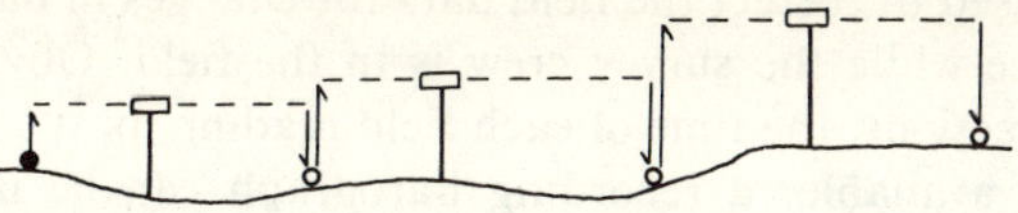

Figure 3.17 Pattern of alternately calculating elevation of sighting plane and elevation of rod station. Solid circle indicates position of first rod station. Open circles indicate positions of subsequent rod stations. Dotted lines indicate sighting planes of level. Upward arrow shows that rod reading is to be added to station elevation to calculate elevation of sighting plane. Downward arrow indicates that rod reading is to be subtracted from elevation of sighting plane to obtain elevation of rod station.

the sighting plane above ground is the height of your eye; however, the readings will be more accurate if the level is rested against a staff. The lines of sight must be kept short with a hand level, or the results will be quite erratic. Hand levels with stadia lines are also available.

When the standards of accuracy are more stringent, the hand level is inadequate and an engineer's level or transit must be used instead. In order to achieve accuracy it is extremely critical that the telescope of the instrument be level when the readings are taken. Coarse leveling of the instrument is done by moving the legs of the tripod. Fine leveling is then done with the knurled screws that bear on the head of the tripod. When the process of leveling the instrument is complete, the bubble should not move off center no matter what direction the telescope is turned.

When a reading is being taken from a leveling rod, the rodman should be instructed to wave the rod slowly toward and away from the level. The rod is read at the smallest graduation appearing at the crosshair during the waving. This ensures that the rod is vertical when the reading is taken.

3.6.4 Elevations by Aneroid Barometer

When the requirements of accuracy are low, an aneroid barometer provides a rapid means of determining elevation. The barometer is sensitive to changes in the atmospheric pressure which decreases with increasing elevation. A good aneroid barometer will show changes in elevation as small as 10 ft. This makes the aneroid suitable for use in preparing rough topographic maps or determining spot elevations.

There are, however, several sources of error in using an aneroid barometer. Changes in temperature will affect the instrument, and corrections for this can be incorporated in the design. More importantly, the barometric pressure at a given elevation fluctuates over a relatively short period of time due to meteorological conditions. Anyone who listens to weather reports should be familiar with this sort of fluctuation. The barometer must be checked periodically at a known elevation, or an assistant must take periodic readings on a second barometer that is kept at constant elevation. The readings on the second barometer are used to correct the field data for changes in barometric pressure which take place while the survey crew is in the field. Obviously in order to make these corrections, the time of each field reading must also be noted. If an assistant is not available, a recording barograph can be used to determine changes in the barometric pressure over time.

3.6.5 Elevations and Heights from Vertical Angles

The utility of leveling methods and barometers is limited to determining elevations. Neither of these techniques is useful for measuring heights of tall objects. Vertical angles provide a more versatile approach which can be used for elevations, heights, and slopes. Instruments for measuring vertical angles will be considered prior to procedures for doing the calculations.

Instruments for Measuring Vertical Angles. Instruments for measuring vertical angles include both hand held and tripod mounted types. Tripod mounted types are the most accurate, but hand held types are far more portable and much quicker to use. The term *clinometer* will be used here as a generic name for hand held models. The reader should be aware, however, that a variety of other terms such as altimeter, inclinometer, hypsometer, and level are also in common use. The most common among the tripod mounted types is the engineer's transit. However, several miniature versions of the transit are also available at considerably lower cost.

Clinometers. Clinometers measure angles in a vertical plane upward or downward from the horizontal. One of two basic principles is employed to define the horizontal reference. One is the level bubble, and the other is the plumb bob or pendulum. The scale may be graduated in any of the various ways of expressing slope, as discussed in Section 3.6.1.

To use a clinometer based on the level bubble, first direct the sighting line along the sloping line to be measured. Holding the instrument steady in this position, move the lever that controls the bubble until the bubble is centered on the hairline. Now relax from the sighting position and read the slope as shown by the position of the indicator on the graduated scale. Typical instruments of this type are the Abney level and the clinometer that is built into the so-called "Brunton Pocket Transit."

There is a greater variety in the ways that clinometers built around a pendulum are assembled and operated. In some of these the scale is weighted so that it moves across a hairline located on the line of sight. In others the scale is fixed relative to the sighting line and a pendulum serves as the indicator. The latter type may be constructed so that the scale is visible in the sighting position and the reading must be made while the instrument is held in the sighting position. Alternatively the scale may not be visible in the sighting position, but a brake is provided to lock the indicator so that the reading can be taken after the instrument is lowered. Typical clinometers based on the pendulum are the Suunto clinometer, the Haga altimeter, the Blume-Leiss altimeter, and clinometer scales incorporated into several compasses. A homemade clinometer of this type can be fashioned by attaching a weighted string to a protractor.

Vertical Angles with a Transit. The transit is typical of tripod mounted instruments for measuring vertical angles. Such instruments are equipped with a graduated vertical circle for reading the inclination of the telescope, and with base levels for assuring that this graduated circle is oriented vertically. Although not present on some less expensive instruments, the telescope of a transit is equipped with a level bubble that is independent of the base level. When this level bubble indicates that the telescope is horizontal, the vertical circle should read zero. Failure of the vertical circle to read zero when the telescope is level is called *index error.* Readings of angles made upward or downward from the horizontal must be corrected for the presence of any index error.

Determining Heights from Vertical Angles. Determining the height on the basis of vertical angles usually involves the measurement of one horizontal distance and two vertical angles. The most common situation is depicted in Figure 3.18. In this case the observer's eye level is somewhere between the base and the top of the vertical object to be measured. The first step is to select a convenient sighting position from which the top and base of the object are both visible. From this position measure the vertical angle between the horizontal and the line of sight to the top of the object (angle Θ of Figure 3.18). An angle measured upward from the horizontal in this manner is also called an *elevation angle.* Next measure the vertical angle between the horizontal and the line of sight to the base of the object (angle ϕ of Figure 3.18). An angle measured downward from the horizontal in this manner is also called a *depression angle.*

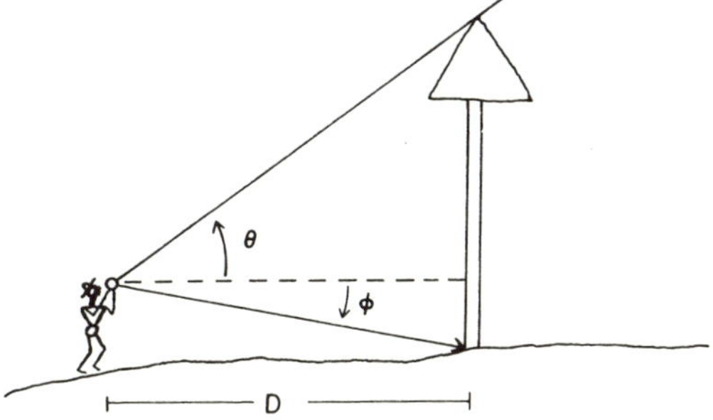

Figure 3.18 Geometry of height measurement by vertical angles when eye level falls between base and top of the vertical object. Dashed line represents horizontal line of sight. Angle θ is elevation angle, and ϕ is depression angle. D is horizontal distance from observer to vertical axis of object.

The distance from the observer's position to the vertical object must also be measured either before or after the angles are measured.

For purposes of doing the computations, we will treat both elevation and depression angles as being positive numbers. A negative sign on the scale of the clinometer serves to distinguish depression angles from elevation angles, but should not be carried into the formulas that follow. The notation of Figure 3.18 is used in the following formulas. If Θ and ϕ are read in degrees, the formula for computing the height is

$$\text{height} = D \times (\tan \Theta + \tan \phi)$$

with the result bearing the same units of measure as the horizontal distance D. For example, an elevation angle of 30° and a depression angle of 10° read from a horizontal distance of 100 ft would give a height of:

$$\begin{aligned}\text{height} &= 100 \text{ ft} \times (\tan 30° + \tan 10°)\\ &= 100 \text{ ft} \times (0.5774 + 0.1763)\\ &= 100 \text{ ft} \times 0.7537 = 75.37 \text{ ft}\end{aligned}$$

If Θ and ϕ are read on a slope scale having a horizontal base of H units (for instance, $H = 100$ for percent slope), the general formula for computing height is

$$\text{height} = \frac{D}{H} \times (\Theta + \phi)$$

which simplifies to

$$\text{height} = \Theta + \phi$$

when $H = D$.

There are two complications that may enter into the measurement of heights from vertical angles. One is that the object being measured may not be truly vertical. In this case the procedure must be modified so that the point on the ground directly below the top of the object is treated as if it were the base of the object. This is necessary because the geometry underlying the formulas is based on the assumption that the object is vertical. This modification is illustrated for an overhanging cliff in Figure 3.19.

The other complication is that the ground may slope so steeply that the observer is below the base of the object or above the top as illustrated in Figure 3.20. If the observer is above the top of the object, both angles will be depression angles. Likewise when the observer is below the base of the object, both angles will be elevation angles.

Given the notation of Figure 3.20, the modified formula for computing the

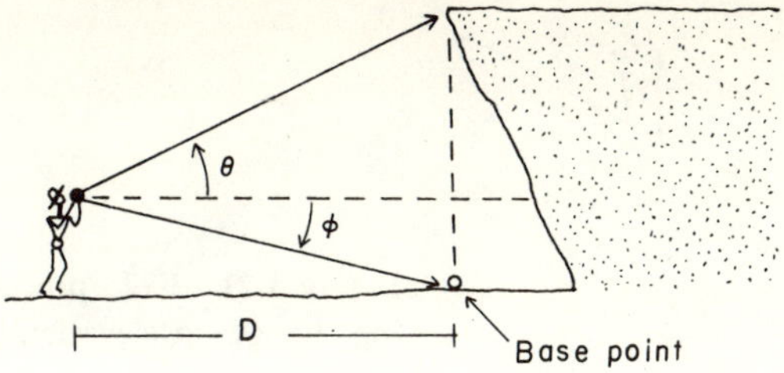

Figure 3.19 Use of base point vertically below top when object is not vertical.

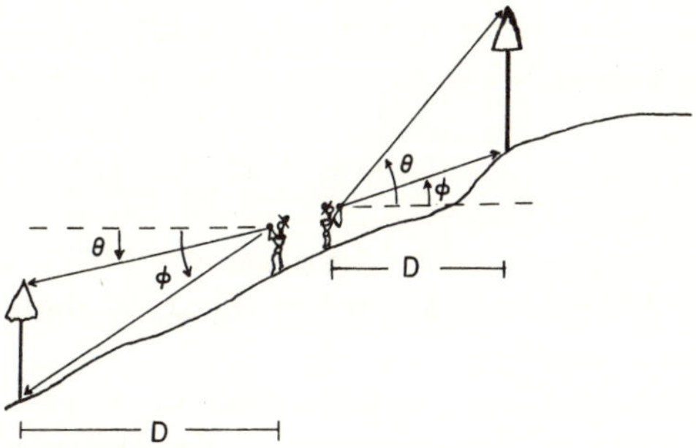

Figure 3.20 Geometry of height measurement when observer's eye level is not between base and top of object to be measured.

height from vertical angles in degrees when the observer's eye level is above the top of the object becomes

$$\text{height} = D \times (\tan \phi - \tan \Theta)$$

When the observer's eye level is below the base of the object, the modification is

$$\text{height} = D \times (\tan \Theta - \tan \phi)$$

Corresponding modifications are required in the formula for computing heights from angles measured in percent, topographic, or other slope units. Note also that the distance *D* is still measured horizontally in these modified formulas.

Determining Elevations from Vertical Angles. The field procedure for determining changes in elevation with a clinometer is illustrated in Figure 3.21. A sighting target is attached to a pole at the same height as the eye of the person who is to use the clinometer. The person using the clinometer stands at a point of known elevation, and an assistant holds the pole over the point for which the elevation is to be determined. The slope angle Θ is determined by sighting on

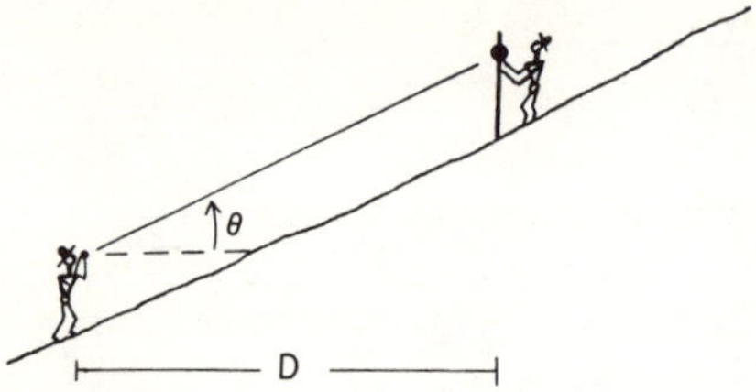

Figure 3.21 Field procedure for determining changes in elevation by clinometer.

the target with the clinometer. The horizontal distance between the two stations is also measured.

If the slope angle is measured in degrees, the formula for the difference in elevation between the two stations is

$$\text{elevation difference} = D \times \tan \Theta$$

in the same units of measure used for the distance D. If the slope angle Θ is an elevation angle, the difference is positive. If the slope angle Θ is a depression angle, the difference is negative.

If the slope angle is measured on some other slope scale having horizontal base H (for instance, $H = 66$ for topographic scale), the corresponding formula is

$$\text{elevation difference} = \frac{D}{H} \times \Theta$$

which simplifies to

$$\text{elevation difference} = \Theta$$

when $D = H$.

When horizontal distances are being measured in chains, the topographic slope scale coupled with a "two-chain trailer tape" provides a convenient way of compensating for the difference between slope distance and horizontal distance. In addition to the two-chain section graduated in the usual way, this type of tape has a "trailer" section which provides the means for compensation. After the tape has been run out the two chains *along the slope*, the ground slope is measured with a clinometer having a topographic scale. This slope reading is then located on the trailer section of the tape, and this additional amount of tape is run out along the slope to achieve a slope distance equivalent to a horizontal distance of two chains. The elevation difference (in feet) between the ends of the line is then equal to twice the topographic scale reading.

3.6.6 Heights and Elevations from Airphotos

The effect of relief in distorting the scale of vertical airphotos was discussed in Section 3.3.3, and the use of this distortion to produce the three-dimensional or stereo effect was covered in Section 3.3.5. In addition to creating the stereo effect, variations in scale due to relief also make it possible to measure heights and elevation differences from airphotos.

Heights from Single Airphotos. Assume that a vertical airphoto is taken from a flying height F over level terrain. Suppose further that two helium filled balloons are tethered from the same stake before the photo is taken. One is tied at ground level, and the other is tied on a long string so that it floats at a considerable height. As long as there is no wind, the balloon above should float vertically over the one on the ground. The geometry that produces the image of the floating balloon is shown in Figure 3.22. Since the shaded triangles are similar,

$$\frac{d_1}{f} = \frac{D_1}{F - H} \qquad \text{or} \qquad d_1F - d_1H = D_1f$$

Likewise the geometry that produces the image of the balloon on the ground is shown in Figure 3.23, from which

$$\frac{d_0}{f} = \frac{D_0}{F} \qquad \text{or} \qquad d_0F = D_0f$$

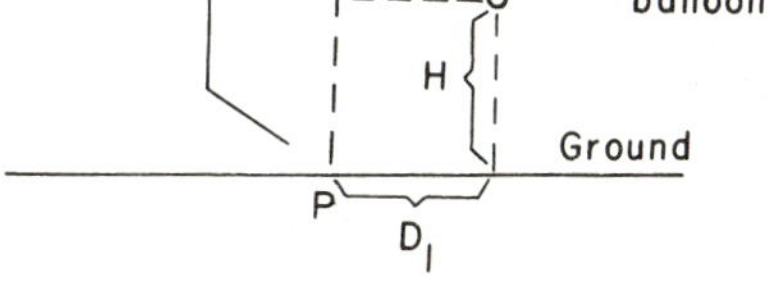

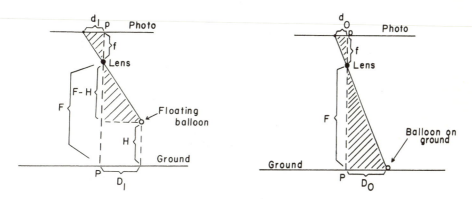

Figure 3.22 Diagram of geometry producing image of floating balloon in a vertical airphoto taken over level terrain. P is terrain point imaged at principal point p of photo. F is flying height; f is focal length; and H is height of balloon. D_1 is horizontal distance between P and balloon; and d_1 is distance on photo between principal point and image of balloon. Shaded triangles are similar.

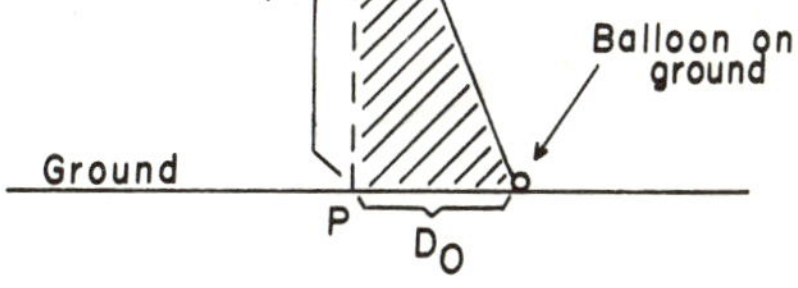

Figure 3.23 Diagram of geometry producing image of balloon on the ground in a vertical airphoto taken over level terrain. P is terrain point imaged at principal point p of photo. F is flying height and f is focal length of lens. D_0 is horizontal distance between P and balloon; and d_0 is distance on photo between principal point and image of balloon. Shaded triangles are similar.

Furthermore $D_0 = D_1$ due to the fact that the floating balloon is directly above the one on the ground. Therefore

$$d_1F - d_1H = d_0F$$

and

$$d_1H = d_1F - d_0F \qquad \text{or} \qquad H = \frac{F(d_1 - d_0)}{d_1}$$

The way that the two ballons would appear in the photo is illustrated in Figure 3.24. The two balloons are analogous to the top and base of any vertical object in the scene, and Figure 3.24 includes the image of a tree for comparison.

In order to find the height of a vertical object in such a photo, then, one can measure the distance d_0 from the principal point to the image of the base and also the distance d_1 from the principal point to the image of the top. The photo geometry is such that the images of the top and base will lie on the same straight line radiating from the principal point. Therefore the difference $d_1 - d_0$ corresponds to the apparent height h of the object as seen in Figure 3.24. The apparent height h is small in comparison to d_0, and better results are obtained if h is evaluated by direct measurement instead of by subtraction. Thus the operational formula for determining heights of vertical objects from a single vertical airphoto taken over level terrain is

$$\text{height} = \frac{F \times h}{d_1}$$

with the result having the same units of measure as the flying height F.

As an example suppose that it is desired to find the height of a tree shown in a vertical photo taken from a flying height of 8000 ft. Assume further that d_1 and h are measured as 2.00 in. and 0.02 in., respectively. Then the height of the tree is

$$\text{height} = \frac{F \times h}{d_1} = \frac{8000 \text{ ft} \times 0.02 \text{ in.}}{2.00 \text{ in.}} = \frac{160 \text{ ft}}{2.00} = 80 \text{ ft}$$

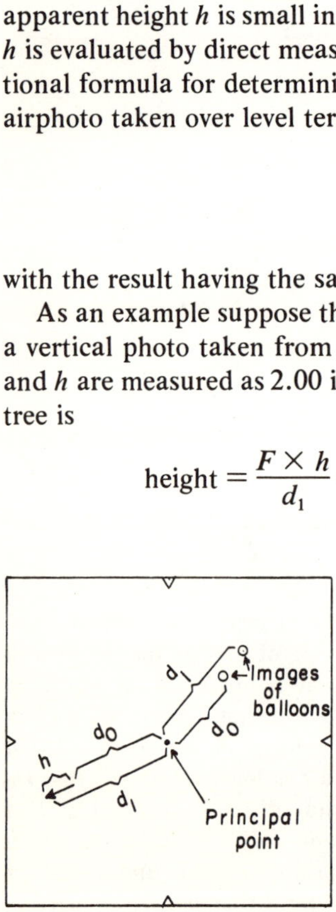

Figure 3.24 Diagram of the way vertical objects appear in an airphoto. Circles represent images of balloons from Figures 3.22 and 3.23. Apparent height of the tree in the photo is $h = d_1 - d_0$.

Heights and Elevations from Stereopairs of Airphotos. The foregoing procedure for determining heights from a single airphoto does have some limitations. There is no relief displacement at the principal point of a vertical photograph, so the method works poorly in the immediate vicinity of the photo center. Also the procedure is applicable only to measuring the heights of vertical objects. An equally important problem is to measure the differences in elevation of points on the ground surface which are not aligned vertically.

These disadvantages can be circumvented by combining the height information from the two photos of a stereopair into a single formula. It is always possible (except for part of a photo at the end of the flight line) to choose a stereopair in which both photos show the object of interest. Since the photos are taken from different positions, the object cannot be near the center of both photos. The problem with lack of relief displacement near the photo center is thus avoided. In the process of developing the procedure it will also be shown that the stereopair method works for finding differences in elevation between two points which are not aligned vertically.

When a stereopair of photos is set up in the proper position for stereo viewing (refer to Section 3.3.5), it is convenient to measure distances parallel to the flight line rather than radially from the principal point. Equivalent results are obtained in terms of formulas by examining the geometry after shifting objects to the flight line along a perpendicular, as shown in Figure 3.25. The photo geometry after projecting features onto the flight line is shown in Figure 3.26. From similar triangles in Figure 3.26,

$$\frac{d}{f} = \frac{D}{F - H} \quad \text{or} \quad d = \frac{fD}{F - H}$$

and

$$\frac{d'}{f} = \frac{G - D}{F - H} \quad \text{or} \quad d' = \frac{f(G - D)}{F - H}$$

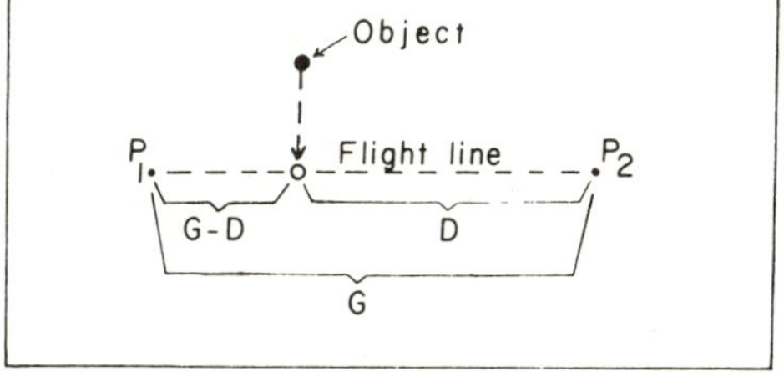

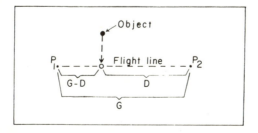

Figure 3.25 Projection of an object onto the flight line. P_1 represents principal point of first photo; and P_2 represents principal point of second photo. G is ground distance between principal points; and D is ground distance between object and P_2 as measured along the flight line.

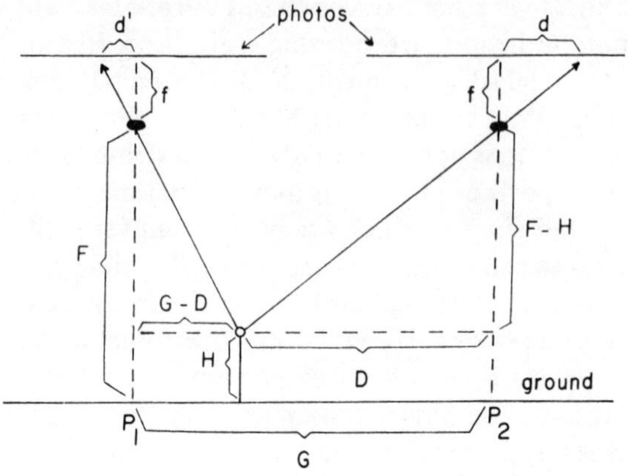

Figure 3.26 Geometry of relief displacement parallel to flight line in a stereopair. P_1 is ground location of principal point for first photo, and P_2 is ground location of principal point for second photo. G is ground distance between principal points of photos; F is flying height above horizontal reference plane; and f is focal length of camera lens. H is height of feature above reference plane; and D is horizontal distance between feature and P_2 as measured parallel to flight line. Distances on photos between image point and principal point as measured parallel to flight line are shown as d and d'.

Therefore

$$d + d' = \frac{fD + f(G - D)}{F - H} = \frac{fD + fG - fD}{F - H} = \frac{fG}{F - H}$$

so

$$d + d' = \frac{fG}{F - H}$$

The quantity $d + d'$ is called the *absolute parallax* of the image point. It is the sum of the distances (measured parallel to the flight line) between the image points and their respective principal points on the two photos, as illustrated in Figure 3.27. Note that the last formula does not involve the distance D of Figure 3.26, which means that the absolute parallax will be the same for all points having the same height above the horizontal reference plane. Thus it is not necessary to use points which are vertically aligned in working with formulas involving the absolute parallax.

Let us use the letter P to denote the absolute parallax of a point lying above the horizontal reference plane. The horizontal reference plane is the plane above which the flying height F is measured. In order to obtain a formula that can be used to compute the height above the reference plane, we also need an

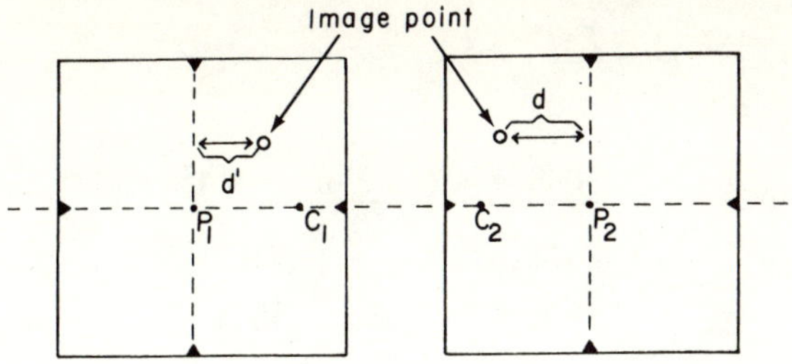

Figure 3.27 Components d and d' of absolute parallax for an image point in a stereopair of photos. Note that the image point lying to the left of principal point in Figure 3.26 is switched to the right of principal point when the photo is turned face up in this diagram, and vice versa. Absolute parallax $= d + d'$. P_1 and P_2 are principal points of photos; C_1 and C_2 are conjugate principal points.

absolute parallax measurement for a point lying on the reference plane. Let us denote the absolute parallax for a point on the reference plane by P_0. Then

$$P = \frac{fG}{F - H} \qquad \text{or} \qquad PF - PH = fG$$

and

$$P_0 = \frac{fG}{F} \qquad \text{or} \qquad FP_0 = fG$$

Equating the two expressions for fG, we have

$$PF - PH = FP_0 \qquad \text{or} \qquad PH = PF - P_0F$$

from which

$$H = \frac{PF - P_0F}{P} \qquad \text{or} \qquad H = \frac{F(P - P_0)}{P}$$

Thus the difference in elevation between any two points can be determined from measurements of the absolute parallax when the flying height is measured from the elevation of the lower point as a reference plane.

For example, suppose that we wish to know the difference in elevation between two points shown in a stereopair of airphotos. Assume that the photos were taken from a flying height of 9000 ft above the elevation of the lower point. Assume further that the absolute parallax of the lower point is measured to be 4.45 in. and that of the upper point to be 4.50 in. Then

$$H = \frac{F(P - P_0)}{P}$$

$$= \frac{9000 \text{ ft} \times (4.50 \text{ in.} - 4.45 \text{ in.})}{4.50 \text{ in.}}$$

$$= \frac{9000 \text{ ft} \times 0.05}{4.50} = \frac{450 \text{ ft}}{4.50} = 100 \text{ ft}$$

One point to note is that our development of the parallax method has been based on the assumption that the object in question is situated between the principal points of the two photos, as in Figure 3.26. This will usually be the case. If the object lies to the left of both principal points or to the right of both principal points, then the absolute parallax is calculated as the difference between d' and d instead of as their sum.

In practice the absolute parallax is usually determined from the differences between span measurements taken across a stereopair that has been taped down in the correct position for stereoscopic study. These span measurements are illustrated in Figure 3.28. Since the same object is being studied in both photos, the perpendicular distance from an image point to the flight line will be the same in both photos. Thus measurements made between corresponding image points are parallel to the flight line as required for determining the absolute parallax. It should be noted, however, that tilt in either photo, a change in flying height between photos, or a failure to line up the flight lines on the two photos will upset this parallelism. Since the positions of the principal points do not change, S_1 need only be measured for the first height determination on a stereopair.

The span distances between images are usually measured with either a micrometer wedge or a parallax bar (also called stereometer or height finder). A

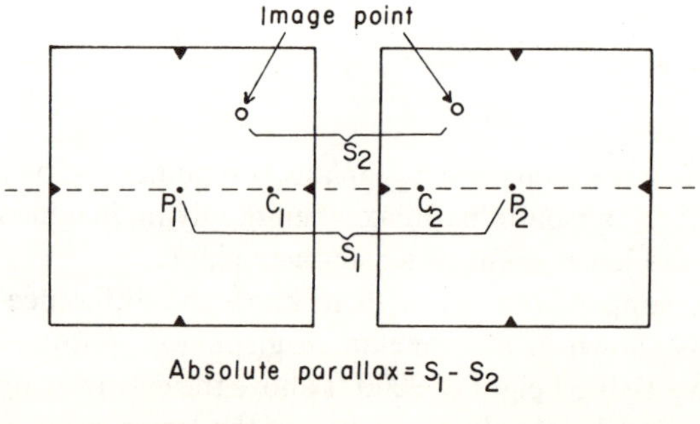

Figure 3.28 Absolute parallax from span measurements when object lies between principal points. P_1 and P_2 are principal points. C_1 and C_2 are conjugate principal points. Absolute parallax $= S_1 - S_2$.

micrometer wedge is a set of two diverging lines printed on heavy transparent overlay material to form a wedge. The distance between lines (measured perpendicular to the bisector of the wedge) is calibrated in intervals of 0.01 in. or less. Such a wedge may also be formed by a series of diverging dots instead of graduated lines. A parallax bar has a dot printed on each of two glass plates. One of the dots is fixed in position, but the other can be moved along the bar by a precision worm gear. The distance between the dots is indicated on the vernier scale of the parallax bar.

The stereoscopic effect provides the best method of using a parallax bar or wedge. When a parallax bar or wedge is properly positioned under a stereoscope, the two lines or dots will appear to fuse and float over the terrain. The closer they are together, the higher they will appear to float. The bar or wedge is read where the dot or crossbar seems to float at the same level as the image for which the span distance is being measured. A parallax bar or wedge can also be used like a ruler without the aid of a stereoscope, but the readings will not be as accurate as those obtained in stereo.

To use a parallax bar without a stereoscope, the distance between dots is adjusted to match that of the corresponding image points and the scale is read. A wedge is laid over the photos with its bisector perpendicular to the flight line and one arm over each photo. The wedge is moved horizontally or vertically (but not canted) until one arm is right over each image point and the scale is read at the image point. Since there are graduations on both sides of the wedge, it is not difficult to tell whether the wedge is in proper position. If the two sides (arms) of the wedge read differently, then the wedge is canted with respect to the flight line and must be straightened.

It should also be mentioned that a slightly different notation is commonly used in presenting the parallax height formula. This notation is

$$\text{height} = \frac{F \times dP}{P + dP}$$

where $F =$ flying height above the plane of the lower point
$dP = (P - P_0)$ in terms of the notation used previously
$P = P_0$ in terms of the notation used previously

In other words, P in this latter notation is the absolute parallax of the lower point, and dP is the parallax difference between the upper and lower points.

When the parallax method is used on photos of level terrain, the absolute parallax for the base of a vertical object will be the same as the base length (length of a line connecting the principal point with a conjugate principal point on the same photo). Because of this, the base length is often used as the denominator of the parallax height formula. In making this substitution, the usual procedure is to measure the base length on both photos and average the two figures which may be slightly different because of small irregularities in the flight pattern.

Heights from Shadows in Airphotos. Parallax measurements are not the only means of determining heights from airphotos. If the angle of the sun's elevation above the horizon at the time of photography can be determined, the lengths of shadows can be used to determine heights. The geometry of a shadow falling on level ground is shown in Figure 3.29. The first step is to measure the length of the shadow image and use the photo scale to find the actual shadow length on the ground. The height is then calculated from the formula

$$\text{height} = \text{shadow length} \times \tan \text{sun's elevation angle}$$

This formula only applies, however, to shadows cast on level ground. The reader is referred to Spurr (1960) for the effect of the ground slope and procedures for calculating the sun angle given the locality and the time of photography.

3.6.7 Depths

In the broadest sense, any decrease in elevation can be considered as a depth. However, methods for determining changes in elevation on the earth's surface have already been covered in previous sections of this chapter. The present concern is with situations where those methods cannot be used, as for instance in determining depths of water bodies or narrow holes in the ground.

Depths by Sounding Lines. Weighted lines, called *sounding lines,* are widely used for determining water depth or the depth of narrow vertical holes. The basic idea is simply to lower the weight until the line goes slack when the weight hits bottom. The line itself may be graduated so that the depth can be read directly, or the line can be measured as the weight is raised. An inexpensive and workable compromise is to mark the line in rather large sections such as 10 ft or 2 or 3 m. The line is measured to the first marker as it is raised, and thereafter one need only count markers.

Sounding in still water over a firm bottom is a relatively straightforward pro-

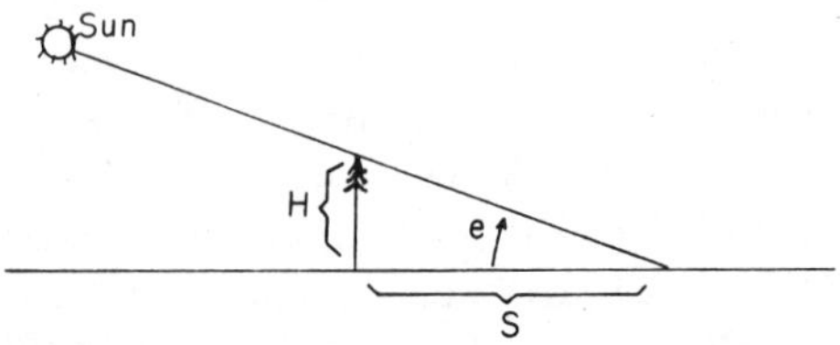

Figure 3.29 Diagram of a shadow cast on level ground. *H* is height; *S* is shadow length; and *e* is elevation angle of the sun.

cess. However, complications arise where there is a strong current or a soft bottom. Strong currents call for a heavy weight of streamlined shape which is not carried easily by the current. Likewise the pressure of the current on the line makes it more difficult to tell the exact point at which the line goes slack. Conversely, soft bottom materials call for a lighter weight of flared shape which will not settle far into the soft substrate. Fortunately the combination of fast current with extremely soft bottom materials is rather rare since such materials are eroded by continued exposure to a strong current.

Depth by Sonar. When a great many depth measurements must be taken in large, deep water bodies, the expense of sonar equipment will probably be justified. Depth measurement by sonar is based on the known rate of propagation of sound waves in water and timing the arrival of echos. Simple versions are marketed for use by sport fishermen. More elaborate equipment is appropriate for hydrographic surveys. Manufacturers' instructions give details of operating procedures. It should be noted, however, that the nature of the bottom materials has an effect on sonar echos.

Depth by Seismic Methods. The depth below the surface of the ground at which changes in the nature of earth materials occur can be determined by seismic methods. Propagation, reflection, and refraction of shock waves generated by explosive charges are dependent on the density and changes in the structure of earth materials. However, the equipment is relatively sophisticated, and the interpretation of seismographs requires a specialized knowledge of the several types of waves involved. Furthermore the handling of explosives is not a matter for amateurs. Therefore the use of seismic apparatus will not be discussed further.

3.7 AREA

Area is the amount of surface lying within a closed boundary on the surface. As with distance, area is usually measured in a horizontal plane rather than along the sloping surface of the ground. Unless otherwise stated, the term "area" will refer to planar surfaces of this nature.

3.7.1 Units of Measure for Area

Any unit of linear measure can be squared to obtain a unit of area measure. Thus 1 sq in. is the area contained in a square having sides 1 in. long. Likewise 1 sq ft is the area in a square having sides 1 ft long. Note, however, that the conversion factor for area is the square of the corresponding factor for linear units. For example, there are $12^2 = 144$ sq in. in 1 sq ft.

In addition to the simple squares of linear units, both English and metric systems have units that are used only for area measure. The basic unit of this type in the English system is the acre. An acre consists of an area equivalent to 10 sq chains or 43,560 sq ft. Therefore 1 sq mi contains 640 acres. The common unit of measure for a land area in the metric system is the *hectare* (ha). One hectare contains the equivalent of 10,000 sq m. The hectare is derived from a less common metric unit of area called the *are* which contains 100 sq m.

3.7.2 Area Measurement by Square Grid

There are several ways of approaching the problem of measuring areas, but most of these methods are designed for use with maps or photographs. The basic idea underlying several of these methods is to lay a transparent square grid over the area to be measured and then count the squares that lie within the area. The approximate area is calculated by multiplying the number of squares by the area represented by a single square. A procedure is needed for handling fractional squares. The usual approach is to count the square if more than half of it lies within the boundary of the area, and to exclude the square if less than half of it lies within the boundary. With squares that appear to be half in and half out, every other one is counted.

The area represented by a single square is calculated from the map or photo scale as follows. First figure what ground distance is represented by the side of a square. As with any other distance determination, this is done by setting up a proportion with the scale. For example, suppose that a grid with squares 0.1 in. on a side is being used on a map which has a scale of $1/20{,}000$. Then the distance represented by the side of a square is calculated as

$$\frac{\text{actual length of side}}{\text{ground distance of side}} = \frac{1}{20{,}000}$$

$$\frac{0.1 \text{ in.}}{\text{ground distance of side}} = \frac{1}{20{,}000}$$

Then

$$\text{ground distance of side} = 0.1 \text{ in.} \times 20{,}000 = 2000 \text{ in.}$$

and 2000 in. = 166.66 ft. Next figure the area of a square on the ground that has sides of this length. For our example,

$$\text{area of square} = (\text{length of side})^2 = (166.66 \text{ ft})^2$$

$$= 27{,}775.6 \text{ sq ft}$$

Converting this to acres, we have

$$\text{area of 1 square} = 27{,}775.6 \text{ sq ft} \times \frac{1 \text{ acre}}{43{,}560 \text{ sq ft}}$$

$$= 0.6376 \text{ acre/square}$$

If 100 squares were counted in measuring a given area with this grid, the area would be

$$100 \text{ squares} \times 0.6376 \text{ acre/square} = 63.76 \text{ acres}$$

The finer the grid (closer spacing of grid lines), the better will be the approximation of the actual area. However, counting squares becomes more tedious with finer grids, unless the small squares are blocked into larger groups by heavy lines.

Area Measurement by Dot Grid. If a dot were placed in the center of each grid square, the square could be counted or not counted according to whether the dot is inside or outside the boundary of the area. Since the dot represents the entire square, there is really no reason to show the square itself. If the grid lines are eliminated leaving only the dots that were in the centers of the squares, the result is a dot grid.

To use a dot grid, one simply counts the dots in the area to be determined and multiplies by the area represented per dot. The area per dot is calculated as before, except that the distance between dots is used in place of the side of a square. Borderline dots are usually handled by counting every other one.

As with square grids, the accuracy of the area approximation increases with closer spacing of the dots in the grid. However, the tedium involved in counting also increases with the density of the dot grid. The counting of dots can be facilitated somewhat by blocking the dots by lines into convenient groups of 100 or so. Dot tally meters shaped like a pen are also available. The tally meter is advanced each time the point is pressed on a dot. With either square or dot grids the best results are obtained by averaging the results of several counts obtained from different positions of the grid.

Area Measurement by Parallel Lines. Another variation of the grid principle is achieved by drawing a line down the middle of each row of squares. Just as the dot in the center represents the entire square, so the length of the line required to cross a single square represents the area of the square. If the sides of the squares are removed, the result is an overlay of parallel lines. The overlay is placed over the area to be measured, and the length of each line within the area is measured. The total measured length of all lines divided by the length per unit area gives the approximate area enclosed by the boundaries. The spacing between lines gives the side of the equivalent square. If overlay material is not available, suitably spaced parallel lines can be drawn directly on the map or photo.

The field version of this parallel line procedure is to run a series of equally spaced parallel lines across the area as illustrated in Figure 3.30. Lines of this type which cross the area in parallel fashion are called *transect* lines. The approximate area of the tract is then calculated from the formula

$$\text{area} = \text{total length of transects} \times \text{transect spacing}$$

For example, suppose that transects are spaced 50 m apart across a tract. Assume further that the total length of the transect lines is 2500 m. Then an estimate of the area is

$$\text{area} = 2500 \text{ m} \times 50 \text{ m} = 125{,}000 \text{ sq m or } 12.5 \text{ ha}$$

3.7.3 Area Measurement by Weight of Cutouts

If a sensitive balance is available, one can take advantage of the fact that the weight of a piece of uniform paper is proportional to its area. First weigh a piece of the map or photo paper having a known area in order to develop a conversion factor from weight to area. Next carefully cut out the areas to be determined from the photo or map. Finally weigh the resulting pieces and use the conversion factor to find the areas.

If the map or photo is not expendable, one can either make a photocopy or trace the areas on overlay material which is then used in place of the original for making the cutouts. Likewise one can compensate for some lack of sensitivity in the balance by pasting the map, photo, or overlay on a uniform piece of heavy backing material before making the cutouts. However, such a backing usually tends to increase the difficulty of making accurate cutouts.

3.7.4 Planimeters

Instruments for measuring areas are called *planimeters*. Planimeter designs range from very sophisticated to very simple. Space does not permit a detailed consideration of all types of planimeters, but a brief description of three types will give an idea of the range available.

Polar Planimeters. This is the most commonly used type of planimeter. The basic components are a point (or pole) which maintains a fixed position on the sheet, a jointed arm which pivots about the pole, a measuring wheel near the joint, and tracing point at the free end of the arm. The pole is fixed in a convenient position outside the boundary of the area to be determined, and the tracing point is moved around the boundary. It is also possible to use the planimeter with the pole position inside the boundary for large areas, but the procedure is more complicated. Some polar planimeters can be set to read

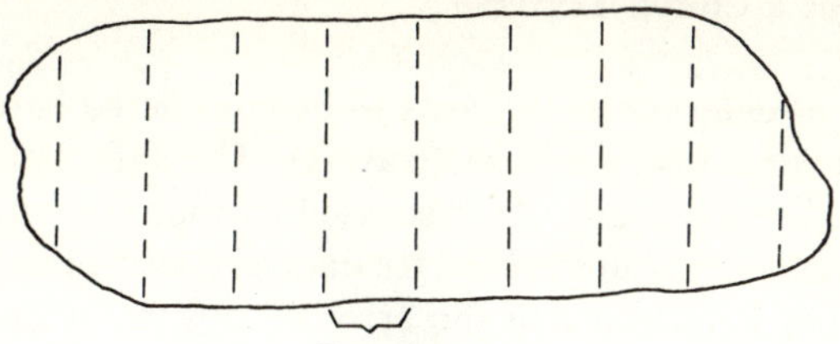

Figure 3.30 Pattern of transect lines run across an area.

directly in map units, while others require a conversion using the map scale. The accuracy of tracing can be checked by repeating the measurements and taking an average. The better polar planimeters are of the "compensating" type. With a compensating polar planimeter the area is measured twice, and the position of the jointed elbow is reversed between the two measurements. The reversal of the elbow causes instrumental errors to be self-compensating in a manner akin to double centering with a transit (see Section 3.4.3).

Hatchet Planimeters. At the inexpensive and simplistic extreme of the planimeter types is the hatchet planimeter. This is a metal bar with both ends bent downward at right angles. One end is sharpened like a pencil point for tracing. The other end is sharpened so that it has a blade edge parallel to the long axis of the bar and slightly curved like a rocker on a rocking chair. The tracing point is placed at the center of the area, and the blade is pushed down on the paper to mark its initial position. The tracing point is moved along a straight line to the near side of the area (side nearest the blade), around the perimeter, and back to the center. The final position of the blade is marked, and the distance moved by the blade is measured. The area is equal to the change in the blade's position times the distance between the point and the center of the blade. Again the tracing can be checked by repetition. The hatchet planimeter is reasonably accurate if the longest distance across the area to be measured is less than half the length of the planimeter (Welch, 1948).

Electronic Planimeters. Electronic planimeters represent the other extreme of accuracy and expense. In essence these planimeters digitize the boundary, and the area is computed electronically. The specific mode of operation varies somewhat, but a common design uses an extremely fine grid of wires in a tracing table to generate the signals as a tracing element is moved around the boundary. Such instruments may give a direct readout, or the output may be recorded on a computer compatible medium to facilitate the input to a larger computer system.

3.7.5 Area of a Closed Traverse

The last method of determining an area to be considered here is one designed for computing the area inside a closed traverse. The nature of a closed traverse was described in Section 3.5.8. As discussed in that section, there is usually some internal inconsistency in the traverse data as collected in the field. This internal inconsistency is reflected in the error of closure. A preliminary step to computing the area of a traverse is to balance out any error of closure so that the figures are internally consistent with the fact that the traverse is closed.

Balancing a Traverse. In order to achieve this consistency, the latitudes of the several lines must be made to sum algebraically to zero, and the same for the departures. There are several methods available for achieving such a balance, but the one that will be described here is known as the *compass rule.* The essence of the compass rule is to distribute both latitude and departure errors in proportion to the length of the lines in the traverse. The following simple three sided traverse will be used to illustrate the process as the explanation proceeds:

Line	Bearing	Distance (chains)	Latitude	Departure
1	S60°E	10.0	−5.00	8.66
2	N30°W	8.0	6.93	−4.00
3	S60°W	5.0	−2.50	−4.33
		23.0	−0.57	0.33

To use the compass rule, the length of each line is expressed as a decimal fraction of the total traverse length. This fraction of the latitude error (algebraic sum of latitudes) is added to or subtracted from the latitude of the line, with the sign being the opposite of the error. The same fraction of the departure error is used to adjust the departure of the line. Thus the adjustments of both latitude and departure are proportional to the length of the line. In formula terms the adjustments are

$$\text{latitude adjustment} = -\left(\frac{\text{length of side}}{\text{total traverse length}}\right) \times \text{algebraic sum of latitudes}$$

and

$$\text{departure adjustment} = -\left(\frac{\text{length of side}}{\text{total traverse length}}\right) \times \text{algebraic sum of departures}$$

The latitude adjustments and the adjusted latitudes for the example are

Line	Latitude adjustment	Adjusted latitude
1	0.248	−4.752
2	0.198	7.128
3	0.124	−2.376
	0.570	0.000

Likewise the departure adjustments and adjusted departures are

Line	Departure adjustment	Adjusted departure
1	−0.143	8.517
2	−0.115	−4.115
3	−0.072	−4.402
	−0.330	0.000

When the traverse has been balanced in this manner, one can proceed directly with the computation of the area. The directions and distances of the traverse lines could be recomputed from the adjusted latitudes and departures as explained in Section 3.5.8, but this is not necessary for computing the area.

Area by Double Meridian Distances. A convenient approach to calculating the area inside a closed traverse is the technique of double meridian distances. The first step in finding an area by double meridian distances is to calculate the latitude and departure for each side of the traverse. This has aleady been done for the example in the process of balancing out the error of closure.

The next step is to calculate the double meridian distance (DMD) and the double area (DA) for each side of the traverse. The DMD of a line is calculated as

$$\text{DMD} = \text{departure of the line} + \text{departure of preceding line} + \text{DMD of preceding line}$$

The double area for a line is

$$\text{DA} = \text{DMD} \times \text{latitude}$$

The DMD of the first line is simply equal to its departure, since the last two terms are zero for that line. A check on the accuracy of the DMD computations is provided by the fact that the DMD of the last line must be numerically the same as its departure, but of opposite sign. The area computation is completed by adding the DAs of all the lines algebraically and dividing this total by 2. If the result is negative, simply drop the sign.

The DMD and DA calculations for the example are as follows:

Line	Latitude	Departure	DMD	DA
1	−4.752	8.517	8.517	−40.473
2	7.128	−4.115	12.919	92.087
3	−2.376	−4.402	4.402	−10.459
				41.155

Therefore the area of the traverse is

$$\frac{41.155}{2} = 20.58 \text{ sq chains or } 2.06 \text{ acres}$$

The double meridian distance technique is equivalent to calculating the areas of triangles and trapezoids formed by projecting each line of the traverse onto a north/south line through the most easterly or westerly point on the traverse. The formula for calculating the area of a triangle is

$$\text{area of triangle} = \frac{\text{base} \times \text{height}}{2}$$

and the formula for the area of a trapezoid is

$$\text{area of trapezoid} = \frac{\text{height} \times \text{sum of bases}}{2}$$

In the double meridian distance technique the division by 2 is postponed until the last step, hence the name "double area." The triangles and trapezoids for the example are shown in Figure 3.31.

3.8 RADIATION, INTERSECTION, AND RESECTION

This section is concerned with three useful techniques for locating the map position in the field. These techniques are ideally suited for use with the plane table and alidade (see Section 3.4.3), but can also be readily adapted for use with other instruments.

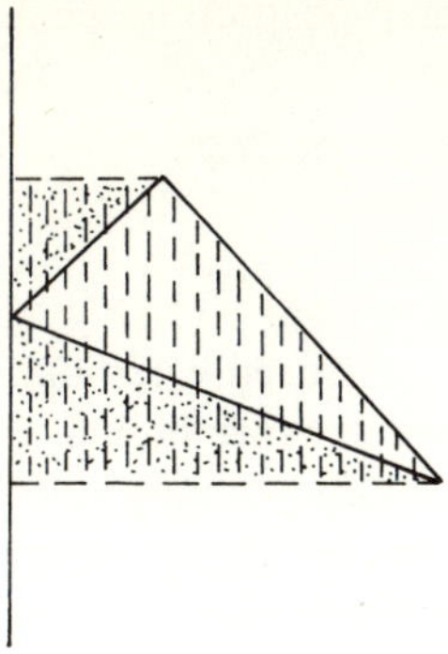

Figure 3.31 Triangles and trapezoids formed by projecting traverse lines on north/south line (meridian) through most westerly point.

3.8.1 Radiation

Radiation is a sophisticated name for a technique that is really very simple. It serves for plotting the map positions of neighboring points when a person is situated at a point of known map position. The points to be plotted must be visible from the known point if this method is to be applied. Suppose a field crew is situated at point A for which the map position is known. It is desired to plot the map position of a nearby point B which is visible from point A. The first step is to determine the direction of the line from point A to B. Next plot a line on the map in this direction from point A. Now measure the distance from A to B on the ground and scale this distance along the plotted line from point A on the map. The map location thus obtained is the correct map position of point B.

3.8.2 Intersection

The name of this technique is quite descriptive. The setting for its application is when a point to be plotted on the map is visible from at least two points of known map position. The advantage of this technique is that no distance measurement is required. Let the letters A and B denote two points of known map position and the letter C a point to be plotted on the map. The field crew moves first to point A and determines the direction of the line from A to C. A line is plotted on the map in this direction from point A as if point C were to be located by radiation. Instead of measuring the distance from A to C, however, the crew moves on to point B. The direction from B to C is then determined, and a line is plotted on the map in this direction from point B. The lines from A and B will cross on the map unless C happens to lie on the same straight line as A and B. The point where the lines cross is the map position of point C.

Intersection works quite well when the two lines cross at nearly right angles. When the angle of intersection is very small or very large, however, it is difficult to decide exactly where the lines cross. Where possible it is desirable to check the plotted location by a line from a third point of known map position to see if all three lines cross at the same point.

3.8.3 Resection

Resection is essentially the reverse process of intersection. It is used when one wishes to plot one's present position on a map given that two or more points of known map position are visible. The directions to the points of known map position are determined, and the backazimuths of these directions are plotted from the respective map points. The plotted lines will cross at the map position of the observer.

The same constraints apply to both intersection and resection. As with intersection, the location determined by resection will be most accurate when the plotted lines cross at nearly right angles. Likewise resection cannot be used if all three points happen to fall on the same straight line. Also, a third line can serve to check the accuracy of resection when three points of known map position are visible.

3.9 GEOGRAPHIC REFERENCING

The ability to locate points in the field and on maps is necessary but not sufficient for assembling integrated data bases. As discussed in Section 2.4.3, some sort of geographic referencing system is required for indexing and merging the multitude of data items that make up a comprehensive information base for ecosystem management and planning. The more common systems of geographic referencing constitute the last topic to be covered in this chapter. Systems designed for use in relatively restricted areas will be considered first, then the scope will be broadened to include systems that are usable over much larger areas.

3.9.1 Public Land Survey Systems in the United States

Two systems are in use for legal descriptions of land parcels in the United States. The older system is known by the name *metes and bounds* because each tract is described solely by its corners and boundaries in relation to local landmarks. The newer system is the standard rectangular survey or General Land Office (GLO) system covering most of the country, which ties each tract of land to a control grid for the area.

Metes and Bounds. The metes and bounds system is used in the eastern parts of the country which were well settled before the rectangular system was put into use. In these parts parcels of land are usually irregular in shape since the boundaries were laid out along convenient natural or manmade features of the landscape. When the tract was laid out, the boundary lines were described and markers were placed at the intersections of the boundary lines, unless a natural marker was already available. Most descriptions of metes and bounds tracts leave a lot to be desired with respect to details usable for retracing the boundaries. However, these descriptions do make quite interesting reading since they contain some rather colorful references to local history.

A copy of the original survey notes is essential for retracing any boundary line, and such copies are usually available from county courthouses. In the case of metes and bounds tracts, however, a copy of the original notes may not be sufficient for retracing the lines. One will often find old-timers who have spent most of their life in the area to be a great help in interpreting the notes and locating the positions of markers. In many cases the original markers will have long since disappeared. It must also be remembered that the equipment used to run the original survey was quite primitive by modern standards. It is a good idea to begin by measuring a line for which the length is given in the notes and end markers can still be located. In this way it is possible to obtain a rough conversion factor for the units given in the notes in terms of current equipment. The metes and bounds system is essentially a nonsystem that is of no use in providing geographic control for modern information systems. Therefore it will not be considered in detail.

Rectangular Survey or GLO System. Most of the United States, including all of the West and Midwest, has been surveyed in the rectangular system. The law providing for use of the rectangular system has been amended several times since its beginning in the 1780s, so the details of the survey may vary slightly, depending on the date that it was conducted. The overall procedure, however, has remained essentially the same.

The locations of all land parcels in a region are tied into the *initial point* for that region. The initial point was located by astronomical measurement and may serve an area encompassing portions of several states. A *principal meridian* was run north and south from the intial point, and a *base line* was run east and west. A straight line started north or south from a point on the earth continues in the same direction to the pole, but the earth's curvature makes a straight line started east or west from a point in the northern hemisphere tend to the south. The base line is corrected every half mile so that it follows a true parallel of latitude (see Figure 3.32).

Standard parallels and *guide meridians* were next run to divide the area into 24 mi square tracts. Standard parallels were run every 24 mi along the principal meridian to the north and south of the initial point. Like the base line, the standard parallels are corrected every half mile for curvature of the earth. A guide meridian begins every 24 mi along a standard parallel and runs north to the next standard parallel. Again because of the curvature of the earth, the guide meridian from the south will strike the standard parallel closer to the principal meridian than the guide meridian that starts toward the north (see Figure 3.33).

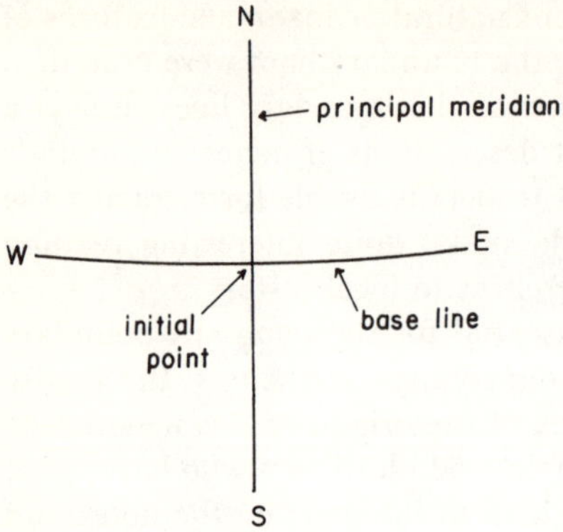

Figure 3.32 Diagram of initial point, principal meridian, and base line.

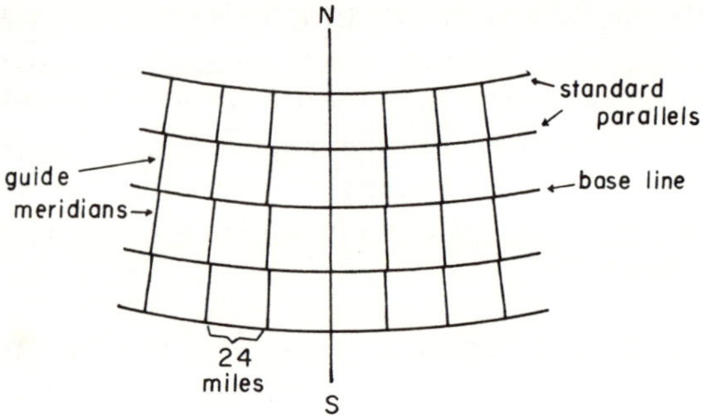

Figure 3.33 Diagram of 24 mi square tracts formed by standard parallels and guide meridians.

The intersection of the meridian from the south with the parallel is called a *closing corner*, and the one for the meridian going north is called a *standard corner*. The standard parallels are numbered with respect to the base line as, for instance, first north, second north, first south, second south, and so on. Likewise the guide meridians are numbered east and west of the principal meridian.

Each of the 24 mi square tracts is divided into 16 *townships*, with each township being 6 mi square. The townships are thus arranged in rows and columns. The term "township" is also applied to a row of township squares. A column of township squares is called a *range*. A township square is located with respect to the initial point by its number of township rows north or south of the base line, and its number of ranges east or west of the principal meridian. In this way of numbering T5N, R2W is the township square in the fifth tier north of the base line and the second column west of the principal meridian (see Figure 3.34).

The townships are in turn subdivided into 36 sections, each approximately 1 mi square. The sections are numbered in a snakelike pattern starting with the

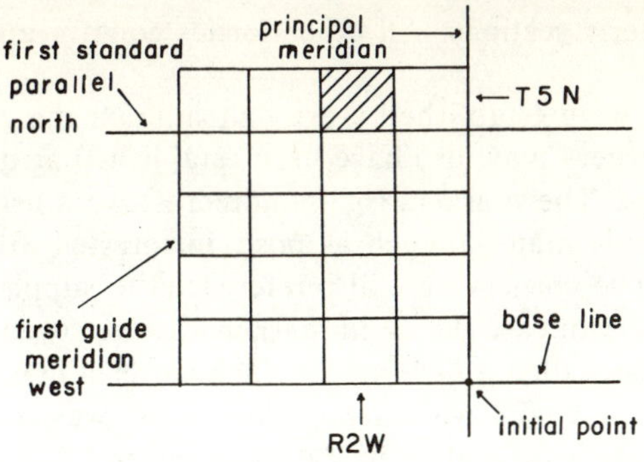

Figure 3.34 Location of T5N, R2W with respect to the principal point.

one in the northeast corner of the township and ending with 36 in the southeast corner (see Figure 3.35).

Due to the convergence of meridians and errors of surveying, townships are not exactly 6 mi square and therefore could not be divided into 36 sections exactly 1 mi square. In order to produce as many sections as possible that are 1 mi square, all the error is thrown into the most northerly and the most westerly quarter mile of the township. Sections 1-7, 18, 19, 30, and 31 may therefore be of odd size.

Sections are subdivided into halves and quarters, with quarter sections being 40 chains on a side. The quarter sections are named by their direction from the center of the section: NE¼, SE¼, SW¼, and NW¼, proceeding in a clockwise direction. The quarter sections are in turn subdivided into quarters or *forties* which are 20 chains on a side and contain 40 acres. The forties are named according to the corner of the quarter section that they occupy. Thus the forty at the extreme northeast corner of a section would be the NE¼ of the NE¼ of the section. Forties are not usually subdivided further in rural areas. In urban areas the forties are subdivided into *lots* of arbitrary shape and size.

The distribution of errors in a township can be restated in terms of forties. The errors are thrown into the most northerly and westerly forties in the most

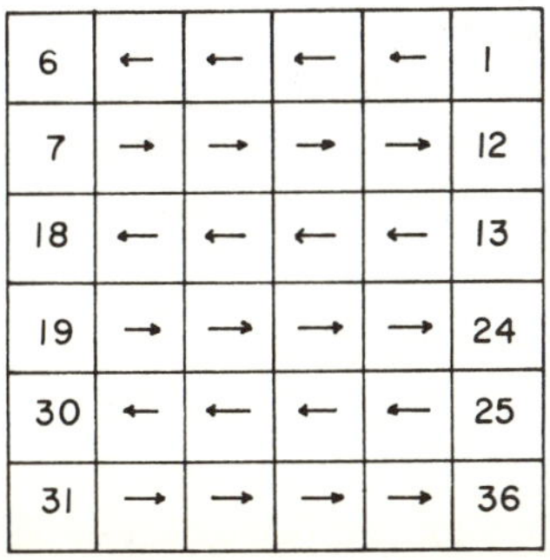

Figure 3.35 Numbering of sections in a township.

northerly and westerly sections. All other forties contain approximately their nominal acreage.

Corner markers were established every half mile on the survey lines in the original survey. Corners may also have been established at quarter mile intervals in later surveys. The markers (or monuments) were usually made of the most readily available material such as posts in forested areas and stones in more open areas. The corners were also referenced to supplementary markers called *witness corners* in case the main marker was destroyed. A tree used as a witness corner is also called a *bearing tree.* Whenever a line crossed a body of water more than three chains wide, a *meander corner* was also established.

The original survey is official, even if it turns out to be somewhat inaccurate upon later investigation. As long as the original markers have not been destroyed, there should be little difficulty in retracing the lines. If the location of a missing marker is known to residents or determinable on other evidence, it should be reestablished in its original position as nearly as possible. A lost corner on a township or more major survey line is located by proportioning the distance to the nearest existing corners on the same line. Lost section corners or corners within sections are located by proportioning the distance to existing corners at right angles as well as on the same line. Because of expected variations in the north and west tiers of forties in a township, however, actual rather than proportional measurements should be used. In any case, a corner should not be considered lost until every effort has been made to locate it. Furthermore legal action in the case of boundary disputes must be based on the work of a registered surveyor.

The systematic structure of nested squares forming the rectangular survey system makes an acceptable basis for geographic referencing in cell oriented information systems. The term "cell oriented" is used here to mean that the area of interest is divided into a set of small area units, and information is stored or retrieved by area unit. These area units can be aggregated to give information for larger areas, but cannot be broken down to give information for smaller areas. The rectangular system also has the advantage of being partially inscribed on the terrain through roads and fence lines. It does, however, have several deficiencies as a general system of geographic referencing. For one thing, the method of allowing for curvature of the earth is rather crude. When this is coupled with the method of handling surveying errors, the result is units of odd size and shape along the north and west boundaries of the townships. Consequently the calculation of distances between widely spaced points is approximate at best. Furthermore initial points are not distributed according to a set geometric pattern, which makes the system awkward for large areas. Finally the system can only be applied in areas which it covers.

3.9.2 State Plane Coordinates

State plane coordinate systems have been set up to provide uniform geographic referencing on a statewide basis. A mathematical transformation (projection) is used to warp the spheroidal surface of the earth so that it corresponds with a planar mapping surface. A rectangular coordinate system on the mapping surface provides a means of uniform geographic referencing throughout the state. Using state plane coordinates, distances and directions between widely spaced points can be calculated accurately.

The *transverse Mercator projection* is used to achieve the warping to a horizontal plane for states that are long in the north/south direction. If such a state is relatively wide, it may also be broken into east/west zones. The *Lambert conformal projection* is used for states that are long in an east/west direction; and states of this type may also be broken into several north/south zones. Most of the more recent USGS "quadrangle" maps provide tick marks showing state plane coordinates. One can construct a state plane coordinate grid over the map by connecting tick marks on opposite sides of the map. If this graphical approach is not sufficiently accurate, one will need to employ mathematical procedures detailed in publications available from the U.S. Government Printing Office.

3.9.3 Geographic Referencing on a Global Scale

Two systems of geographic referencing on a global scale will be outlined here. One is the latitude/longitude system of spherical coordinates which most people have already encountered briefly in a geography or social studies course. The other is a rectangular grid system similar in nature to state plane coordinates, which is used extensively by the U.S. military.

The Latitude/Longitude Graticule. A series of circles around the earth perpendicular to its axis of rotation provide a system of north/south coordinates called *parallels of latitude.* The largest of these circles is the equator, which is taken to be 0° latitude. The north pole is taken to be 90° north latitude, and the south pole is taken to be 90° south latitude. Any other parallel of latitude is assigned a value in degrees according to its relative position between the equator and the pole. For example, the parallel situated halfway between the equator and the north pole is 45°N latitude. The Greek letter phi (ϕ) is often used to denote the latitude coordinate of a point on the earth. A degree of latitude is 60 nautical miles, or approximately 69 statute miles. Degrees are further divided into minutes and seconds. A nautical mile covers 1 minute of latitude.

A series of semicircles connecting the poles provides a system of east/west coordinates called *meridians of longitude*. The meridian of longitude passing through Greenwich, England, is arbitrarily designated as the zero or *prime meridian.* The meridian lying on the side of the earth opposite the prime meridian is designated as the 180° meridian. As one moves along a given parallel of latitude, the longitude increases uniformly toward either east or west of the prime meridian until the 180° meridian is reached. A degree of longitude along the equator is 60 nautical miles, but it decreases in size progressively toward either pole. The Greek letter lambda (λ) is often used to denote the longitude coordinate of a point on the earth. It should also be noted that not all countries recognize the same prime meridian. Some other countries recognize an important point within their own boundaries as defining the zero meridian.

The lines of latitude and longitude taken together form a geographic referencing network called the *graticule*, which is sufficient to fix the location of any point on the earth's surface or to compute directions and distances based on solid geometry. The major drawback of this network is that either the latitude lines or the longitude lines or both become curved when the graticule is projected onto a map sheet. Working with a curved coordinate axis tends to be a little awkward.

The UTM Grid. The military has adopted a global geographic referencing system that provides rectangular mapping coordinates similar to the state plane coordinate system. It is called the *universal transverse Mercator* (or UTM) *grid*.

The UTM grid applies to the portion of the earth between 80° south latitude and 84° north latitude. Grids for the polar areas are developed differently, and will not be considered here. This part of the earth is divided into 60 *zones*, each containing 6° of longitude. This division begins at the 180° meridian and extends eastward around the globe. The zones are numbered 1 through 60 in this same order. Thus zone number 1 spans longitudes from 180 to 174°W (see Figure 3.36). For quick reference to large blocks, the zones are divided from south to north into 8° segments of latitude, except that the most northerly segment in a zone covers 12° of latitude. These segments are lettered from south to north, starting with C and skipping both I and O (which could be confused with numbers). The letters A, B, Y, and Z are used for polar grid designations. Thus a combination of a zone number and a letter is sufficient to designate any segment of any zone.

Within each zone a transverse mercator projection is used to construct a square grid having its origin at the intersection of the equator with the central meridian for the zone. The west/east coordinate of the grid is called the *easting*, and the south/north coordinate is called the *northing*. The origin of the grid is assigned an arbitrary easting of 500,000 m and a northing of 0 m for the portion of the grid lying in the northern hemisphere (see Figure 3.37). For the portion of

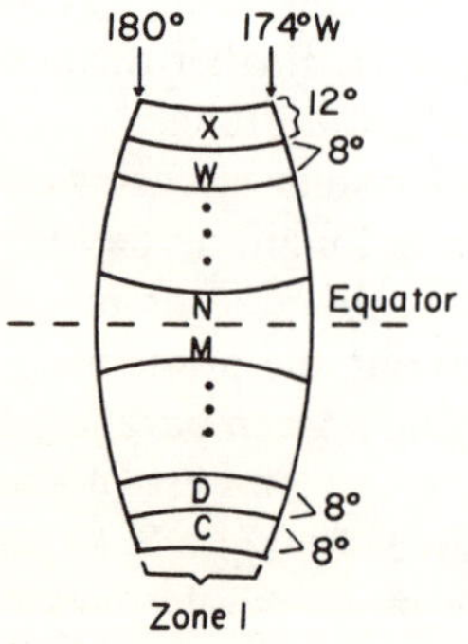

Figure 3.36 Major divisions of UTM grid as illustrated by zone 1.

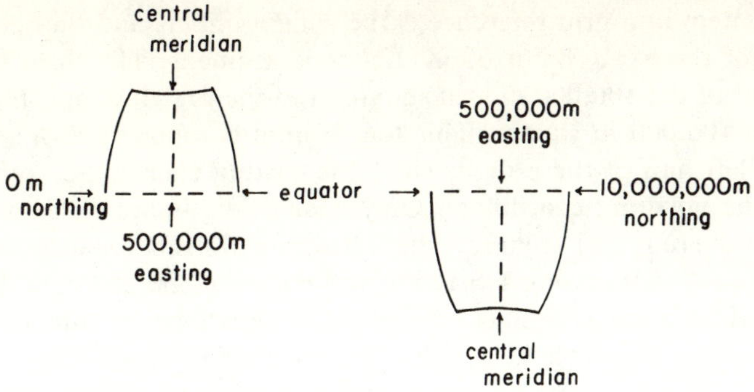

Figure 3.37 Easting and northing of origin in a UTM zone.

the grid zone lying in the southern hemisphere, the easting of the origin is again 500,000 m, but the northing is 10,000,000 m. Eastings increase toward the east (right) and decrease toward the west. Northings increase toward the north (up) and decrease toward the south, except for the break at the equator.

The grid in each zone has a coarse mesh of 100,000 m squares, which are given letter designations as well as easting/northing coordinates. At the equator there are six full squares and two partial squares across the width of a zone. Lettering of these squares along the easting axis begins at the 180° meridian and proceeds continually eastward (that is, it does not begin again at A for the second or third zone). Since the letters I and O are omitted from the sequence to avoid possible confusion with numbers, the available letters are exhausted after three zones, and the sequence begins again at A for the next three zones. The lettering of 100,000 m squares along the northing axis goes from south to north in cycles of A through V, again omitting I and O. However, the lettering is offset between odd numbered zones and even numbered zones. In odd numbered zones the row of squares immediately north of the equator is lettered A. In even numbered zones this row is lettered F. Mesh lines subdividing the 100,000 m squares go only by easting and northing coordinates and are not lettered.

The reason that the origin of a grid zone is assigned the easting 500,000 instead of zero is to avoid negative coordinates. For this reason it is called a "false easting." For the same reason the origin for the southern hemisphere is assigned a false northing of 10,000,000 m.

The north direction on the central meridian is taken as *grid north* for all points in the zone. Thus the true north and the grid north coincide only on the central meridian of a zone.

The scale factor for the grid varies slightly within a zone. It is accurate for lines 180,000 m east and west of the central meridian. The scale decreases toward the central meridian from these lines, being reduced by a factor of about 0.0004 at the central meridian. Likewise the scale increases in the outward directions, being inflated by a factor of about 0.0001 near the boundaries of the zone along the equator.

The first item in a grid reference is the zone number, and the second is the letter code of the 6° × 8° or 6° × 12° zone segment. The third item is the easting letter of the 100,000 m grid square, and the fourth item is the northing letter of the 100,000 m square. This fourth item is followed by a sequence of digits. The left half of these digits show the easting coordinate, and the right half show the northing coordinate. Only enough digits are used to locate the point within the required accuracy, the remaining digits being dropped. For example, the point with easting 495,125 m and northing 550,250 m in the 100,000 m square ML in zone 2, segment N, would be located to within 10,000 m on either axis by the reference 2NML95. The last four digits of both easting and northing were dropped because the desired accuracy was only to within 10,000 m.

As with state plane coordinates, the easiest way to work with the UTM grid is to obtain a map with UTM ticks and complete the grid by connecting tick marks. The alternative is to use a mathematical approach as described in U.S. Army technical manuals TM 5-241-1, TM 5-241-8, and TM 5-241-11.

3.10 SUGGESTIONS FOR INDIVIDUAL STUDY

A few suggestions for individual study were offered at the beginning of this chapter. We shall reiterate those and add a few more at this point.

If the reader is not already familiar with the use of keys for identifying things, it would be worthwhile to spend a little time becoming acquainted with them. Numerous keys for the identification of plants, animals, rocks, and minerals are available. One or more such keys for features of interest to the reader should be acquired and used until familiarity has been gained with their structure and application. Having become familiar with keys in this manner, the reader should then attempt to construct a few keys to objects of personal interest.

The fact that the reader has come in contact with this book should imply enough interest in field oriented activities to acquire a hand held orienteering compass. Field navigation by compass and pacing should be practiced until reasonable proficiency is achieved. At this point one should undertake the preparation of simple maps for several areas of personal interest.

When one has become familiar with the basic principles of map preparation and use, one should order topographic maps and airphotos for one's home vicinity and neighboring areas.

3.11 PART I IN REVIEW

Hopefully the materials presented in Part I of this book have provided a reasonably firm foundation on which we can build in Part II. The reader should have emerged from Chapter 2 with an understanding of the components that make up a survey information system and how these components must function in concert to produce an effective information delivery service. Chapter 2 should also have provided similar insight to the survey planning process. Chapter 3 has been devoted to the nature of numerical data, the measurement of physical dimensions, and methods for locating and describing the position of features on the terrain. These are key concerns regardless of the particular ecosystem components on which a survey may focus. The chapters that comprise Part II deal with the use of these and other techniques for collecting and presenting data on major components of ecosystems.

3.12 BIBLIOGRAPHY

Adams, O., and C. Claire. *Manual of Plane Coordinate Computation.* Washington, D.C.: U.S. Government Printing Office, U.S. Coast and Geodetic Survey Special Publication 193.

Adams, O., and C. Claire. *Manual of Traverse Computation on the Lambert Grid.* Washington, D.C.: U.S. Government Printing Office, U.S. Coast and Geodetic Survey Special Publication 194.

Adams, O., and C. Claire. *Manual of Traverse Computation on the Transverse Mercator Grid.* Washington, D.C.: U.S. Government Printing Office, U.S. Coast and Geodetic Survey Special Publication 195.

Amidon, E., and M. Whitefield. *Length and Area Equivalents for Interpreting Wildland Resource Maps.* Berkeley, Cal.: U.S. Department of Agriculture, Forest Service, Pacific Southwest Forest and Range Experiment Station, Research Note PSW-190, 1969.

Avery, T. E. *Interpretation of Aerial Photographs,* 3rd ed. Minneapolis, Minn.: Burgess, 1977.

Baker, W. *Elements of Photogrammetry.* New York: Ronald, 1962.

Bouchard, H., and F. Moffitt. *Surveying,* 4th ed. Scranton, Pa.: International Textbook Co., 1961.

Brinker, R., Ed. *Elementary Surveying,* 5th ed. Scranton, Pa.: International Textbook Co., 1969.

Chamberlin, W. *The Round Earth on Flat Paper.* Washington, D.C.: National Geographic Society, 1950.

Compton, R. *Manual of Field Geology.* New York: Wiley, 1962.

Greenhood, D. *Mapping.* Chicago: University of Chicago Press, 1964.

Kjellstrom, B. *Be Expert with Maps and Compass.* New York: American Orienteering Service, 1955.

Mitchell, H., and L. Simmons. *The State Coordinate Systems, A Manual for Surveyors.* Washington, D.C.: U.S. Government Printing Office, U.S. Coast and Geodetic Survey Special Publication No. 235.

Moffitt, F. *Photogrammetry.* Scranton, Pa.: International Textbook Co., 1964.

Reeves, R., A. Anson, and D. Landen, Eds. *Manual of Remote Sensing,* Vols. 1;2. Falls Church, Va.: American Society of Photogrammetry, 1975.

Smith, J., Jr., and A. Anson, Eds. *Manual of Color Aerial Photography.* Falls Church, Va.: American Society of Photogrammetry, 1968.

Spurr, S. *Photogrammetry and Photo-interpretation, with a Section on Applications to Forestry,* 2nd ed. New York: Ronald, 1960.

Stamp, L., Ed. *Glossary of Geographical Terms.* New York: Wiley, 1961.

Steers, J. *An Introduction to the Study of Map Projections.* London: London University Press, 1962.

Stevens, S. "On the Theory of Scales of Measurement." *Science,* **103,** 677–680 (1946).

Strandberg, C. *Aerial Discovery Manual.* New York: Wiley, 1967.

Thompson, M., et al., Eds. *Manual of Photogrammetry,* Vols. 1;2, 3rd ed. Falls Church, Va.: American Society of Photogrammetry, 1966.

U.S. Army. *Universal Transverse Mercator Grid.* Washington, D.C.: U.S. Government Printing Office, U.S. Department of the Army Technical Manual TM 5-241-8, 1958.

U.S. Army. *Universal Transverse Mercator Grid Tables for Latitudes 0°–80°, Clarke 1866 Spheroid (Meters) Coordinates for 7½-minute Intersections.* Washington, D.C.: U.S. Government Printing Office, U.S. Department of the Army Technical Manual TM 5-241-11, 1959.

U.S. Army. *Grids and Grid References.* Washington, D.C.: U.S. Government Printing Office, U.S. Department of the Army Technical Bulletin TM 5-241-1, 1967.

U.S. Army. *Map Reading.* Washington, D.C.: U.S. Government Printing Office, U.S. Department of the Army Field Manual FM 21-26, 1969.

Wanless, H. *Aerial Stereo Photographs for Stereoscope Viewing in Geology, Geography, Conservation, Forestry, Surveying.* Northbrook, Ill.: Hubbard, 1969.

Welch, P. *Limnological Methods.* New York: McGraw-Hill, 1948.

PART II–SURVEYING ECOSYSTEM COMPONENTS

CHAPTER 4–THE FOUNDATION OF ECOSYSTEMS: TOPOGRAPHY, SOILS, AND GEOLOGY

The physical environment is a complex of several major factors and a multitude of more minor factors. Some of these factors are measurable to a usable degree of accuracy with equipment that is relatively simple and inexpensive. Other factors require analytical systems that are both sophisticated and costly.

The form in which the quantification is expressed may be quite easy to interpret and translate into implications for problems of applied ecology, as when a contour map is used to quantify topography. Conversely, the initial quantification may take the form of a taxonomic classification that does not translate in an obvious way to implications for ecosystem management. In the latter case, discipline specialists must prepare interpretive tables to help the user translate taxonomic categories into applied terms before the survey information becomes directly useful. The "genetic" classification scheme employed by agricultural soil scientists furnishes an example of this need for intermediate interpretation.

Since physical factors produce a set of constraints that limit both human activities and the development of biological communities, data concerning the physical environment provide the starting point for most analyses of ecosystems.

4.1 THE LAY OF THE LAND: TOPOGRAPHY AND DRAINAGE

Topography and drainage are key factors in both manmade and natural communities. Relatively level and well drained areas are required for most types of construction. Lack of such areas does not necessarily preclude development entirely, but it does imply much greater costs of construction for grading and drainage as well as extensive alterations of and impacts on natural systems. Any recurring disturbance of the vegetative cover on steep slopes carries a high risk of accelerated erosion which can destroy natural communities, agriculture, foundation support of structures, reservoirs, waterways, and esthetics. The conformation of watersheds affects the rate of runoff, ground water recharge, and the likelihood of flooding. Locating habitations and related structures on floodplains is a gamble that can only lead to human tragedy. Likewise topographic relief is an important feature of natural ecosystems since it creates a variety of microenvironments which lend diversity to the ecosystem.

Topography arises from a complex of internal geological processes which produce volcanic activity, upheaval, or subsidence, coupled with the erosional forces of wind, water, and ice which remove surface material from some areas and deposit it in other areas. Animal species, especially man, produce some direct effects by moving earth materials. However, their greatest influence arises from the disturbance of the vegetative cover which exposes surface materials to accelerated erosion.

Survey information on topography and drainage can be presented in either graphical or numerical form. The essence of the graphical display problem is to depict a three-dimensional system on a two-dimensional sheet of paper. The problem in numerical description is to strike a balance between the need to summarize a confusing mass of data so that it is more easily interpretable, and the need to preserve details which may be important to the user of the survey information.

4.1.1 Sources of Topographic Data

There are two basic ways of obtaining topographic data. One is the obvious approach of going into the field with instruments for determining direction, horizontal distance, and change in elevation. There are many combinations of the instruments, as discussed in Chapter 3, which will serve the purpose, depending on the nature of the terrain and the requirements of accuracy. The basic operations are to locate the horizontal position on a map, determine the elevation of the point, and then plot the elevation reading in its correct map position. One should be certain to obtain readings at all points where an obvious change in slope of the ground occurs, and also to make notes regarding the nature of the changes. Additional readings can be taken between these key points until the desired density of elevation data is reached.

The second basic way of obtaining topographic information is through stereopairs of vertical airphotos. As long as the photos are vertical, taken from the same flying height, and the ground is not hidden from view by vegetation or deep snow, stereo viewing provides a three-dimensional scale model of the terrain common to the two photos. Procedures for determining directions, horizontal distances, and elevations from such photos were covered in Chapter 3. The elevations of a few locations in the scene must be known to serve as control points for parallax measurements. If the interest centers primarily on mapping slopes in broad classes, acetate overlays are available with which the interpreter matches the ground slope to a calibrated series of successively steeper lines as seen in the stereoscope.

Although the original data are seldom available, topographic maps already exist for most areas. These topographic maps are usually of the contour line type discussed in the next section. The most likely source for such maps is the U.S. Geological Survey (USGS) through the National Cartographic Information Center (NCIC). The USGS produces two series of topographic maps. In the most complete series each map covers a 15 minute quadrangle of latitude and longitude. More detailed maps in the second series cover 7$^1/_2$ minute quadrangles, but this series is less complete.

4.1.2 Graphical Description of Topography

As mentioned in the introductory paragraphs of Section 4.1, elevation information can be presented to the user in several ways. Graphical formats will be considered first because they are used more commonly than numerical formats.

Elevation Contour Maps. The elevation contour map comes close to being a universal language for presenting topographic information. It is simple in concept, visually oriented, and capable of transmitting relatively detailed information on the conformation of the earth's surface. After a little time spent studying contour maps, a person soon develops the ability to visualize the terrain shown on the map in three dimensions.

The basic idea behind a contour map is simply to draw a series of lines, with the elevation being constant along any given line. Each such line is called a *contour line*. By convention the difference in elevation between adjacent contour lines is a constant preset value throughout the map. This *constant difference in elevation* between adjacent contour lines is called the *contour interval*. Every fifth contour line on the map (or some other convenient interval) is drawn more heavily and labeled with its elevation. In some cases it is a little difficult to tell at a glance whether the elevation is increasing or decreasing as you move in a given direction. Short hatch lines are sometimes placed on the downhill side of contour lines in depressions to ease this problem in more difficult areas.

Perhaps the best analogy for visualizing the structure of a contour map is the one used by Greenhood (1964). Imagine a huge dike constructed around the area of interest, an unlimited supply of water, and pumps of tremendous capacity. Temporarily raise the water level in a series of equal steps. For each successive water level draw a map of the shoreline. The series of map lines tracing the shorelines would constitute elevation contour lines.

Another way to gain a feeling for the nature of contour maps is to use one in constructing the scale model of a terrain from plywood or similar material and a filler such as papier-mache. Find the contour line representing the highest elevation. With a razor blade or scalpel cut along this line to remove the area that it encircles. Position the resulting hole in the map conveniently on a sheet of plywood or similar material and draw a line on the plywood along the inside edge of the hole. Cut out the piece of plywood encircled by the resulting line, and paste the cutout from the map on the plywood block. Now move to the next lower contour line and cut along it. Prepare a plywood block as before. Paste the ring of paper from the map in position along the edge of the plywood block; then place it under the previous block so that the edge of that block fits within the ring of map paper. Continue this process until the whole map has been incorporated into the model. The vertical scale of the model is determined by the contour interval and the thickness of the plywood sheet. If ¼-inch plywood is used with a contour interval of 10 ft, each vertical inch in the model would represent a rise of 40 ft on the terrain. Thus the vertical scale would be 1:480. At this stage the model has a stairstep appearance as seen from the side, and it looks like the original map as seen from directly above. The final step in constructing the model is to fill in the stairsteps with papier-mache or similar material so that the steps are smoothed out into more natural slopes.

If one is to interpret contour maps effectively, several important properties of contour lines must be understood. First, a contour line must either close back on itself within the area of the map or end at the edge of the map. Contour lines do not just come to an end within the map itself, since this would require a landform shaped something like the peak of a roof with slanting end panels and yet large enough to show at map scale. Likewise overhanging cliffs and caves are the only landforms that would cause contour lines to cross; and even then the lines would appear to merge rather than cross unless the scale was very large. Second, the direction of the steepest slope always lies at right angles to the contour lines. Third, uniformly spaced contour lines represent a steady slope or incline. Closely spaced contour lines indicate a steep slope, whereas widely spaced contour lines indicate a gentle slope. There will be no contour lines crossing an area that is truely level. Fourth, contour lines will take V-shaped form where a stream cuts into a hillside, with the arms of the V-shape opening in a downstream direction.

Elevation profiles are simple to construct from contour maps, and an illustration of the process may also help to give a better feel for the nature of contour maps. Figure 4.1 is a crude contour map of a hypothetical area that includes a small watershed. A *watershed* is an area from which the surface runoff feeds a particular stream system. The boundary of a watershed is called the *drainage divide*. Outside the drainage divide, runoff water will flow into other

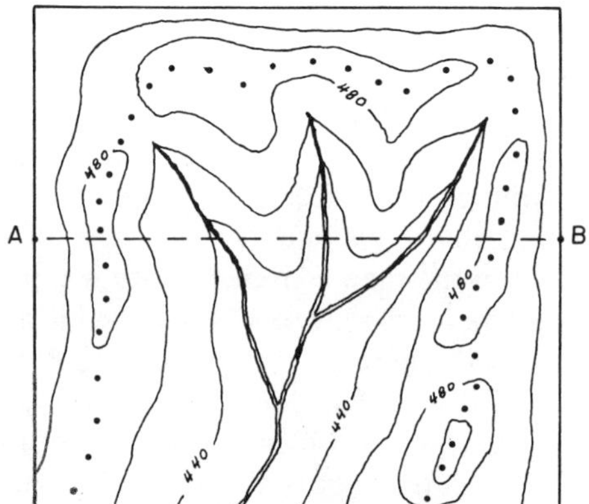

Figure 4.1 Contour map of hypothetical watershed. Solid lines are elevation contours. Dotted line is drainage divide. Dashed line from A to B is profile line for Figure 4.2.

stream systems. The drainage divide in Figure 4.1 is shown as a dotted line. This figure will be used to illustrate the construction of elevation profiles.

The first step is to draw the line on the map for which the profile is to be made. The dashed line from A to B in Figure 4.1 will serve this purpose. It should be noted that the profile line does not necessarily have to be straight over its entire length. For example, the profile line might be constructed along a streambed.

Choose a scale for the vertical axis (elevation) of the profile graph and graduate it accordingly; then provide a scale for the horizontal (distance) axis. The two scales need not be the same; in fact, it is usually advantageous to exaggerate the vertical axis considerably.

Start at one end of the profile line and measure the distance along the profile line to the point where the first contour line crosses. Scale this distance along the horizontal axis of the profile graph and plot the elevation of the contour above this point. Repeat this process for each successive intersection of a contour line with a profile line.

When all intersections have been plotted, then sketch a line connecting the plotted points. The relative spacing between contour lines can be used as a guide for sketching this line. Figure 4.2 is a profile graph for the line from A to B in Figure 4.1. Profile graphs are essential for construction projects that involve earthwork and are extremely helpful in a variety of other situations.

The preparation of contour maps from spot elevations determined in the field and plotted on a base map calls for interpolating the position of lines between the known points. A little practice is required before one develops the ability to do this interpolation rapidly and accurately. It is very helpful in doing the interpolation if a *form line sketch* is prepared when the elevations are determined in the field. This is simply a sketch in which the positions of the contour lines are estimated visually instead of by actual measurement. As mentioned earlier, one

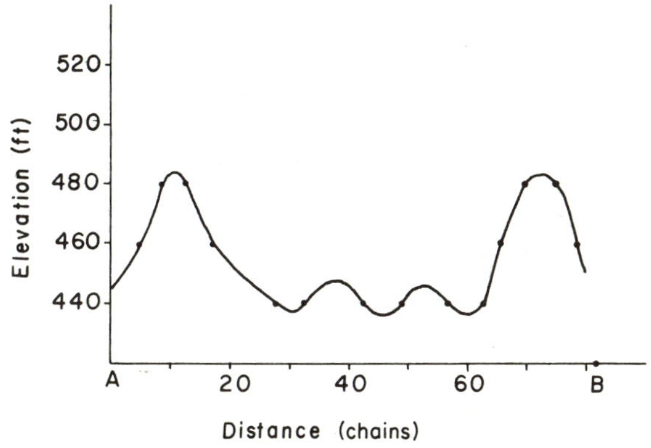

Figure 4.2 Profile graph for dashed line from A to B in Figure 4.1.

should also be certain to take spot elevations and make appropriate notations at points where major changes in slope occur.

Drawing contour lines from stereo airphotos requires less interpolation since one can set a parallax measuring device for constant elevation and trace the line along which it makes contact with the terrain as seen through the stereoscope.

Slope Maps. In many cases the steepness of the ground slope is a limiting factor which restricts use of the land. The usual procedure in such cases is to define slope classes that correspond to changes in the feasible use of the land, and then to prepare maps showing those slope classes. For example, the Cooperative Soil Survey in Michigan typically uses classes of 0-2%, 2-6%, 6-12%, 12-18%, 18-25%, and greater than 25% for the classification of soils according to feasible management practices. Usually colors or shading patterns are used to show the slope classes on a map, but the classification itself can be done in several ways.

One way of classifying areas according to the ground slope is to make direct measurements in the field with a clinometer. A second way already mentioned is to use a parallax device on stereo airphotos. The parallax device creates a graded series of successively more steeply sloping lines against which the interpreter matches the ground slopes to make a classification.

If contour maps are available, they offer a third convenient approach. As described earlier, the spacing between adjacent contour lines is determined by the slope of the ground. Take for instance a ground slope of 10% as shown on a topographic map with a scale of $1/24{,}000$ and a contour interval of 20 ft. A 10% slope implies a vertical rise of 100 ft for every 1000 ft of horizontal distance. On a map of scale $1/24{,}000$ 1 in. corresponds to 2000 ft of horizontal distance on the ground. The vertical rise over 2000 ft horizontal distance shoud be 200 ft. Each space between contour lines represents a 20 ft rise, so there will be 10 spaces between contour lines per inch on the map. Thus a 10% slope on this map is indicated by a spacing between contour lines of 0.1 in. A slope classification by this method goes quite rapidly when an acetate overlay is prepared with contour line spacings that correspond to boundaries of the desired classes.

Hachure and Shadow Maps. Two other maplike forms are sometimes used to convey general impressions of relief. In a *hachure map* short lines drawn downslope are used to show relief. Where confusion between hills and valleys is likely to occur, an arrowhead can be added to a hatch line every now and then to show the direction of the slope. The hachure system can be improved somewhat by using different line weights to represent slope classes, but the results are still not very satisfactory.

The other approach is the artist's technique of using shadings to mimic shadow patterns of hills. Maps of this latter type can be pleasing to the eye, but are qualitative at best in terms of topographic information.

Elevation–Area Graphs. Elevation–area graphs are useful for summarizing the topography of an area without showing the elevation of each point. One of the variables in such a graph is the elevation or height above some base level. The other variable can be the area between successive elevation contours or the area above (or below) successive elevation contours. Elevations and areas can be stated in either absolute or percentage form.

A workable approach is to use the format of histograms and distribution graphs as discussed in Section 2.6.3 with the elevation variable plotted on the X axis and the area taking the place of frequency.

However, geographers, geologists, and hydrologists usually prefer to have the elevation variable plotted on the vertical (Y) axis and the area variable plotted on the horizontal (X) axis. A graph showing the percent of relief above base level plotted on the Y axis against the percent of area with higher elevation plotted on the X axis is called a *hypsometric curve.* Figure 4.3 is a hypsometric curve for the hypothetical watershed in Figure 4.1.

Other varieties of elevation–area graphs also have special names, but these will not be discussed at length. The important thing is that the graph be well designed and clearly labeled so that the reader may understand its information content with a minimum of study. The construction of elevation–area graphs requires the measurement of the area between successive contour lines by one of the methods described in Section 3.7.

4.1.3 Numerical Description of Topography

There is often a need to summarize the topographic characteristics of an area that constitutes some sort of ecological unit, as for example a watershed which feeds a stream, river, or lake. The shopping list of such descriptors which have been proposed is quite extensive, so this discussion will be limited to the more commonly used parameters. For the origins of these parameters and information on more exotic ones the reader should consult hydrology and geomorphology texts such as Wisler and Brater (1959) or Ruhe (1975).

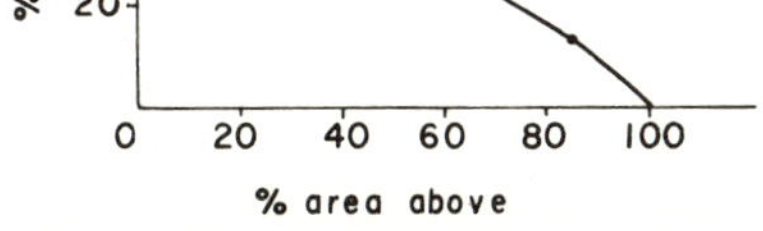

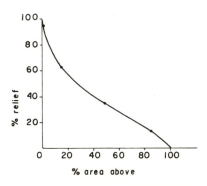

Figure 4.3 Hypsometric curve for watershed of Figure 4.1

Elevation Range. The maximum elevation, the minimum elevation, and their difference are all important topographic characteristics. The fact that the climate varies with elevation should come as news to very few. Higher elevations are cooler and often receive more precipitation than lower elevations. The composition of flora and fauna changes in response to these climatic changes. Even within the elevation range of a species, developmental rates change with altitude. Thus it is important to know both the highest and the lowest elevation in absolute terms. Likewise the difference or relief gives an indication of the variability in conditions and also shows the span over which gravity can act on surface water to produce erosive forces and stream currents.

Median and Mean Elevation. As discussed in Section 2.6.4, one must be aware of both average conditions and variability. Average conditions in terms of elevation are usually expressed as either *mean elevation* or *median elevation.* In either case the average is calculated with respect to the area.

Median Elevation. The median elevation is that elevation for which 50% of the area is higher and 50% is lower. This is most easily determined from an elevation-area graph in which the area above or below a given elevation is expressed as a percent of the total area. Given such a graph, one projects the 50% point of the area to the curve and thence to the elevation axis.

Mean Elevation. The mean elevation may be determined in two ways. One is to place a dot grid of suitable density over a topographic map, and then calculate the mean of the elevations as read at the points where the dots fall. The other method is to measure the area between each pair of successive contour lines. For each pair multiply this area by the midelevation. Sum these products, then divide by the total area. For both methods the accuracy increases with a decreasing size of contour interval. The accuracy of the first method is also determined by the density of the dot grid. The accuracy of the second method is affected by the care taken in measuring the area between pairs of contour lines.

The comparative advantages of means and medians have already been discussed in Section 2.6.4.

Mean Slope. The average slope within an area is usually expressed by the mean slope, which is determined from the formula

$$\text{mean slope} = \frac{\text{total length of contours} \times \text{contour interval}}{\text{total area}}$$

The reader is referred to Wisler and Brater (1959) for the derivation of this formula. The length of the contours is most easily determined by tracing with a map measurer (opisometer). The total area can be determined by any of the several methods described in Section 3.7. One should be careful to attach the proper units of measure to the mean slope. For example, suppose that the contour interval is expressed in meters, the length of the contours is expressed in kilometers, and the total area in square kilometers. The resulting units of measure for the mean slope are

$$\frac{\text{km} \times \text{m}}{\text{sq km}} = \text{m/km}$$

In order to convert this to percent slope it would be necessary to divide by 10. Likewise with the contour interval in feet, the length of the contours in miles, and the area in square miles, the units of the mean slope would be feet per mile. In order to convert feet per mile to percent slope, one would divide by 52.8.

Orientation of Slope. The direction in which a slope faces is called its *aspect*. This is the direction from which sunlight can fall most directly on the surface. It can be expressed in any of the ways used to describe directions such as azimuth, bearing, and so on. The aspect is important ecologically since north facing slopes tend to be cool and moist while south facing slopes tend to the warm and dry (in the northern hemisphere). This difference is often so pronounced that the composition of vegetation is quite different on opposite sides of a ridge.

Shape. The Shape of a topographic unit is particularly important with respect to hydrology. Since many storms are quite small in terms of area covered, a compact watershed is more likely to be covered in its entirety by a single storm than is one with an elongate or lobed shape. This tends to increase the variability of stream flow and the tendency toward flooding. There are exceptions, of course. Storms also tend to come from a particular direction because of prevailing winds. An elongate watershed that is oriented along a storm track will also tend to have high peak flows. Similar considerations also apply to influences other than rainfall. Environmental alterations in a compact unit are likely to be quite noticeable over most of the unit unless there is considerable relief to mask the view. At least partial concealment is easier to achieve on an elongate or irregularly shaped area.

The more common indexes of shape involve a comparison with a circle through relationships between perimeter and area. The *compactness coefficient* (Wisler and Brater, 1959) is the ratio of the actual perimeter to the circumference of a circle having equal area, that is,

$$\text{compactness coefficient} = \frac{\text{actual perimeter}}{\text{circumference of equal area circle}}$$

The circumference of a circle is computed from its area as

$$\text{circumference} = \sqrt{\pi \times \text{area} \times 4}$$

Since a circle has the smallest perimeter in relation to an area of any plane geometric form, the minimum value of the compactness coefficient is 1, and larger values indicate an increasing departure from circularity.

The values for the *circularity ratio* (Ruhe, 1975) tend in the opposite fashion since

$$\text{circularity ratio} = \frac{\text{actual area}}{\text{area of circle with same perimeter}}$$

With this latter index the maximum value is 1, and lower values indicate a departure from circularity.

Comparison with a circle may also be achieved through maximum length in relation to diameter of an equal area circle (Ruhe, 1975). This index is

$$\text{elongation ratio} = \frac{\text{diameter of equal area circle}}{\text{maximum length of unit}}$$

Like the circularity ratio, this index has a maximum of 1, and smaller values indicate an increasing departure from circularity.

The form may also be compared with a square through relationships between length or width and area, but these indexes are somewhat more difficult to interpret. Likewise indexes can be constructed from ratios of width to length which do not involve comparison with a standard form. There are several possible ways of measuring length and width. The various possibilities will not be discussed here, except to indicate that an average width is obtained by dividing the area by some measure of length.

4.1.4 Classification of Drainage Patterns

Ground slope and surface materials combine to produce characteristic stream drainage patterns which provide important clues to the geology of an area. Streams flowing over relatively uniform surface materials in level to rolling terrain will form a pattern which resembles the branches of a tree and is therefore called *dendritic*. Streams that follow lines of weakness in the geological materials often merge at nearly right angles forming an *angular* drainage pattern. Streams flowing into a valley from several directions will form a *radial* pattern.

These are just a few examples from a fairly large taxonomy of stream patterns which has been developed. Since this taxonomy of drainage patterns and relations to geological structure is covered in most textbooks on geology and photointerpretation, the subject will not be discussed in detail here. However, its importance should not be minimized.

Delineation of stream courses from aerial photos constitutes the first step in a basic nine-step method of terrain analysis set forth by Strandberg (1967). Way (1973) gives an excellent coverage of relations between drainage pattern and geological structure as well as implications for human ecology.

It should also be noted that an analysis of drainage patterns is better done from airphotos than from maps. The scale and generalization in maps is often such that many details of the drainage network are eliminated or obscured.

4.1.5 Numerical Description of Drainage Networks

The importance of water supply and flood control for human ecology, coupled with the difficulty of representing large drainage nets on a map of reasonable scale, has led to the development of several types of numerical descriptors for drainage networks. Further details regarding the development and application of numerical methods covered here are available in many textbooks dealing with physical geology, hydrology, and water resources. Linsley, Kohler, and Paulhus (1975); Ruhe (1975); Strahler (1964); Strandberg (1967); and Wisler and Brater (1959) constitute a sample of such references.

Stream Orders. Order numbers are used to show the degree of branching in streams. The Strahler system of numbering is most useful for the analysis of restricted areas. In this system streams are numbered from source to mouth. A stream with no tributaries is designated as being of order 1. When two streams of order 1 join, they form a stream of order 2. When two streams of order 2 join, they form a stream of order 3, and so on. Order does not increase unless two streams of the same order come together. Thus a stream of order 2 could receive a tributary of order 1 without changing order.

A system of numbering in the reverse direction is often used in regional analyses. In this system a stream with its mouth at the ocean is of order 1. A tributary to an order 1 stream is of order 2, a tributary to an order 2 stream is of order 3, and so forth.

Stream Frequency. Stream frequency (sometimes called stream density) is the number of streams per unit area. The first step in its determination is to measure the area under consideration by one of the methods described in Section 3.7. The next step is to assign order numbers to the streams. Under the Strahler system the total number of streams is obtained by adding the numbers of streams of different orders. This is equal to the number of order 1 streams plus the number of confluences where the stream order increases. The final step is to compute

$$\text{stream frequency} = \frac{\text{number of streams}}{\text{area}}$$

Likewise the stream frequency may be computed separately by order.

It should also be noted that the stream frequency according to an earlier system of ordering (Horton, 1945) is equal to the frequency of Strahler order 1 streams.

In general a greater stream frequency implies better drainage. There may be exceptions to this, however, because the stream frequency does not take direct account of the stream length.

Drainage Density. Drainage density is the average length of streams per unit area. It is determined by first measuring the total area, then measuring the total length of the streams with a map measurer (opisometer), and finally computing by the formula

$$\text{drainage density} = \frac{\text{total length of streams}}{\text{area}}$$

Again greater drainage density implies better drainage of the area. In this case, however, the extent of branching in the drainage network is not shown explicitly. Therefore the best indication of drainage is obtained by using the drainage density and the stream frequency in combination.

4.1.6 Description of Channels and Lake Basins

Since stream channels and lake basins are just submerged sections of topography, it should not be surprising that their morphology is quantified by minor variations of the methods already discussed. The terminology is somewhat different, but the techniques are essentially the same.

Hydrographic Maps. Depth contours are used to map the morphology of channels and basins in the same way that elevation contours are used to map the topography. The only difference is that field data for drawing the depth contours are collected by sounding lines or sonar, and the interpolation of depth contour positions must be done without the help of a form line sketch. Likewise the process of constructing depth profiles is entirely parallel to that for constructing elevation profiles as discussed in Section 4.1.2.

One added feature of profile analysis in stream channels is that the cross-sectional area is often required. It may be obtained from the profile graph by any of the area determination techniques applicable to maps. The concave nature of stream channels, however, often makes it convenient to compute the cross-sectional area from the slices formed by pairs of contours. If straight lines are used to approximate the slope of the bank between depth contours, each pair of contours forms a trapezoid. The area of such a trapezoid is easily computed by averaging the lengths of the upper and lower contours and then multiplying by the contour interval. The total area is calculated as the sum of the trapezoidal slices.

Hypsographic Curves. Like contour maps, elevation-area graphs translate directly to analyses of lake basins where depth is substituted for elevation. Since this type of graphical display has been discussed in Section 4.1.2, it will not be reviewed here.

Stream Gradient. The rate of fall of a streambed is important since it affects the speed with which water will move through the channel once it reaches the stream. The average rate of fall between two points along the channel is called *stream gradient*. The first step in determining the gradient is to find the difference in elevation between the two points in question. The second step is to measure the length of the stream along its course between the two points. Then the gradient is computed as

$$\text{gradient} = \frac{\text{elevation difference}}{\text{length}}$$

Volume of Basin. The volume of a lake basin is an important characteristic which does not have a direct parallel in the analysis of topography. The volume is important for determining concentrations that will result from dumping a given quantity of soluble material into the lake, and also for determining the relationship between heat input and rise of water temperature. Likewise volume is a crucial consideration in working with the water storage of reservoirs.

A unit of volume which may be unfamiliar to many readers is often used in this type of analysis, the *acre foot*. An acre foot is a volume of water which would cover an acre to a depth of 1 ft, or 43,560 cu ft.

It would be quite feasible to compute the volume as the product of mean depth and surface area, with the mean depth being determined from a map of depth contours and a dot grid in the same manner as the mean elevation (see Section 4.1.3). However, this is not the approach usually taken. The usual approach is to compute the volume in slices formed by adjacent depth contours and then add the slices together. The volume of such a slice is computed under the assumption that the slices take the form of a frustum of a cone (Welch, 1948). The formula for the volume of a frustum of a cone is

$$\text{volume} = \frac{h}{3} \times (a_u + \sqrt{a_u a_l} + a_l)$$

where a_u = area of upper surface
a_l = area of lower surface
h = height or depth

Dimensions of Basin. The obvious dimensions for describing a lake basin are length, width, and depth, but each of these dimensions may be determined in several ways (Welch, 1948).

The length is usually measured as both *maximum length* and *maximum effective length*. Maximum length is essentially the longest distance of travel by boat between points on the shore, with the travel path lying approximately midway between the left and right hand shores. However, the line of travel is allowed to cross small islands, which a boat obviously could not do. If the lake has a curving form, as for example an oxbow lake, then the line for the measurement of the maximum length is also curved. In contrast, the maximum effective length is the longest straight line distance across open water. The latter distance is more important with respect to the development of waves and cooling due to wind action.

The width is measured in three ways. *Maximum width* and *maximum effective width* are measured at approximately right angles to the corresponding lengths and in the same manner. *Mean width* is obtained by dividing the maximum length into surface area.

The *maximum depth* is simply the deepest point in the lake as determined from sounding lines, sonar, or a map of depth contours. The *mean depth* is usually obtained by dividing the surface area into volume. However, it could also be determined from a map of depth contours and a dot grid in the same way as the mean elevation of a watershed or other topographic unit (see Section 4.1.3).

Mean Slope of Basin. The *mean slope* of a lake basin can be determined in the same manner as the mean slope of a watershed or other topographic unit, that is,

$$\text{mean slope} = \frac{\text{total length of contours} \times \text{contour interval}}{\text{surface area}}$$

However, a frequent modification (Welch, 1948) is to reduce the sum of contour lengths by one-half the combined lengths of the surface and the deepest contours. This takes into account the fact that the top of the basin coincides exactly with the surface contour. Since the surface contour is the longest, the reduction by half its length has some effect on the result. In contrast, the deepest contour is usually short, and the correction for it could normally be omitted with little effect on the result.

There is also an occasion to find the mean slope of the lakebed between a pair of contour lines. The appropriate formula (Welch, 1948) is

$$\text{mean slope} = \frac{\text{sum of contour lengths} \times \text{contour interval}}{2 \times \text{area between contours}}$$

Orientation of Basin. The extent of cooling, wave action, and related effects generated by wind action will depend to some extent on the coincidence between the wind direction and the maximum effective length of the basin. Therefore it is important to know the direction in which this axis is oriented.

Shape of Basin. As with watersheds or other topographic units, several indexes have been developed for expressing the shape of a lake basin. The term *shoreline development index* is used instead of compactness coefficient when describing the shape of a lake, but the computational procedure is the same, that is,

$$\text{shoreline development index} = \frac{\text{actual length of shoreline}}{\text{circumference of equal area circle}}$$

The *volume development index* is the ratio of the actual volume of the lake to the volume of a cone having the same maximum depth and surface area. Welch (1948) gives the shortcut formula:

$$\text{volume development index} = \frac{3 \times \text{mean depth}}{\text{maximum depth}}$$

Other indexes of shape (Welch, 1948; Strandberg, 1967) are the ratio of mean depth to maximum depth and the ratio of maximum depth to the square root of the surface area. The former is simply one-third the volume development index. The latter takes the value 1 for a cubical basin.

4.2 THE ENVIRONMENTAL SUBSTRATE: SOILS

Having dealt with the shape of the land surface in the foregoing sections, we consider next the soils which carpet that surface. The nature of the soil exerts a major influence in determining the feasibility of land uses and the composition of natural communities. One function of the soil is to provide the base support for human structures and rooted plants. Another is to supply mineral nutrients, air, and water for plants. A third is to collect, transport, and store groundwater. Soil constitutes the primary environment for a vast array of microorganisms, and it also provides shelter for larger animals that have a burrowing habit. The chemical and biological processes in the soil serve as a gigantic disposal system for the waste products of man, other animals, and plants. Since we walk upon the soils, pave them, build upon them, obtain our food from plants that grow in them, and bury our waste in them, the importance of having survey information on soils should be fairly apparent.

Soils are formed by the interaction of climate, plants, and animals with parent geologic materials exposed at the surface of the earth. The parent materials may be either fragments from the underlying geologic structure or transported from another location by erosive forces. Soil is a dynamic entity subject to surprisingly rapid alteration by the physical and chemical influences of man, other animals, and plants. The physical and chemical properties of soils are quite variable, making different types of soil better for some purposes than others. The first order of business is to review these properties and ways by which they may be determined.

Although space does not permit a detailed discussion of apparatus, reagents, and procedures, the general nature of soil tests and conditions for their application are described. Several references are cited in which details of the tests can be found, and most regions have one or more soils laboratories which will perform tests on a commercial or service basis. A variety of field test kits such as LaMotte, Hellige-Truog, and Simplex are also available commercially for making approximate determinations rapidly. These kits are designed for easy use by persons with little previous experience in making soil tests. The set of two volumes on *Methods of Soil Analysis* published by the American Society of Agronomy (1965) is an excellent reference for all types of soil analysis.

A caution is in order regarding the utility of soil tests. Existing interpretations of soil tests have been developed largely for agricultural crops. Relatively little information is available upon which to base interpretations for natural ecosystems. Such tests can, however, provide a basis for exploring the possible reasons for observed differences within and between natural ecosystems.

4.2.1 Soil Properties

The intensity of the influence by climate and organisms decreases with the depth below the surface. The result is a series of changes in soil appearance and properties with increasing depth. Although they are gradational, fairly distinct zones can often be identified as one digs downward from the surface. These zones are called *horizons*, and together they make up the *soil profile*.

Soil Profiles. The major soil layers or horizons which one normally encounters in a profile are given letter designations, and numbers are used for a first subdivision of these major horizons. It should be understood, however, that actual profiles encountered in the field may have some horizons modified or missing from the typical sequence depending on the developmental history of the soil in that location. In some cases the parent material may not have been exposed to soil forming processes long enough for a layered structure to be recognizable. In other cases layers may have been altered or removed by the activities of man or erosion. Variations in the parent material also modify the typical sequence. In any case a certain amount of judgment usually enters into deciding where one horizon ends and the next begins.

O Horizon. This is a surface layer of organic matter from plant and animal debris. In the upper portion of the layer, designated O1, decay by microorganisms is slight. In the lower portion, designated O2, decay has progressed to the point where it is difficult to recognize the original nature of the debris.

In older terminology the O1 layer was often called the L (litter) layer, and the O2 was called the F (fermentation) layer.

A Horizon. The first layer below the O horizon in which mineral matter predominates is called the A horizon. The upper portion of the A horizon, designated A1, is characterized by an accumulation of organic matter from the O horizon above. The central portion of the A horizon, called A2, is characterized by some leaching away of soluble and fine materials. The lower portion of the A horizon, designated A3, is transitional with the B horizon below, but is more like A than B.

B Horizon. The B horizon is characterized by deposition of soluble and fine materials leached from the A horizon above. The upper portion, designated B1, is transitional with the A horizon, but more like B than A. The B2 is the main zone of deposition or accumulation of solubles and fines. The lower, or B3, portion is transitional with the C horizon below, but more like B than C.

C Horizon. The C horizon is the deep zone in which soil forming processes have not caused appreciable layering. This zone is not usually subdivided by numbers except for the convenience in describing any variations that may occur in the C horizon materials. If such variations occur, they are not primarily the result of soil forming processes (influences of climate, plants, and animals).

R Horizon. This designation is used for the rock underlying the soil horizons.

Additional numbers (such as B11, B12) may be used for convenience in describing minor variations within horizons. If there is a major change in the materials from which the soil is formed, this is indicated by a Roman numeral prefix (such as IB2) for horizons located in the new type of material. Horizons after the first change carry the prefix I, those after the second change carry the prefix II, and so on. Small letters are also used as modifiers to indicate specific types of variation in the profile. For example, A1p is used to designate a layer in the upper portion of the profile in agricultural areas where thorough mixing has taken place from plowing. Furthermore combinations of capital letters (such as AB) may be used when transitional layers between horizons (such as A3 and B1) cannot be readily separated. For additional details on the description of soil horizons the reader should consult Soil Survey Staff (1962 and 1975) or Buol, Hole, and McCracken (1973).

In working with soils the various horizons should be described individually, including thickness, variations in thickness, and distinctness of boundaries between horizons. Likewise the major horizons should be sampled separately for laboratory testing of soil properties.

Color. Color is one of the more convenient properties for visual separation of soil horizons since it reflects several more fundamental properties of the soil. Common usage of color names, however, allows so much latitude for interpretation that its utility is largely lost.

A consistency in the description of colors can be achieved by the use of color charts such as those produced by the Munsell Color Company, Inc., of Baltimore, Md. Munsell charts are arranged according to the three color variables *hue, value,* and *chroma*.

The term *hue* is used to characterize the dominant wavelength of reflected light, or spectral color. In the Munsell color notation the various spectral colors are broken down into ranges, and each range is given a letter code, such as Y for yellow. Each letter range is covered by a numerical scale of 0–10. A given hue is expressed by its numerical position in the range followed by the code letter(s) for that range (such as 2.5Y).

The term *value* is used to characterize the relative lightness or darkness as determined by the amount of light reflected. The value is scaled 0–10 according to increasing lightness. In Munsell color notation, the value number follows the letter code of the hue.

The term *chroma* is used to characterize color purity or the extent to which the color differs from neutral gray of the same lightness. Chroma is scaled 0–20 in order of decreasing grayness. In Munsell color notation, the chroma number follows value number, separated from it by a slash.

Munsell color chips are arranged on pages, with holes to facilitate comparison with specimens. Each page contains a given hue with values increasing up the page and chroma increasing from left to right. Common names of colors are also given on the sheets. Since soils do not cover the entire range of hues, a subset is available that is appropriate for use with soils.

Since color changes with the moisture content in some soils, it should be noted whether the soil is wet or dry when the color observations are made. Additional terminology has been adopted for describing patterns of soil coloration (Soil Survey Staff, 1951), which will not be described here.

Particle Size Distribution. Although small amounts of organic matter from plant and animal residue are usually intermixed, most soils are composed primarily of mineral particles. *Texture* is the term used to characterize the size distribution of mineral particles in the soil. The major textural classes in the order of decreasing particle size are *gravel, sand, silt,* and *clay*. The term *loam*, with appropriate modifiers, is used for mixtures of sand, silt, and clay. There is a general, but not complete, agreement among the various groups concerned with the analysis of soils as to where the divisions between the size classes should be drawn.

Depending on the classification scheme, the lower limit of the gravel class ranges from 4.0 mm diameter to 2.0 mm diameter; the lower limit of the sand class ranges from 0.074 mm to 0.053 mm; the lower limit of the silt class ranges from 0.005 mm to 0.002 mm, and there is no set lower limit for the clay size. The four major classes may also be broken down into subclasses carrying descriptive labels such as very coarse, coarse, medium, fine, and very fine. The clay fraction is of special significance because fine clays have colloidal properties.

A quantitative analysis of the particle size distribution is done by standard soil laboratory testing procedures. Separations within the gravel and sand classes are usually accomplished by shaking through a nest of standard soil sieves and then determining the percentage of material retained on each sieve by weighing.

The analysis of the silt and clay classes is accomplished by using Stokes' law of settling to determine the size classes of the materials which remain in water suspension as a function of the time that the suspension is allowed to stand. For maximum accuracy the suspension is sampled at specific time intervals with a pipette (Day, 1965). More commonly, however, a hydrometer is used to measure the change in specific gravity of the soil suspension over time. See Bowles (1970), Day (1965), Lambe (1951), or MacIver and Hale (1970) for details of these testing methods. Wilde, Voigt, and Iyer (1972) give a less accurate hydrometer method that can be performed rapidly with portable equipment.

A convenient method for displaying the results of a particle size analysis is to plot the weight percent finer against the log of the particle diameter. An alternative for the particle size axis, which has some desirable statistical properties (Dapples, 1959), is to use a phi scale, where

$$\phi = \log_2 \left(\frac{1}{\text{diameter}} \right)$$

with the diameter expressed in millimeters.

The numerical presentation of grain size data usually takes the form of weight percent by size class. In addition the *coefficient of uniformity* and the *coefficient of concavity* are often computed as an aid to soil classification for engineering purposes. The formulas are

$$\text{coefficient of uniformity} = \frac{D_{60}}{D_{10}}$$

and

$$\text{coefficient of concavity} = \frac{(D_{30})^2}{D_{10} \times D_{60}}$$

where D_n is the diameter for which n % of the material is finer. Large values of the coefficient of uniformity indicate heterogeneous material with respect to particle size. The coefficient of concavity is a product of two D_n ratios,

D_{30}/D_{10} and D_{30}/D_{60}. Therefore the coefficient of concavity will be 1 whenever D_{30}/D_{10} and D_{60}/D_{30} are equal.

A rough determination of the soil texture can be done in the field by applying pressure to moist soil with the hand and fingers (Soil Survey Staff, 1951). Most clays will form a flexible ribbon when moist soil is squeezed between the fingers. Clay loam will also form a ribbon when moist soil is squeezed between the fingers, but the ribbon is fragile rather than flexible. Coarser soils will not form a ribbon, but can be formed into a cast when moist soil is squeezed by the handful. The feel of soil between the fingers coupled with the durability of the cast is used to judge the texture. Sand is very gritty when rolled between the fingers, and a moist cast crumbles when touched. Sandy loam also feels very gritty, but a moist cast can be handled gently without breaking. Loam feels somewhat gritty when squeezed, and a moist cast bears considerable handling. Silt and silt loams feel quite smooth to the touch and form durable casts.

Structure and Consistence. The individual soil particles may be aggregated together to form secondary units. Units of this type are called *peds*. Soil *structure* is the term used to characterize the nature of the peds. This is an important soil property since structured soil may behave quite differently from unstructured soil, especially when considerable amounts of clay are present. The four major types of structure are *platy, prismlike, blocklike,* and *spheroidal* (Soil Survey Staff, 1951), where the names relate to the shape of the peds.

In a platy structure the peds are more or less flat and oriented horizontally, In a prismlike structure the peds have relatively straight faces and are oriented more or less along vertical lines. The prismlike type of structure is broken down into two subtypes according to the nature of the ends of the peds. The term *columnar* is used for prismlike peds with rounded ends, and the term *prismatic* is used if the ends are not rounded. The term blocklike is fairly descriptive of peds in that type of structure. A blocklike structure is broken down into *angular blocky* and *subangular blocky* according to the angle of intersection between faces of the ped and the degree of rounding along the edges. The spheroidal type is broken down into *granular* for tightly packed peds and *crumb* for loosely packed peds.

The size of the peds is described as very fine (or thin), fine (or thin), medium, coarse (or thick), and very coarse (or thick). See Soil Survey Staff (1951) for the size ranges of peds in each type of structure. The term *grade* is used to characterize the amount of aggregation between peds. Grade is usually described as structureless, weak, moderate, or strong according to the difficulty involved in separating peds from one another.

Consistence is a term used to characterize the physical behavior of soil in a rather general way. Consistence arises from texture and structure. Important aspects of consistence in wet soil are stickiness and plasticity. The consistence of moist soil is described by adjectives such as loose, friable, or firm.

The consistence of dry soil is described by terms like loose, soft, or hard. The degree of cementation is another aspect of consistence. It is usually more important to note the consistence of soil in a wet condition that when moist or dry.

Bulk Density, Specific Gravity, and Porosity. Soil is a composite of particulate matter and voids (spaces) between the particles. The spaces between particles are important both biologically and for engineering purposes since they are occupied by air and/or water. The load bearing capacity of many soils decreases with increasing water content, particularly so for fine textured soils. Plant roots and soil organisms must have adequate supplies of air and water, both of which must come from the void spaces. The percentage distribution of the total soil volume between solid and voids can be determined from the specific gravity of soil solids and bulk density.

Bulk Density. *Bulk density* is the weight of dry soil materials per unit total volume of soil. The essential steps in determining the bulk density are to obtain a known volume of soil, dry the soil materials, and then weigh the dry materials. The bulk density is then computed as weight divided by volume. The volume used in determining the bulk density must be the volume occupied by the sample before removal from the field. If the soil is cohesive enough to be removed without crumbling, a cylindrical sampler of known volume can be used. For loose soil materials the original volume can be determined by the sand cone method or rubber balloon method (Blake, 1965; Bowles, 1970). Drying is achieved by placing the sample in a constant temperature oven set in the range of 105 to 110°C until its weight no longer changes from the evaporation of moisture. Weights should be determined carefully using a sensitive balance. Alternatively, backscattering of gamma radiation can be used for the field determination of bulk density. This involves adding a gamma radiation source and a gamma detector to the apparatus to be described later for measuring the soil moisture content by neutron scattering.

Specific Gravity. *Specific gravity of the soil solids* is the weight of solids relative to the weight of water. This is the ratio that would be obtained from weighing a given volume of soil solids (which contained no void spaces) and dividing it by the weight of an equal volume of water. The specific gravity is determined by using a *pycnometer*, which is a carefully calibrated volumetric flask. The basic steps are to determine the weight that the pycnometer has when filled with water alone (W_w) the weight of the pycnometer when filled with a mixture of soil and water (W_m), and the weight of the soil when removed from the pycnometer and dried (W_s). Then the specific gravity of soil solids is

$$\text{specific gravity} = \frac{W_s}{W_w + W_s - W_m}$$

However, the temperature must be controlled and air must be removed from both the water and the soil-water mixture in order to get accurate results. See Blake (1965), Bowles (1970), Lambe (1951), or MacIver and Hale (1970) for details of procedure and apparatus. Unless otherwise stated, the term *specific gravity* is used in this discussion to mean the average specific gravity of soil solids as determined in this manner.

Porosity. The fraction (or percentage) of total soil volume occupied by voids is called *porosity*. The porosity is usually calculated indirectly by first finding the solid fraction and then subtracting from 1 (or 100%). The solid fraction in terms of bulk density and specific gravity is

$$\text{solid fraction} = \frac{\text{bulk density}}{\text{specific gravity} \times \text{unit weight of water}}$$

and

$$\text{porosity} = 1 - \text{solid fraction}$$

Voids may also be expressed by *void ratio*, where

$$\text{void ratio} = \frac{\text{void fraction}}{\text{solid fraction}}$$

Moisture Content. As mentioned above, the void spaces are occupied by water, air, or part of each. Unless soil is submerged (below the water table), water drains from the larger voids (macropores) under the force of gravity. Capillary forces hold water in the smaller voids (micropores) until it is removed by plant roots or drying. Thus air is supplied primarily from the macropores and water is supplied from the micropores. The finer the particles that make up the soil, the greater the amount of micropores relative to macropores, and the greater the water holding capacity of the soil. Thus clays tend to be waterlogged and sands tend to be droughty. Since they affect both engineering properties and the suitability for growing plants, the moisture relations of a soil are very important. A prerequisite for investigating soil moisture relations, however, is to have a way of determining the moisture content.

Gravimetric Method. The simplest (but not always the most convenient) method of determining the soil moisture content is by loss of weight on drying. This is called the *gravimetric method*. The procedure consists of obtaining a sample of moist soil, weighing the moist sample, oven drying the sample, and weighing the oven dry sample. The moisture content as a percentage of the dry weight is then calculated as

$$\%\ \text{moisture} = \frac{\text{weight loss on drying}}{\text{oven dry weight}} \times 100$$

A few cautions are necessary in applying the gravimetric method. One is to use an airtight sample container to avoid evaporation before first weighing or absorption after oven drying. Another is to remember that the weight of the empty container (tare weight) must be subtracted in determining oven dry weight of the sample. If desired, the moisture content on a dry weight basis can be converted to a volume basis by multiplying by the factor (bulk density/unit weight of water). Wilde, Voigt, and Iyer (1972) give a modified method based on weighing a sample in water, which avoids the necessity of oven drying. However, the specific gravity must be known in order to apply their method.

Neutron Scattering Method. Moisture determinations can be made in the field without extracting samples by using the capacity of the soil water for slowing the speed of fast neutrons. An access tube is implanted in the soil through which a fast neutron source (such as radium-beryllium) is lowered along with a slow neutron detector (such as boron trifluoride). A portable scaler (for example, Nuclear-Chicago model 2800) is used to compare the count rate in the soil against that from a standard neutron absorber in the shield. The moisture content on a volume basis is then read from a calibration curve. Since the access tube can be left in place for later use, this method is particularly convenient when reasonable accuracy is required and a number of determinations must be made in the same location over a period of time. See Gardner (1965) and Wilde, Voigt, and Iyer (1972) for details of equipment and procedure.

Electrical Conductivity of Porous Blocks. The electrical conductivity of porous blocks buried in the soil can also be used for repeated determinations of the moisture content without extracting samples. However, there are more sources or error in this method, so the results are only approximate and frequent checking against gravimetric determinations is necessary. A porous block made from gypsum which contains imbedded electrodes is buried in the soil with wire leads running up to the surface. The block absorbs moisture from the soil, and the electrical resistance of the block changes with the moisture content. A meter is attached to the wire leads to measure the electrical resistance of the block, and the moisture content is read from a calibration curve.

The tension with which water is held in the soil determines the absorbance by the block. However, the water tension at a given percentage moisture content varies with the soil type. Clay soils exert greater tension at a given moisture content than coarser soils. Therefore a given type of block must be calibrated separately for each different type of soil. Furthermore the plot of tension against moisture is different when the soil is drying than when it is being wetted (called hysteresis effect); and substances dissolved in the water can also alter the calibration curve. Due to these several sources of error, the porous block method should be regarded as providing an index of the moisture content (actually moisture tension) rather than an accurate measurement.

Tensiometers. Tensiometers provide an index of the moisture status similar to that obtained with the electrical conductivity of porous blocks. A tensiometer is a porous cup with a vacuum gauge attached which can be filled with water and sealed. When the tensiometer is imbedded in the soil, the tendency for water to be pulled through the porous walls of the cup into dryer soil exerts a suction which is registered on the vacuum gauge. As with porous blocks, a hysteresis effect occurs between wetting and drying. However, tensiometers are not affected by dissolved substances in the soil water. The use of tensiometers is restricted to relatively moist soils. Often they are used to provide an index of the need for irrigation in agricultural crops. They are well suited to this purpose since an accurate determination of the soil moisture content is not required and the soil is never allowed to become very dry. See Richards (1965) for a more complete discussion of the utility and limitation of tensiometers.

From these discussions it should be apparent that the gravimetric and neutron scattering methods provide the most satisfactory determinations of the soil moisture content. The gravimetric method is applicable when samples can be removed from the field for laboratory analysis. The neutron scattering method is applicable for in-place moisture determinations. The porous block and tensiometer methods are inexpensive but approximate techniques which provide an index of the soil moisture status as reflected by moisture tension.

Soil Moisture Relations. As stated earlier, the soil moisture has marked effects on both the engineering properties and the suitability as a medium for growth of plant roots and other soil organisms. The engineering aspects of soil moisture will be considered first, followed by the biological relationships.

Atterberg Limits and Indexes. As moisture is added, the soil changes physical states. Depending on the moisture content, the soil may behave as a solid, semisolid, plastic, or liquid. Atterberg (1911) proposed tests to determine the limiting moisture contents of the several physical states. Although Atterberg's tests are rather crude, they have been widely adopted in the analysis of soils for engineering purposes.

The moisture content at which a soil changes from plastic to liquid behavior is called the *liquid limit*. The liquid limit is determined by graphing on semilog paper the number of jarring blows required to close a standard groove against the moisture content. The moisture content is plotted on the arithmetic axis, the blow count on the log axis. Liquid limit is the moisture content at which 25 blows cause the standard groove to flow shut over ½ in. of its length. This point is interpolated from the semilog graph. The conditions of the test are standardized by the use of a jarring machine and a grooving tool.

The moisture content at which the soil changes from semisolid behavior to plastic behavior is called the *plastic limit*. The plastic limit is determined by altering the moisture content until the soil just crumbles when rolled into a thread 1/8 in. thick.

The moisture content at which the soil changes from semisolid to solid behavior is called the *shrinkage limit*. As the term implies, no further shrinkage of the soil occurs when the moisture is reduced below the shrinkage limit. The shrinkage limit is determined by using mercury displacement to determine the change in volume of a soil pat upon drying from a moisture content in excess of the liquid limit. The reduction in water content from the initial value to the shrinkage limit is equal to the loss in volume of the soil pat during drying.

Only that fraction of the soil passing the #40 sieve is used in determining Atterberg limits. See Bowles (1970), Lambe (1951), MacIver and Hale (1970), or Sowers (1965) for details of the procedure and equipment involved in the Atterberg limit tests.

Several useful indexes are also computed from the Atterberg limits. The *plasticity index* is the difference between the liquid and the plastic limits. It shows the range of moisture contents over which the soil has a plastic behavior. The *flow index* is the difference between the moisture content at which a liquid limit test groove closes in 10 blows and that at which it closes in 100 blows. This is determined from the plot of the moisture content against the log of the blow count in the liquid limit test. The *toughness index* is the ratio of plasticity index to flow index. The *coefficient of linear extensibility* (abbreviated COLE) is a common index of the shrink-swell potential in soils. This is the linear shrinkage upon drying expressed as a fraction of the dimension at the shrinkage limit.

Optimum Moisture Content for Compaction. Compaction is the least expensive method of improving the suitability of soil materials as a structural base. For a given compactive energy the maximum density that can be achieved in compaction depends on the moisture content at which compaction takes place. Therefore it is of considerable practical interest to know the optimum moisture content for compaction and the density that can be achieved if compaction is done at this moisture content. Proctor (1933) proposed laboratory tests for determining the optimum moisture content and maximum density of compaction. Modifications of Proctor's tests are widely used in engineering practice. The tests involve the use of standard hammers for compacting samples over a range of moisture contents. The maximum compacted density and the moisture content at which it was achieved can be determined by plotting compacted density against moisture content. See Bowles (1970), Felt (1965), Lambe (1951), and MacIver and Hale (1970) for details of compaction tests.

Permeability. Permeability and hydraulic conductivity are terms used to describe the ease with which water moves through a soil. It is an important consideration in engineering applications that involve the movement or retention of water, such as construction of water impoundments or drain fields. The permeability is expressed by the *coefficient of permeability* which relates the volume of flow per unit time to the hydraulic gradient and cross-sectional area through which the flow takes place. The units attached to the coefficient of permeability are length of flow per unit time. The larger the coefficient of permeability, the greater the rate of flow. However, the flow rate also depends on temperature since the viscosity of water changes with temperature. Thus the coefficient of permeability must be for a stated temperature. A temperature of 20°C is frequently used in this respect.

There are two main types of laboratory methods for testing the permeability, known as the constant head and the falling head methods. See Bowles (1970), Lambe (1951), and MacIver and Hale (1970) for details of these and related tests as performed with standard laboratory apparatus. Klute (1965) gives simplified versions of laboratory permeability tests. Since the void ratio affects permeability, the results of permeability tests are usually presented by plotting the coefficient of permeability against some function of the void ratio.

It should be noted, however, that laboratory tests may not be a true reflection of the conditions in undisturbed soil because of a variety of reasons such as the alteration of the soil structure in preparing samples and a possible stratification of soil materials as they occur in nature. Boersma (1965) describes two methods for use in the field below the water table level. One is based on the rate of water rise in an open auger hole. The other is based on an auger hole that is lined by a pipe in the upper portion and open in the lower portion to form a piezometer (pressure measuring device). He also describes a double tube system for use above the water table.

Infiltration. Infiltration refers to the rate at which water enters the soil from the surface. This is important with respect to the recharge of groundwater supplies, surface runoff, and flooding. Devices for measuring the infiltration rate are called *infiltrometers*. A rather wide variety of infiltrometer designs are in use (Bertrand, 1956; Wisler and Brater, 1959). All infiltrometers employ some means of isolating a small area of known size to which water is applied. In some models the confined area is flooded with a known amount of water, and the time required for the water to be absorbed into the soil is determined. In other models water is sprinkled onto the confined area at a known rate, and runoff is permitted in an attempt to mimic the natural situation more closely. The runoff is collected and subtracted from the total quantity applied to determine the amount absorbed.

Because of the closer resemblance to natural rainfall, the sprinkling type gives more meaningful results. However, it also introduces the additional variable of the rate of application into the problem of interpreting the results. It should be noted that infiltration is greatly influenced by the nature of surface litter as well as the subsurface character of the soil. This is especially true for infiltrometers that simulate natural rainfall.

Field Capacity. A soil is saturated when all voids are filled with water. When a saturated soil is allowed to drain under the force of gravity, water will drain from the macropores, but remain in the micropores. When the macropore water which drains under the force of gravity has been removed, the soil is at *field capacity*. Moisture conditions for plant growth are ideal at field capacity. The micropores are full of water, and water in the larger of these micropores is easily absorbed by plant roots. The air supply is also adequate in the soil at field capacity since drainage from the macropores leaves them filled with air. The volume of air in a soil can always be determined by subtracting the moisture content on a volume basis from porosity.

The moisture content at field capacity can be approximated by saturating intact soil samples and then allowing them to drain for about 24 hours on a moist blotter. If a pressure chamber or tension table apparatus is available, previously saturated soil samples can be allowed to equilibrate at ⅓ atmosphere (11 psi) in order to approximate field capacity.

Availability of Water for Plant Growth. Soil water at field capacity is held against gravity by capillary forces at a tension of about ⅓ atmosphere (Foth and Turk, 1972). This low tension is easily overcome by plant roots in the process of absorbing water from the soil. As soil moisture is progressively depleted by drying and root absorption, the capillary tension on the remaining moisture increases, thus making it more difficult for plant roots to remove the moisture. Beyond a tension of about 15 atmospheres plants can no longer remove water fast enough to maintain their turgidity (Foth and Turk, 1972). This level of moisture tension and the moisture content at which it occurs is often called *permanent wilting point*. The relationship between moisture tension and moisture content varies with soil texture and structure.

Moisture tension levels less than 0.8 atmosphere (Richards, 1965) can be monitored directly in the field with relatively inexpensive tensiometers as discussed earlier in this section. Therefore tensiometers are useful for controlling irrigation systems to maintain the soil moisture in the optimum range for plant growth. Moisture tensions in excess of 1 atmosphere are best studied in relation to the soil moisture content by using laboratory pressure chamber or tension table apparatus which will not be considered here. Wilde, Voigt, and Iyer (1972) also give a procedure for approximating the moisture content at the permanent wilting point with a simple suction flask.

The moisture available to plants between field capacity and the permanent wilting point is called *available moisture capacity* and is often expressed as inches of water/inch of soil.

Load Bearing Capacity. The load bearing capacity of soils is a major concern in most engineering projects. It has already been stated that the moisture content affects the load bearing capacity directly and also influences the compaction operations designed to improve the load bearing capacity. This is taken into account in laboratory tests either by plotting the measures of strength against the moisture content or by reporting the strength for the least favorable moisture content which is likely to be encountered in a field situation. As a general rule coarse textured soils have a greater load bearing capacity than fine textured soils, although the mineral composition of the soil grains may cause exceptions. Critical tests of the load bearing capacity usually require laboratory facilities including power supplies and relatively heavy equipment for applying known loads to samples under controlled conditions. General descriptions of the more common tests are given in the following paragraphs.

California Bearing Ratio Test. The California bearing ratio test, usually abbreviated CBR, is probably the simplest strength test for laymen to understand and interpret. CBR is the load required to drive a standard piston into the test material expressed as a percentage of the load required for well-graded crushed stone. The standard piston has an area of 3 sq in., and the standard loads for crushed stone are taken to be 1000 psi for 0.1 in. penetration or 1500 psi for 0.2 in. penetration. The test is usually made at the optimum moisture content for compaction as determined by the Proctor test. See Bowles (1970) and Goodwin (1965) for details of the test and its interpretation.

Direct Shear Tests. Direct shear tests are performed by applying opposing forces to the parts of a specimen in a split box, thus causing the parts to slide along the plane of separation in the box. Depending on the specifics of the test, the specimen may also be loaded perpendicular to the plane of failure. Procedures differ somewhat for cohesive and noncohesive soils. Since the plane of forced failure may not represent the plane of greatest weakness in the sample, inferences to field situations are somewhat uncertain. Bowles (1970), Lambe (1951), and Sallberg (1965) give details of the tests. Results may be presented in the form of stress-strain curves, or in a tabular format including peak load during failure and other indices of strength.

Triaxial Compression Tests. For triaxial tests a cylindrical test specimen is encased in a flexible membrane and placed in a pressure chamber. Major loading takes place from the ends of the cylindrical specimen, but the chamber can also be filled with liquid under pressure to exert a lateral force. The triaxial apparatus does not restrict failure to a single plane like direct shear machines. The membrane seal around the specimen coupled with the facility for applying lateral pressure gives much better control of test conditions than does direct shear. By opening or closing valves, drainage can be allowed or prevented, either before or during heavy loading. Likewise balancing the end and lateral pressures coupled with a variable rate of loading gives control of preconsolidation in the specimen. Again there are procedural differences for cohesive and noncohesive soils, especially with respect to the preparation of specimens. See Bowles (1970), Lambe (1951), MacIver and Hale (1970), or Sallberg (1965) for methods of conducting the tests and analyzing the results.

Unconfined Compression Test. This might be considered a special case of triaxial testing in which there is no lateral membrane or confining pressure. Therefore its use is limited to cohesive soils which will hold a cylindrical shape when not under stress.

Consolidation Tests. Consolidation tests are used to determine volume changes that may be expected when a soil is placed under load. These tests are run on saturated soil with a provision for water to escape as consolidation takes place. The soil is placed in a ring which confines it laterally, and the load is applied vertically. Excess water escapes through porous base and top plates. The load is applied in incremental steps, with the load being doubled each time. Adequate time is allowed for the volume to stabilize before the load is increased. Two types of lateral retaining rings are used for consolidation tests. With a *fixed ring* the ring rests directly on the base so that all movement takes place from above. With a *floating ring* both the base and the top fit inside the ring so that movement may occur from either above or below. Frictional effects are greater in a fixed ring, but this type has the advantage of allowing permeability to be determined in the course of the consolidation test. See Bowles (1970), Holtz (1965), Lambe (1951), or MacIver and Hale (1970) for details of procedure and analysis.

Penetrometers. Penetrometers are field instruments that can be used to approximate some of the results that would be obtained from laboratory tests mentioned above. A penetrometer consists of a probe and some means of gauging the force required to insert the probe manually into the soil. There are several types available (Davidson, 1965) such as the pocket, Proctor, and cone penetrometers. The various models differ mainly in the size of the probe and the accuracy of the gauge used to measure the force required for insertion. The Proctor type has several interchangeable probes. In order for penetrometer readings to be of much value they must be correlated either with past experience or with the results of laboratory tests. Also, penetrometers are of little use in situations where rocks are encountered during insertion of the probe.

Soil Reactivity. Soil water dissolves chemical substances from the surfaces of particles comprising the solids until equilibrium is reached. Water movement through the soil leaches away these dissolved substances, upsetting the equilibrium and causing more material to be dissolved. Depending on the composition and diversity among the solid particles from which substances are dissolved, various reactions may take place in the soil solution. Molecules and ions of water and other substances may also form weak bonds at the surface of soil particles in response to electrical charges on these surfaces. The extent of bonding and the rate at which the substances enter solution depend on the amount of surface area exposed on the particles. Since smaller particles have larger surface areas for a given weight of solids, there is greater chemical activity in fine textured soils than in coarse textured soils. Microorganisms in the soil are also responsible for some of the chemical activity.

Plants obtain their chemical nutrients from substances dissolved in the soil water or adhering to surfaces of particles, so chemical properties constitute a major determinant of soil fertility. Chemical properties also control the amount of corrosion that takes place in buried metal or concrete components of structures and conduits. The general aspects of chemical reactivity in soils will be covered here. Suggested approaches for an analysis of specific chemicals will be covered in Section 4.2.6.

Reaction. The strength of acidic or basic tendencies in the soil moisture solution provides an index to many chemical properties of the soil. This tendency is called the *soil reaction* and is usually expressed on a pH scale. It was originally intended that the pH measurement obtained from a soil should be the negative logarithm of the hydrogen (hydronium) ion concentration (moles/liter) in the soil solution. However, interferences within the soil solution cause the apparent concentration as reflected by pH measurements to differ from the true concentration. Furthermore pH measurements only reflect the materials dissolved in the soil solution. Typically there are additional acids or bases adsorbed on the soil solids. The adsorbed materials will also dissolve and become active if those already in solution are neutralized. Therefore pH measurements made in soils are best viewed simply as a relative index of active acidity or basicity.

There is an inverse relationship between acidity and basicity. Values of pH near 7 indicate neutrality; values decreasing from 7 indicate greater acidity, and values increasing from 7 indicate greater basicity. The pH of most soils ranges between 4 and 10 (Foth and Turk, 1972).

The solubility of many plant nutrients is greatly reduced under strongly acid or strongly basic conditions. Although total acidity and conductivity are more critical, a strongly acid reaction also indicates corrosivity for buried concrete and unprotected metal components of structures or conduits. Foth and Turk (1972) draw an analogy between determination of soil pH and taking the temperature of an animal. Both can be very useful for diagnosing problems, but neither gives very specific information.

The soil reaction can be measured either with color indicators or with battery operated pH meters. Color indicators undergo color changes according to the pH of the sample being tested. Color indicator pH test kits are available commercially from several suppliers. Details of the procedure for mixing indicator solutions with the soil sample and making color comparisons vary somewhat between kits. It is usually more expedient to use a commercial kit and follow the enclosed instructions carefully, rather than try to assemble one's own kit. The type of kit used should be specified when reporting results.

The most common type of pH meter is based on calomel and glass electrodes which are immersed in a soil extract, suspension, or paste. A potentiometer gives a readout on a dial graduated in pH units. The meter must be checked against buffer solutions of known pH before use and at intervals during a series of determinations. The electrodes must be rinsed after each determination and kept moist by being lowered into a reservoir of distilled water when not in used. Differences may arise from the method used to prepare the soil sample, so the method of preparation should be stated when reporting results. Battery operated meters are sufficiently portable for use in the field, but they are more bulky and expensive than indicator kits. If the directions for the use of color indicator kits are followed carefully, the results should agree with electrometric determinations within 0.5 pH unit or less. See Peech (1965) along with the manufacturers' instructions for additional information on the care and use of pH meters.

Cation Exchange Capacity. Silicate clays found in temperate regions carry a negative electrical charge on the surface of the particles. This causes positively charged ions (cations) of other substances to adhere to these surfaces. These cations are held rather loosely, so an exchange of position between adsorbed cations and substances in solution takes place quite readily. The total quantity of such exchangeable cations that a soil will absorb is called the *cation exchange capacity* and is usually expressed in milliequivalents/100 grams of soil. A milliequivalent is 0.001 gram equivalent weight of an ion. The cation exchange capacity (abbreviated CEC) is an important property because many of the cations involved are plant nutrients. Thus a soil with a large cation exchange capacity is capable of storing large quantities of plant nutrients.

There are several methods for determining the CEC of a soil, but most of them share a common principle. The soil is leached with liquid containing a cation that is capable of displacing the other adsorbed cations until the total CEC is occupied by a single type of cation. The amount of this cation held by the soil is then determined by chemical analysis. Ammonium acetate, sodium acetate, or barium chloride with triethanolamine are often used for the leaching process. Each of these has certain advantages for particular types of soils. An alternative approach often used for soils having an acidic reation is to make separate chemical determinations of the several groups of cations that are likely to be adsorbed and then calculate the CEC as the sum of these. See Chapman (1965) for procedures and a comparison of methods.

Since there is not always good agreement between methods for a given type of soil, the method used should be specified when reporting results. It should also be noted that the surface charge of oxide clays characteristic of the humid tropics is dependent on the pH (Foth and Turk, 1972), so the situation is not the same as with silicate clays of temperate regions.

Base Saturation and Total Acidity. The exchangeable cations on a silicate clay are of two broad types. The group that includes the major plant nutrients consists of cations from the alkali metal and alkaline earth families in the periodic table of elements such as Na^+, K^+, Mg^{2+}, and Ca^{2+}. Cations of this group are called *exchangeable bases.* The other group consists of H^+ and Al^{3+} cations, which are called *exchangeable acids.*

Although this is not the complete picture, the acid or base terminology for the two groups can be viewed in terms of their effect on the pH of the soil solution when they move from an adsorbed position on the surface of a soil particle into the solution surrounding the particles. Keep in mind the inverse relation between acidity and alkalinity, with less alkaline being synonymous with more acidic. The hydroxyl ion (OH^-) is primarily responsible for basic or alkaline reactions in the soil solution. The exchangeable base cations are typically replaced on the soil particle by H^+ which comes from the dissociation of a water molecule into H^+ and OH^- ions. The H^+ adheres to the soil particle, leaving the OH^- ion in solution. The OH^- ion makes the soil reaction more alkaline or basic. When an H^+ ion moves from an adsorbed position into solution, it neutralizes an OH^- ion to form a water molecule. Thus the solution becomes less alkaline, or more acidic. When an adsorbed Al^{3+} ion goes into solution, it pulls apart three water molecules to form insoluble aluminum hydroxide, and the residual parts of the three water molecules (H^+ ions) make the solution more acidic. In summary, replacement of the exchangeable bases makes the soil solution more alkaline, whereas replacement of the exchangeable acids makes the solution less alkaline (more acidic).

The sum of exchangeable bases and exchangeable acids equals the CEC of the soil. Since plant nutrients come from the base fraction and the acid fraction must be neutralized with lime if one wishes to raise the pH, it is of interest to know how the CEC is partioned between bases and acids. The percentage of the CEC accounted for by bases is called *percent base saturation.*

There are several approaches to determining the amount of exchangeable bases (in milliequivalents/liter), exchangeable (or total) acids (in milliequivalents/liter), and base saturation (%). One approach is to analyze the several bases separately, total them to get exchangeable bases, and subtract from the CEC to get exchangeable acids. Another approach is to selectively displace either the bases as a group or the acids as a group. See Chapman (1965), Peech (1965), and Wilde, Voigt, and Iyer (1972) for analytical methods and relative merits.

Conductivity. The electrical conductivity of the soil reflects its salinity. This is important for both agricultural and engineering uses of the soil. High concentrations of salts inhibit the germination and growth of most plants. Likewise, high salt concentrations indicate corrosivity for buried concrete and unprotected metal components of structures and conduits. There is a reciprocal relationship between electrical conductivity and electrical resistivity. The standard unit for electrical resistance is the ohm, and the unit for conductance is the reciprocal ohm or mho. Since the conductivity of soil pastes is relatively low, measurements of conductance are expressed in millimhos, with 1 millimho being 0.001 mho.

Conductivity determinations are made with a unit consisting of a conductivity cell and an alternating current type of Wheatstone bridge. The sample is prepared as a soil paste, but different ratios of soil to water are used. Therefore the method of preparing the sample should be stated when reporting results. See Bower and Wilcox (1965) or Wilde, Voigt, and Iyer (1972) for details of equipment and procedure.

Organic Matter. Mineral soils contain varying amounts of organic matter, usually not exceeding a few percent by weight. However, some soils such as peat and muck are composed primarily of organic matter. Since decomposed organic matter is highly colloidal, the organic matter fraction has a very significant effect on the soil properties. It increases the water and plant nutrient holding capacity of the soil, thus increasing soil fertility for plant growth. It also helps form water stable aggregates of clay particles, thus improving the soil structure. When present as a small percentage by weight in mineral soils, organic matter does not adversely effect engineering properties to a noticeable degree. However, soils composed primarily of organic matter have very poor engineering properties.

The amount of organic matter in a soil is determined by either wet or dry combustion. The dry combustion procedure employs direct heat, whereas the wet combustion procedure is based on oxidation by chromic acid. Details of both approaches vary considerably depending on the accuracy required. Measurements may involve loss in weight on ignition, collection of carbon dioxide evolved, titration, or indicator solutions. Procedures also vary according to whether or not the soil contains carbonates. The presence of carbonates can be detected by effervescence upon the addition of hydrochloric acid. See Allison, Bollen, and Moodie (1965), Allison (1965), Broadbent (1965), and Wilde, Voigt, and Iyer (1972) for details of methods and relative merits.

4.2.2 Classification of Soils

It would be impractical to consider taxonomic systems for all major ecosystem components in this book. The classification of soils, however, does require some discussion because there are several systems in use. The system adopted by the USDA Soil Conservation Service (SCS) is most widely used in soil surveying and mapping for agricultural purposes. The SCS system is based primarily on factors of soil formation, and is therefore called a genetic approach. In contrast, systems popular with engineers are entirely utilitarian and designed so that the classes translate easily into implications for engineering practice. Although there are several systems oriented to the needs of engineers, only the two most commonly used systems will be discussed here. One is known as the AASHTO (American Association of State Highway and Transportation Officials) system, and the other is the Unified system.

USDA Soil Conservation Service System of Soil Classification. As already mentioned, the SCS system is based mainly on factors of soil formation. Since a given set of soil forming factors will produce similar soils irrespective of location, soils in the same category of the SCS system all have similar properties. However, the category names in the SCS system give little information regarding properties to a person who is not a soil scientist. Therefore soil survey reports must include a comprehensive discussion of the properties to be expected in the various soil categories shown on accompanying maps. When adequately interpreted through the report, SCS soil surveys provide useful information for most applications in ecosystem analysis and engineering practice. The higher categories of the SCS system have been revised in recent years to reflect increasing knowledge of soils. Fortunately the few revisions in the lower categories actually shown on maps have little effect on users. The earlier system that controlled many of the existing surveys is described by Soil Survey Staff (1951). The most recent version is described by Soil Survey Staff (1975). This most recent version is outlined briefly as follows.

Soil Orders. Ten soil orders form the highest level of the system. *Alfisols* are characterized by a clay accumulation in the B horizon with high base saturation and good moisture availability. *Aridosols* are desert soils with little moisture or organic matter. *Entisols* are soils in which there is little evidence of soil forming processes. *Histosols* are highly organic soils. *Inceptisols* are young soils with some but not extensive alteration by soil forming processes. *Mollisols* are soils with a dark upper layer that typically form under grasses in areas where moisture is seasonally scarce. *Oxisols* are highly weathered soils of humid tropical areas. *Spodosols* are characterized by an accumulation of organic matter and aluminum in the B horizon giving this layer a black or reddish color and pH dependent cation exchange capacity. *Ultisols* are soils of warm, moist areas that have clay accumulation in the B horizon like Alfisols, but base saturation and natural fertility are low. *Vertisols* are clay soils that develop deep cracks on drying. Water washes surface soil into the cracks and wetting of lower layers through the cracks causes soil heaving. The net result is vertical mixing within the soil.

Lower Categories of the SCS System. The SCS system is hierarchical or nested, with each lower category being subdivided sequentially. The characteristics of an order can be produced by different combinations of soil forming processes, and these variations constitute the basis for *suborders* within orders. *Great groups* form the third level of subdivision in the system. In great groups the soil is considered as an assemblage of horizons. Great groups are further broken down into *subgroups* according to variations in horizons that indicate transitional forms between great groups (intergrades) or modifications of typical forms (extragrades). *Families* constituting the fifth level in the system are utilitarian categories used to designate soils within a subgroup which have similar properties affecting their responses to management. The lowest level in the hierarchy is the soil *series*. A series is a set of soils within a family which are similar with respect to some important property or set of properties affecting management. In the older version of the SCS system, *types* within series were recognized on the basis of texture in the A horizon. Types are no longer recognized in the revised system.

Soil Phases, Complexes, and Associations. The term *phase* is not category specific, that is, phases can be defined within any category to reflect certain variations that are significant to soil management. The rocky phase and the eroded phase are two examples of such variations. The terms *soil complex* and *soil association* are used mainly for convenience in mapping. A soil complex is an intricate mixture of different kinds of soils within a small area which would be impractical to show separately at scales commonly used for preparing detailed soil maps meant for a variety of uses. The term soil association is used in generalized soil maps. An association is a typical pattern of occurrence of several soils in a landscape. The difference between a complex and an association is primarily a matter of pattern and area. Whereas a complex is an intricate mixture with each soil covering a small area, the pattern in an association is not necessarily intricate, and individual soils often cover considerable areas of the landscape as contiguous units.

AASHTO System of Soil Classification. This system of soil classification is widely used in highway engineering. It originated with the Bureau of Public Roads and has been revised several times. It has been adopted as a standard by the American Association of State Highway and Transportation Officials (AASHTO), whence comes the name. Until recently the word *Transportation* was not part of the society name, so both the society and the classification system were designated as AASHO. The system is designed to classify soils according to their suitability for subgrades. The system has eight major categories, designated A-1 through A-8, with suitability for subgrade decreasing as the category number increases. A sieve analysis along with liquid and plastic limit determinations for the fraction passing the #40 sieve form the basis for classification. Subcategories are defined to reflect the situation where the coarse fraction has favorable characteristics for subgrade, but properties of the fine fraction detract from overall suitability.

The system also incorporates a *group index* which is used for comparing soils in the same category. The group index is determined from the following formula (Portland Cement Association, 1962):

$$\text{group index} = 0.2A + 0.005AC + 0.01BD$$

A and B are obtained from a sieve analysis, whereas C and D come from Atterberg limits. A is found by deducting 35 from the percentage passing the #200 sieve. If the result is less than 1, A is assigned the value 1. Likewise,

A is assigned the value 40 if the subtraction gives a result greater than 40. *B* is found by deducting 15 from the percentage passing the #200 sieve and is constrained to be between 1 and 40 in the same manner as *A*. *C* is found by deducting 40 from the liquid limit and is constrained to fall between 1 and 20. *D* is found by deducting 10 from the plasticity index and is constrained to fall between 1 and 20. *A, B, C,* and *D* are all rounded to whole numbers, and the group index itself is also rounded to a whole number. The higher the group index, the less suitable is the soil for subgrade. The group index is placed in parentheses after the category designation. For example, A-2-7(3) denotes a soil with a group index of 3 which is placed in major category A-2 on the basis of its coarse fraction, but with fine materials similar to soils in the A-7 category.

The AASHTO system is summarized in the form of a dichotomous key as follows:

1a.	Organic soils (peat or muck)	*AASHTO A-8*
1b.	Mineral soils	2
2a.	35% or less passes #200 sieve	7
2b.	More than 35% passes #200 sieve	3
3a.	Liquid limit 40 or less	4
3b.	Liquid limit greater than 40	5
4a.	Plasticity index 10 or less	*AASHTO A-4*
4b.	Plasticity index greater than 10	*AASHTO A-6*
5a.	Liquid limit 10 or less	*AASHTO A-5*
5b.	Liquid limit greater than 10	6
6a.	Plasticity index less than (liquid limit − 30)	*AASHTO A-7-5*
6b.	Plasticity index greater than (liquid limit − 30)	*AASHTO A-7-6*
7a.	Nonplastic soil	8
7b.	Soil somewhat plastic	10
8a.	50% or less passes #40 sieve	10
8b.	More than 50% passes #40 sieve	9
9a.	10% or less passes #200 sieve	*AASHTO A-3*
9b.	More than 10% passes #200 sieve	10
10a.	50% or less passes #10 sieve	11
10b.	More than 50% passes #10 sieve	14
11a.	30% or less passes #40 sieve	12
11b.	More than 30% passes #40 sieve	14
12a.	15% or less passes #200 sieve	13
12b.	More than 15% passes #200 sieve	14
13a.	Plasticity index 6 or less	*AASHTO A-1-a*
13b.	Plasticity index greater than 6	17
14a.	50% or less passes #40 sieve	15

14b. More than 50% passes #40 sieve	17
15a. 25% or less passes #200 sieve	16
15b. More than 25% passes #200 sieve	17
16a. Plasticity index 6 or less	*AASHTO A-1-b*
16b. Plasticity index greater than 6	17
17a. Liquid limit 40 or less	18
17b. Liquid limit greater than 40	19
18a. Plasticity index 10 or less	*AASHTO A-2-4*
18b. Plasticity index greater than 10	*AASHTO A-2-6*
19a. Plasticity index 10 or less	*AASHTO A-2-5*
19b. Plasticity index greater than 10	*AASHTO A-2-7*

The reader should consult Portland Cement Association (1962) for an introduction to management properties of soils as categorized in the AASHTO system.

Unified System of Soil Classification. The Unified soil classification system originated with Arthur Casagrande during World War II and has undergone several revisions since. Like the AASHTO system, it is based primarily on texture and Atterberg limits. Emphasis in the system is on its suitability for roads, airfields, embankments, and foundations. It has been adopted by the U.S. Corps of Engineers and the U.S. Bureau of Reclamation and is also widely used in commercial engineering practice. The designation for most categories consists of two letters. The letters used are G for gravel, S for sand, M for silt, C for clay, O for organic fractions in mineral soils, W for well-graded, P for poorly graded, L for low liquid limit, and H for high liquid limit. Peat and muck, which consists primarily of organic matter, are coded Pt. Soils that have properties reflecting a mixture of two classes are given a double code with a hyphen separating the parts (such as, CL-ML). The Unified system is summarized in the form of a dichotomous key as follows:

1a. Soils primarily composed of organic matter	*Pt*
1b. Soils primarily composed of mineral matter	2
2a. Less than 50% of material passes #200 sieve	3
2b. 50% or more passes #200 sieve	16
3a. Of the material not passing #200 sieve, less than 50% passes #4 sieve	4
3b. Of the material not passing #200 sieve, 50% or more passes #4 sieve	10
4a. Less than 5% of material passes #200 sieve	5
4b. 5% or more passes #200 sieve	6
5a. Coefficient of uniformity greater than 4 *and* coefficient of concavity between 1 and 3	*GW*
5b. Not as in 5a	*GP*

6a. More than 12% passes #200 sieve	7
6b. 5 to 12% passes #200 sieve—*borderline case, use double code*	
7a. Plasticity index less than 0.73 (liquid limit − 20)	*GM*
7b. Plasticity index greater than 0.73 (liquid limit −20)	8
8a. Plasticity index less than 4	*GM*
8b. Plasticity index 4 or greater	9
9a. Plasticity index 7 or less—*Borderline case, use double code*	
9b. Plasticity index greater than 7	*GC*
10a. Less than 5% of material passes #200 sieve	11
10b. 5% or more passes #200 sieve	12
11a. Coefficient of uniformity greater than 4 *and* coefficient of concavity between 1 and 3	*SW*
11b. Not as in 11a	*SP*
12a. More than 12% passes #200 sieve	13
12b. 5 to 12% passes #200 sieve—*Borderline case, use double code*	
13a. Plasticity index less than 0.73 (liquid limit − 20)	*SM*
13b. Plasticity index greater than 0.73 (liquid limit − 20)	14
14a. Plasticity index less than 4	*SM*
14b. Plasticity index 4 or greater	15
15a. Plasticity index 7 or less—*Borderline case, use double code.*	
15b. Plasticity index greater than 7	*SC*
16a. Liquid limit less than 50	17
16b. Liquid limit 50 or greater	19
17a. Little organic matter	18
17b. Abundant organic matter	*OL*
18a. Fine materials predominantly silt	*ML*
18b. Fine materials predominantly clay	*CL*
19a. Little organic matter	20
19b. Abundant organic matter	*OH*
20a. Fine materials predominantly silt	*MH*
20b. Fine materials predominantly clay	*CH*

The Unified codes are more suggestive of potential uses than the AASHTO codes. The suitability for engineering uses (other than sealing impoundments) usually decreases from coarse to fine textures. Likewise well-graded coarse materials are more favorable than poorly graded coarse materials; and fine materials with low Atterberg limits are more favorable than fine materials with high Atterberg limits. The interested reader is referred to U.S. Army Corps of Engineers (1960) for further information on the Unified system. Additional information on classification systems oriented toward engineering is given by Portland Cement Association (1962).

4.2.3 Soil Survey Procedures

In discussing procedural aspects of soil surveys it is helpful to think in terms of three levels of soil surveys with respect to the intensity or detail of information collected. In the least intensive type of survey, called a *reconnaissance* survey, general patterns of soil occurrence are mapped. At the intermediate level of intensity individual units of soil, as opposed to soil patterns, are mapped. Surveys of this type are usually called *detailed* soil surveys. The most intensive type of survey takes place within localized areas in preparation for construction projects. Surveys of the latter type are often called *site investigations* and involve mapping of all variations in soils that will affect the construction effort. Procedures for implementing surveys differ appreciably for the three levels of intensity.

Reconnaissance Surveys. Soil reconnaissance surveys are typically large area operations, although they may also be carried out as a preliminary to site investigations. The first phase of a reconnaissance survey involves breaking the area down into groups of landscape units such that the landscapes within a group are similar in most respects. This is done on the presumption that similar sets of soils will have similar natural vegetation, support similar kinds of land uses, and also be topographically similar. Airphotos are usually employed for grouping landscape units, and remote sensing from satellites may even be used to good advantage.

In the second phase of a reconnaissance survey a sample of landscape units is selected from each group, and relatively detailed field work with supporting laboratory analysis of soil samples is carried out to determine the pattern of soil occurrence within the landscape units. If all the sample landscape units within a group have essentially the same set of soils occurring in similar spatial patterns, then the landscape units comprising the group can tentatively be considered to represent sections of a natural association of soils which is the mapping unit of a reconnaissance survey. If not, the criteria used in setting up the groups must be reexamined. The final step is to do a more cursory check of other landscape units in each group to see whether they fit the pattern worked out from landscape units studied more intensively.

The delineation of soil boundaries in a detailed examination of the sample landscape units is conveniently done on copies of the most recent black and white airphotos available. This makes for easy placement of lines relative to the terrain with a minimum of compass work and distance measurement. The gradational nature of the changes in soils usually does not justify a highly accurate map placement of soil boundaries anyhow.

Transects usually provide the most efficient layout for field travel when the spatial distribution of an environmental feature is under investigation. Transects are simply a series of parallel and equally spaced lines with data collection taking place either continuously or at regular intervals along the lines. As applied in soil surveys, observations on soil are made at intervals along the transect lines, with the interval depending on the complexity of the soil pattern. The observations consist of using an auger or core sampler to examine the soil to a depth of 5 ft or more.

There are two main types of augers. Both types have a cutting and collecting element at the end of a rod which is twisted into the ground by means of a T-shaped handle. In one type the cutting and collecting element resembles a screw. In the other type a pair of cutting blades feeds the loosened soil into a small bucket. There are also variations on the type of bucket. The bucket type has the advantage of giving a more localized sample. Many augers are constructed so that sections of rod may be added below the handle for boring to greater depths. A disadvantage common to all augers is that the soil is disturbed in the process of removal.

Core samplers have the advantage of extracting a more or less intact column of soil. A core sampler is basically a tube with a cutting edge and a handle which is forced into the ground and then removed full of soil. The tube may be modified in several ways to facilitate removing the core. Typical modifications are to have the tube partially open on one side or to use a split tube arrangement. However, stones in the soil make core samplers difficult to use.

Descriptions of the soils are developed in the field from the borings, and samples are taken for laboratory analysis as needed. At less frequent intervals soil pits are dug for observing complete profiles and taking undisturbed samples of lower horizons. Likewise full advantage is taken of roadcuts and other excavations for observing soils at depth.

Results of reconnaissance soil surveys are conveniently mapped at scales of 1 in. to the mile or smaller, with larger scale supplementary maps covering the areas that were examined in more detail. Enlargements can be made to match scales of base maps, but one should carefully avoid giving a false impression of accuracy in the enlargements.

Detailed Soil Surveys of Large Areas. The standard type of soil survey performed by the USDA SCS in cooperation with state agricultural experiment stations covers a county, with individual soil units (series and phases of series) serving as mapping units. However, soil complexes may be used as mapping units where soils occur in a very intricate pattern within a small area.

The procedure for conducting surveys of this kind is essentially the same as that outlined previously for studying patterns of soil occurrence in sample areas of reconnaissance surveys, except that the detailed field study is extended over the entire area. Soil boundaries are delineated on black and white aerial photos. In this case, however, the definition and delineation of soil associations is a secondary concern accomplished by aggregating information from the completed field sheets. Procedures for conducting soil surveys of this type are covered in detail by Soil Survey Staff (1951).

Typical scales for presenting results of such detailed soil surveys range from 3 in. to the mile to 5 in. to the mile. The soil information is usually presented most effectively by retaining the airphoto base for final maps, but showing photo detail in subdued tones. This allows the user to relate soil information to other terrain features easily.

Site Investigations. It is more difficult to generalize with respect to procedures for conducting site investigations because methods will vary according to the purpose. For example, the emphasis will be quite different for constructing a sewage lagoon than for a building foundation. Nevertheless copies of existing general purpose soil survey reports can be a great help in planning site investigations.

Soil borings in site investigations will usually be much more closely spaced than for general purpose surveys. Small inclusions of variant types of soils are permitted in mapping units for general purpose surveys, but such inclusions can have significant implications for construction work. Likewise drilling operations as opposed to shallow soil borings are often needed in site investigations because the depth to the bedrock is a frequent concern as well as the conditions that will be encountered in cut and fill operations.

Soil samples taken for laboratory tests need to be large in order to accommodate the variety of tests required. Compositing of soil samples from different locations is not recommended because the variation in physical and chemical properties is important. Field tests of percolation rates and similar properties are often required.

Hough (1969) devotes a chapter of his book on *Basic Soils Engineering* to site investigations, and guidelines for specific kinds of site investigations are prepared by various engineering societies and similar organizations. Although somewhat outdated, Hvorslev (1949) is a useful reference in this regard.

4.2.4 Soil Maps and Their Interpretation

Maps prepared from detailed soil surveys for general purpose use are usually presented at scales of 1/20,000 or 1/15,840 on a subdued airphoto base. However, both larger and smaller scales are also used. The airphoto base makes it easy to locate areas of interest and to relate soil information to other features of the environment.

As mentioned earlier, however, neither the names of the soil units nor the map symbols used to represent them carry much meaning for a person who is not a soil scientist. Therefore the usefulness of soil maps for a broad range of purposes depends on the amount of interpretive information furnished in the survey report that accompanies the maps. If the interpretive information is scanty, the user is faced with the time consuming and expensive task of developing his or her own correlations between map units and suitability of soils for a given purpose. The USDA SCS is devoting major efforts to the translation of soil names and map symbols into behavioral properties of soils that are of direct concern to prospective users of the maps. The following discussion centers on the results of these interpretive efforts.

Interpretive Tables. One way of placing soils information in a user context is through interpretive tables. Modern soil survey reports contain a series of tables in which each category of soil shown on the map is rated in terms of properties and limitations affecting major categories of use. Included are a set of engineering tables, a set of farm management tables, a table of limitations for recreational uses, and a table of suitability for elements of wildlife habitat and kinds of wildlife.

One of the engineering tables contains estimated ranges of values for the depth to the seasonal high water table, percentages passing standard sieves, permeability, available water capacity, reaction, and shrink-swell potential; along with the probable texture according to the depth below surface and likely classifications in the AASHTO and Unified systems.

A second table of engineering interpretations has columns giving the suitability as a source of construction materials; features affecting highway location, foundations for low buildings, and winter grading; the degree and type of limitations for septic tank disposal fields; and the corrosion potential for buried conduit. The ratings are based on a combination of laboratory testing, past experience with the soil, and a comparison with similar soils. Terms such as not suitable, poor, fair, good, slight, moderate, severe, low, and high are used to give general guidance without implying that the ratings are a substitute for on-site investigations.

One of the farm management tables gives predicted yields of crops under different levels of management. A companion table gives interpretations of features affecting farm uses such as drainage, irrigation, terraces and diversions, grassed waterways, and farm ponds.

The table of limitations for recreational uses considers features such as service buildings play areas, camp areas, picnic areas, and paths or trails.

The wildlife tables include columns for habitat elements such as food plants, cover, and water along with overall suitability ratings for animals that inhabit openland, woodland, and wetland.

Capability or Suitability Classifications. A slightly different approach to the interpretation problem is through capability or suitability classes. This approach is used for interpreting soil survey information in terms of agricultural cropping patterns and woodland management.

SCS Land Capability Classification System. SCS uses a nationwide system for classifying land according to its capability for agricultural production. The system contains three levels of grouping. The eight categories in the first level, called land capability classes, are designated by Roman numerals.

Class I, II, III, and IV lands have few, moderate, severe, and very severe limitations, respectively, restricting their use for general agricultural production. As the class number increases through these first four classes, so also does the need for a careful choice of crop species and conservation practices. Classes V, VI, and VII have limitations that restrict their uses to pasture, range, forestry, or wildlife management. These three classes differ in nature and severity of the limitations. Class V land, for example, does not have an erosion hazard. Class VIII lands have limitations so severe that they cannot even be used for commercial grazing or timber harvesting.

Subclasses constitute the second level of grouping. Subclasses are designated by a letter that indicates the major limitation. The letters are *e* for erosion, *w* for water, *s* for shallow soil, and *c* for climate. Class I land has no subclasses since there are no major limitations. Likewise there is no *e* subclass in class V since there is no erosion hazard in that class.

The third level of grouping is made up of *capability units* within subclasses. Lands in the same capability unit are essentially the same with respect to feasible uses and the intensity of management required. The capability unit is indicated by a number separated from the subclass letter by a hyphen. Thus a complete land capability code might be IIIs-3. Each soil category shown on the map is assigned a capability classification in the accompanying report.

Woodland Suitability Groups. SCS uses a system of woodland suitability groups which is somewhat similar to the land classification system. However, major forest types are quite different in different parts of the country, so woodland suitability groups are set up on a statewide basis rather than nationwide. The woodland suitability classification consists of three parts.

The first part is a number that serves as a rating of the productive capacity for the tree species to which the land is best suited. The potential productivity decreases as the number increases. Thus the number 1 is used for the most productive areas. The rating of the productive capacity is based on the *site index* which is the height that dominant trees in the stand can be expected to attain at a certain age called the *index age.*

The second part of the suitability classification is a small letter that indicates limitations in woodland management. The letter *o* indicates lack of major limitations. The letter *s* indicates an area that is excessively sandy and dry. The letter *w* is used for wet areas.

The third part of the woodland suitability classification is a number that serves to separate groups differing in respects not covered by the first two parts of the classification. For example, the number 3 might indicate soils with one or more severe management problems that are best suited for conifers;

whereas, the number 6 might signify soils with one or more severe management problems that are best suited to hardwoods.

Soil Management Groups. Another system of grouping soils for interpretive purposes is used in Michigan (Mokma, Whiteside, and Schneider, 1974; Tilmann and Mokma, 1976), but not on a nationwide basis. This system is based on the fact that soils similar in texture, natural drainage, and slope also respond in a similar way to management.

The first part of the designation for a soil management group consists of a number that reflects the dominant texture in the soil profile. The number increases with the coarseness of texture. For instance, 0 indicates fine clay, whereas 5 indicates sand. Both decimal and fractional numbers are permitted. Decimal numbers are used for a more detailed breakdown of texture, and fractions are used to represent profiles having layers of contrasting texture. Also the capital letters G, L, R, and M are used in place of texture numbers to denote gravel, alluvial materials, rock, and muck or peat, respectively.

The texture code is followed by a small letter reflecting the natural drainage of the soil. The letter *a* indicates well and moderately well drained soils, the letter *b* indicates somewhat poorly drained soils, and the letter *c* indicates poorly and very poorly drained soils. The combination of texture code and drainage code constitutes a *soil management group.*

Slope class codes may be used to refine soil management groups a step farther. The slope classes are 0-2%, 2-6%, 6-12%, 12-18% 18-25%, and 25+%. These slope classes are designated by capital letters A-F. When a slope class is added to a soil management group, the result is called a *soil management unit.* The primary virtue of this system lies in reducing the number of categories as compared to series and phases while preserving a great deal of useful information on the response of soils to management.

Color Codes for Severity of Limitation. Bauer (1966) and Quay (1966) describe the use of interpretive soil maps specially prepared from conventional soil maps. Each interpretive map gives an effective visual display of soil limitations for a single type of land use. Limitations for the proposed use are rated as none, slight, moderate, severe, and very severe; or suitability is coded in the converse way as good, fair, poor, and so on. These ratings are then transferred to a copy of the map or to an overlay by using shading patterns or colors.

Shadings or colors should be selected for visual gradation from fewest limitations to most limitations. Shadings can become increasingly dense with the degree of limitation. Likewise, colors can grade from white through increasingly "hot" colors like yellow and orange on up to red.

The articles by Bauer and Quay mentioned above are part of the proceedings of a symposium on applications of soil surveys in land use planning sponsored by the Soil Science Society of America and the American Society of Agronomy. Papers presented at that symposium describe applications of soil survey information for such diverse purposes as regional planning, urban planning, highway construction, recreation, zoning, and tax assessment. The experiences shared by the various authors should suggest a whole host of potential uses to the alert reader.

4.2.5 Erosion Potential

Land capability classifications, woodland suitability groups, and interpretative tables incorporated in soil survey reports prepared by the USDA SCS in cooperation with state agricultural experiment stations, all contain ratings of the susceptibility to soil erosion. These ratings are usually adequate for a general assessment of the erosion potential. However, a more quantitative approach is sometimes needed for making numerical estimates of potential erosion rates (tons of soil loss/acre/year) under specific soil management regimes. The universal soil loss equation (Wischmeier and Smith, 1965) has been developed for this purpose. Although details of procedures for applying the universal soil loss equation are beyond the scope of this book, an outline of the system can serve as a point of departure for further study.

The universal soil loss equation contains six multiplicative factors which are in turn composites of the basic variables that control erosion. A rainfall factor, denoted by R, incorporates the climatic variables of rainfall intensity, duration, and erosive force. The influence of soil properties such as texture is expressed through a factor denoted by K. Two factors in the universal soil loss equation relate to the influence of topography. One of these factors is the average slope length denoted by L, and the other is the average slope gradient or steepness denoted by S. A fifth factor reflects the protective influence of the vegetative cover and is denoted by C. The final factor reflecting the effectiveness of erosion control practices is denoted by P. Given estimates of these six composite factors, the estimated soil loss A (in tons acre/year) is calculated as

$$A = R \times K \times L \times S \times C \times P$$

Given estimates of five factors and a target rate of acceptable loss, the equation can be solved for the required value of the sixth factor. The required level of erosion control practices P can be determined in this fashion.

The interested reader should consult Beasley (1972), Guy (1970), Leopold (1968), Wischmeier (1974), Wischmeier and Smith (1965), Wischmeier, Johnson, and Cross (1971), and Wischmeier and Meyer (1973) for further information. Tilmann and Mokma (1976) illustrate the use of the universal soil loss equation for evaluating regional soil losses and loss from construction sites in Michigan.

4.2.6 Chemical Composition and Nutrient Status of Soils

Soil reaction, cation exchange capacity, base saturation, total acidity, and conductivity have already been introduced as indexes to the chemical properties of soils. These indexes are useful for classifying soils and rating their suitability for particular uses. However, plants require certain chemical nutrients in specific forms and quantities. Furthermore these requirements for chemical nutrients vary between plant species. Therefore modern agricultural practice necessitates more detailed information on soil chemistry than general indexes such as soil reaction and conductivity can provide.

An analysis of the chemical nutrient status in soils is complicated by the fact that nutrient elements may be present in considerable quantities but in forms that cannot be utilized by a particular type of plant. Thus the *availability* of nutrients for plant growth is more important than the total quantity present. Furthermore the capacity of a plant to utilize a particular nutrient also depends on the availability of other types of nutrients. Therefore the *balance* between nutrients must be considered along with the availability of individual nutrients. An additional consideration is that nutrients can become toxic to plants when present in excessive amounts of readily available forms.

Useful information about the nutrient status of soils can be obtained at several levels of analytical sophistication. Simple tests can often be conducted directly in the field to obtain an indication of gross nutrient imbalance. Due to the lack of precision and control in field tests, however, laboratory analysis is usually necessary for determining corrective treatments to be applied. If the amount of analysis to be performed is not large, suitable tests can usually be conducted with relatively simple laboratory equipment. As the analytical load increases, it soon becomes necessary to acquire more sophisticated equipment. Although the initial investment for such sophisticated equipment is large, advantages include greater accuracy, lower cost per sample, and greater versatility of analysis.

At this point we should also repeat our earlier warning that most of the background data available for interpreting soil tests apply to agricultural ecosystems. Little information of this nature is available for natural ecosystems, so soil tests normally must be used for comparative purposes when dealing with native plants.

Analytical Approaches for Field Use. Most results of soil tests are usually expressed on a dry weight basis, such as percent by oven dry weight or quantity per 100 grams of oven dry soil. Since accurate weighing is difficult to accomplish in the field, this poses an immediate limitation for most field tests. It would be feasible to carry a small beam balance sensitive on the order of 0.1 gram into the field, but even this would be rather bulky and awkward. Therefore field tests are usually started by filling a small spoon with soil that has been passed through a suitable sieve. Variations in void ratio and moisture content make a translation from this volume basis to a dry weight basis very approximate. Consequently results of field tests usually take the form of a more or less qualitative index of availability.

The second operation normally involves the use of an extracting solution to remove the nutrient under study from the soil sample to a degree that approximates the plant's ability to remove the nutrient. This is a weak link in most soil tests for nutrient availability. Plants vary in their ability to extract nutrients from soils, and no extracting solution mimics the removal by plants in a completely satisfactory manner. Since laboratory tests share this limitation which cannot be overcome by precise measurement, laboratory tests are little better than field tests as an actual measure of nutrient availability to a specific plant. The real value of the test lies in the correlative work which has been done in relating plant growth to test results. Very little such correlative information is available for many native plants comprising natural ecosystems.

Depending on the extent to which soil materials interfere with subsequent steps in the test, it may be necessary to separate soil and extract by filtration. Lack of a suction apparatus may make filtration a little slower in the field than in the laboratory, but this is not a major problem.

Colorimetric Tests. A very common type of field test for the chemical analysis of soils involves adding a reagent to the extract which produces a color reaction. The results of the test are judged on the intensity of the color produced. A color comparator is used to judge the intensity of the color. The color comparator may take several forms ranging from colored chips on a sheet to a tube holder for liquid standards. Color comparison with such a comparator is necessarily subjective, which constitutes a source of possible error. Subjectivity is avoided in the laboratory by using a colorimeter or spectrophotometer.

Titrimetric Tests. Another type of test that can be used in the field is *titration.* In this approach a solution containing a known concentration of one reagent (called the *titrant*) is added in small increments to the test solution until an end-point indicator shows that a reactant in the test solution has been consumed. The endpoint indicator is usually a color change. The amount of reactant in the test solution is then calculated from the amount of titrant used in reaching the endpoint. A graduated tube with a stopcock, called a *burette,* is used to dispense the titrant. The amount of titrant used is determined as the difference between the initial and final titrant levels in the burette. Titration is less frequently used in the field than colorimetry because it requires more skill and care on the part of the analyst, and the equipment is also more bulky.

Electrometric Tests. The use of a battery operated pH meter for determining the soil reaction has been mentioned earlier. This type of meter operates by detecting the difference in electrical potential between special electrodes immersed in the test solution. Electrodes are also available that show high specificity for other types of ions. These specific ion electrodes should be used with due caution to eliminate possible interferences which depend on the type of ion being detected.

Cleanliness and contamination are concerns in all types of chemical analyses. Although field tests are generally not expected to be as accurate as laboratory tests, cleanliness is still important; particularly so since there are many opportunities for contamination, and washing is more bothersome than in the laboratory.

Neatly packaged kits for making a variety of soil tests in the field are available commercially from several suppliers. In addition to compactness, these kits have the advantage of simple instructions for use by those without chemistry backgrounds along with convenient guides for interpreting the results of the tests. Wilde, Voigt, and Iyer (1972) give tests designed for use in the field by those who prefer to assemble their own apparatus and reagents.

Analytical Approaches for Simple Laboratories. If the chemical analysis of soils is to be done with any regularity, it is essential to have at least simple laboratory facilities available. Such facilities are assumed to include hot and cold running water, gas, compressed air, fume hood, cabinets for storing chemicals, and standard laboratory glassware. The glassware must include distillation apparatus. A controlled temperature oven is necessary for drying soil samples. Analytical balances are needed for dispensing dry chemicals and performing gravimetric analyses such as the determinations of the moisture content. A nest of standard soil sieves, preferably with mechanical shaker, is required for particle size separations. A mixer of the type used for making malts and shakes from ice cream is used in hydrometer analyses, and a mechanical stirring or shaking device for flasks containing a solution is also extremely helpful. Likewise a small grinding mill is useful if samples of plant tissue are to be analyzed. Determinations of pH are required so frequently that a pH meter will soon pay for itself in time saved.

Colorimetric determinations are so common in the chemical analysis of soils that a colorimeter becomes a practical necessity. A colorimeter measures the absorbance or transmission of a light beam passing through the solution. A set of standards (solutions of known concentration) are used to plot a graph of concentration against colorimeter reading. The concentration of an unknown is then determined by taking a colorimeter reading and using the graph to find the concentration. The wavelengths comprising the light beam in the colorimeter can be controlled by the use of filters.

Nutrient elements required by plants in relatively large amounts are nitrogen (N), phosphorus (P), potassium (K), calcium (Ca), magnesium (Mg), and sulfur (S). Since they are required in relatively large quantities, these elements are called *macronutrients*. Nutrients needed by plants in only trace amounts are manganese (Mn), copper (Cu), zinc (Zn), boron (B), iron (Fe), chlorine (Cl), and molybdenum (Mo). Since they are required in only trace amounts, the latter elements are called *trace elements* or *micronutrients.* Since macronutrients are required in larger quantities, they are most frequently limiting for plant growth. Consequently they are also the most frequent targets for chemical analysis. Total carbon (C) is also of interest because the ratio of carbon to nitrogen in soil affects the availability of the nitrogen. Conventional methods of analysis for macronutrients and carbon which can be performed with simple laboratory apparatus are indicated below. The point also bears repeating that the choice of the extracting solution can have a decided effect on the test results. Therefore the test procedure should be reported along with the results.

Total Nitrogen. Nitrogen follows a rather complex cycle of interconversions between nitrogen gas, ammonia gas, ammonium ion (NH_4^+), nitrate ion (NO_3^-), nitrite ion (NO_2^-), and nitrogen incorporated into organic molecules. This nitrogen cycle is of considerable importance in the dynamics of ecosystems. The conventional method of analysis for nitrogen in the organic and ammonium forms is the Kjeldahl method. This begins with digestion using a sulfuric acid solution which converts nitrogen to the ammonium form. The nitrogen is then distilled off as ammonia gas and collected in boric acid (Winkler modification). The ammonium borate is titrated with sulfuric acid. Although the Kjeldahl method is usually labeled as total nitrogen, it does not include the nitrate and nitrite forms. Furthermore gaseous nitrogen normally present in the soil atmosphere is not part of the analysis. The latter is of no consequence since the gaseous form is not utilized by plants anyhow. See Bremner (1965) or Wilde, Voigt, and Iyer (1972) for details of the Kjeldahl procedures.

Nitrate Nitrogen. Nitrogen in the form of nitrates can be determined by extracting with a copper sulfate solution followed by colorimetric determination with phenoldisulfonic acid. However, nitrates along with nitrites and ammonium are subject to rapid changes. Therefore it may be necessary to freeze the samples if they cannot be analyzed shortly after collection. See Bremner (1965) or Wilde, Voigt, and Iyer (1972) for details of this method.

Nitrite Nitrogen. Nitrogen in the form of nitrites can be determined by the Griess-Ilosvay colorimetric method, involving diazotization to form colored azo compounds. See Bremner (1965) for details.

Ammonium Nitrogen. Exchangeable ammonium can be determined by distillation from a potassium chloride solution with either colorimetric Nesslerization or titration by sulfuric acid. See Wilde, Voigt, and Iyer (1972) for details.

Phosphorus. The extraction of phosphorus may be accomplished either with dilute sulfuric acid or with a solution of ammonium fluoride and hydrochloric acid. The extraction is followed by a colorimetric determination of molybdenum blue developed from acid molybdate with reducing agents. See Olsen and Dean (1965) or Wilde, Voigt, and Iyer (1972) for details.

Potassium. Exchangeable and water soluble potassium may be determined by extraction with ammonium acetate followed by precipitation with cobaltinitrite and titration with potassium permanganate. See Pratt (1965) for details.

Magnesium. Exchangeable magnesium may be determined by extraction with ammonium acetate followed by titration with the chelating agent EDTA (ethylenediaminetetracetic acid). This same basic procedure is used for determining calcium, so steps must be taken to separate the two. See Heald (1965) for details.

Calcium. Use the same basic procedure as for magnesium above, with steps being taken to separate the two.

Sulfates. Since sulfates are readily available to plants, an extraction of sulfates with ammonium acetate provides an index of the availability for sulfur. Barium sulfate developed from treatment with barium chloride produces turbidity. A colorimeter is used to measure the turbidity. See Bardsley and Lancaster (1965) for details.

Chlorides. Although chlorine is one of the micronutrients, problems of toxicity can arise from chlorides used for removing ice from highways. Such chlorides can be determined by extracting with distilled water and titrating with silver nitrate in the presence of sodium bicarbonate and potassium chromate. See Wilde, Voigt, and Iyer (1972) for details.

Total Carbon. Both wet and dry combustion methods are commonly used for the analysis of total carbon in soils. Since an accurate determination by dry combustion requires a muffle furnace, only wet combustion methods would be usable in a laboratory lacking a furnace. Wet combustion methods involve oxidation with chromic acid and measurement of the carbon dioxide evolved. See Allison, Bollen, and Moodie (1965) for procedural details.

The second part of the comprehensive manual on *Methods of Soil Analysis* sponsored by the American Society of Agronomy (ASA) and edited by Black et al. should be consulted for the analysis of micronutrients and other chemical constituents of soils. With the exception of Wilde, Voigt, and Iyer (1972), the citations given above for analytical methods also come from this manual. Some of the techniques covered in the ASA manual can be used in simple laboratories, but many require more sophisticated equipment.

It should also be noted that the problem of assessing the nutrient status in soils can be approached by examining plants growing in the soils. Living plants exhibit deficiency symptoms of varying types and degrees depending on the species and the nutrient in short supply. Likewise a chemical analysis of plant tissue can be used to study variations in chemical composition which reflect nutrient deficiencies. Neither of these approaches is as simple as one might think. The nature and degree of the deficiency symptoms vary with the species, age of the plant, and position on the stem. Similarly the various plant species have different patterns of nutrient uptake, and different nutrients vary with respect to ease of transport within the plant. The most effective approach to the assessment of the nutrient status is through a combination of soil analysis, observation of deficiency symptoms, and analysis of plant tissues.

Analytical Approaches for Sophisticated Laboratories. Although the level of chemistry and physics underlying much of the sophisticated instrumentation is beyond the scope of this book, brief descriptions can provide a general idea of the capabilities and applications for these systems.

Muffle Furnaces. In many analyses there is a need for very high temperatures coupled with controlled entry and exit of gases. Electric muffle furnaces serve this purpose. In comparison with some of the other systems mentioned below, a muffle furnace is a relatively inexpensive piece of equipment. Muffle furnaces are standard in well equipped general laboratories.

Centrifuges. A centrifuge is used to spin tubes of solution at high speed. This causes the heavier materials in the tubes to collect at the bottom. A centrifige is very helpful in separating finely divided precipitates from the mother liquor. Like the muffle furnace, the centrifuge is relatively standard for well equipped laboratories.

Spectrophotometers. A spectrophotometer is similar to a colorimeter, except that a diffraction grating permits the selection of any desired wavelength of light without the use of filters. Spectrophotometers with manual wavelength selection and readout are almost as common in laboratories as colorimeters. However, spectrophotometers that automatically scan a spectral range and plot the results on graph paper are also available. This latter type of unit is considerably more expensive.

Autoanalyzers. Many procedures for chemical analysis can be automated by proper selection and integration of the system components. The resulting system is called an autoanalyzer for the particular kind of determination it is designed to perform. Autoanalyzers greatly accelerate the processing of large batches of samples. Some are even designed to operate largely in the unattended mode, except for a periodic restocking of chemical supplies and recording charts.

Flame Photometers. When a material is vaporized in a flame, the resulting atoms or molecules are excited to higher energy states. They return to lower energy levels by giving off light at wavelengths that are characteristic of the material. A spectral analysis of the emitted light permits the identification of materials and the determination of concentrations. Whether vaporization produces molecular units or atomic units depends on the temperature of the flame. This type of analysis is also called flame emission spectroscopy.

Atomic Absorption Spectroscopy. Using an analogy with photography, atomic absorption spectroscopy might be described as the negative of flame emission spectroscopy. In absorption spectroscopy the materials are vaporized with a lower temperature flame since excitation is not necessary. A light beam of known spectral characteristics is passed through the vapor. The vaporized atoms absorb light at characteristic wavelengths permitting identification and quantification.

Infrared Absorption Spectroscopy. This type of analysis can be considered a variation of the atomic absorption spectroscopy principle using infrared radiation.

X-ray Emission Spectroscopy. The basic idea in this technique is similar to that of flame photometry. Instead of a flame, X rays are used to excite sample materials. Secondary X rays emitted by the sample materials are then analyzed. This type of analysis does not require vaporization.

Fluorometry. The principle of fluorometry is similar to other methods which involve excitation with a subsequent emission of characteristic radiation. In this case the excitation is accomplished by light energy. Each type of material has characteristic wavelengths at which it absorbs light. Absorbed energy is then emitted (fluorescence) at longer wavelengths which are also characteristic of the substance and can be used for analysis.

Nephelometry. Nephelometry is based on the scattering of light by particles suspended in liquids. The scattered light is used to determine the amount of suspended material.

Chromatography. Chromatography encompasses several techniques used to separate mixtures of materials. The mixture is dissolved in a carrier, which is then passed over an absorbant material. The mixture of substances will be separated because of differences in the attraction of the absorbant for components of the mixture. The weaker the attraction, the more rapidly the substance will be washed along the absorbant surface by the carrier. In gas-liquid chromatography a gas serves as the carrier, and absorption takes place on a resin column wetted with an appropriate liquid. In paper chromatography a liquid serves as the carrier, and a strip of paper acts as the absorbant. In thin layer chromatography a liquid serves as the carrier, and absorption takes place in a thin layer of material coated on a glass plate. The name chromatography presumably comes from early uses to separate colored substances.

Mass Spectrometry. Mass spectometry also serves to separate different types of materials, but in a very different way than chromatography. In this case the material is vaporized and then ionized by a stream of electrons. Electrical forces are used to accelerate the positive ions, and magnetic forces are used to direct the ions through an exit slit according to their mass and electrical charge.

Radiometry. Fluorometry, flame photometry, and X-ray emission spectroscopy are all based on secondary radiation resulting from the excitation of test materials. The difference lies in the use of successively higher energy levels for excitation. Increasing the energy of excitation still further can produce radioactivity in the test materials. Neutron activation analysis, for example, makes use of the radioactivity produced by bombarding test materials with neutrons. Alpha, beta, and gamma rays produced by radioactive decay can be measured with standard Geiger-Muller tubes and scintillation counters linked to appropriate scalers.

Articles included in the book edited by Walsh (1971) provide a further introduction to the application of instrumental methods for analyzing soils and plant tissues. Skoog and West (1971) provide a more theoretical introduction to a wide variety of techniques for instrumental analysis.

4.3 THE ENVIRONMENTAL UNDERPINNING: SUBSURFACE GEOLOGY

As evidenced by the frequent references to other sources for further details, a great deal more could be said about soil surveys. Although soils are very important in ecosystem analysis, they comprise only a thin layer blanketing the subsurface geologic structure. This subsurface structure also constitutes a major determinant of the environment, and space limitations require that attention be turned in this direction.

Save for mountains, gorges, and bluffs, subsurface geology often goes unnoticed by the casual observer. Likewise the fact that it is usually subject to little change in the short run may lead to an attitude of indifference. Nevertheless subsurface geology indirectly controls broad patterns of development for natural ecosystems and has major implications for human uses of the land.

The surface conformation (topography and drainage) is in large part a reflection of subsurface geology. Volcanic activity and other large scale earth movements create mountain ranges, fault scarps, and other major landforms. The hardness and cementation of rock materials are expressed topographically through their influence on the rate of erosion. Likewise the pattern of drainage channels on the surface conforms to lines of weakness in the underlying materials. Soils develop from parent materials exposed at the surface, and the composition of these materials strongly influences, fertility, texture, depth, and other important soil properties. The location and characteristics of aquifers (materials permeable to water) determine the availability of groundwater in an area. The dependence of extractive industries on subsurface geology should not require elaboration. Also obvious are the dangers for human development associated with earthquakes. However, there are many less obvious hazards arising from a failure to take account of the geologic structure. These are set forth clearly by Keller (1976) in his book on environmental geology. Issues of site development for major geological features are also considered in detail by Way (1973). Typical hazards are landslides, flooding, coastal erosion, groundwater depletion, groundwater pollution, deposition in waterways, structural collapse, and so on.

First on the agenda in this section is a brief review of the building blocks comprising the geologic structure, and this is followed by a discussion of survey information on geology.

4.3.1 Geologic Structure

Minerals are the basic geologic building blocks and also the target of most extractive industries. Minerals combine to form rocks. Rocks in turn comprise geologic units such as formations. Geologic formations, then, are altered over time by various natural forces.

Minerals. Although they contain impurities to varying degrees, minerals consist largely of a single chemical substance at the molecular level. Properties commonly used to identify minerals in the field are as follows:

Crystal Form. Many minerals occur as crystals large enough to be seen with the naked eye or a hand lens. There are six major types of crystal form with respect to symmetry, and each major type has several variations with regard to the number and arrangement of the faces on the crystal. In the *isometric* form there are three mutually perpendicular axes of symmetry, with crystal dimensions being the same along all axes. In the *tetragonal* form there are three mutually perpendicular axes of symmetry, but only two of the axes have equal crystal dimensions. In the *orthorhombic* form there are three mutually perpendicular axes, but no two axes have equal crystal dimensions. In the *monoclinic* form there are three axes of symmetry, two of which are perpendicular and the third slanted. Crystal dimensions are not equal on any two axes. In the *triclinic* form all three axes of symmetry are slanted, and no two have equal crystal dimensions. The axes of symmetry in the *hexagonal* form are arranged like a wheel and axle. There are four axes in this latter form. One axis forms the axle of the wheel, and the other three form spokes.

Cleavage. Cleavage is the term used to describe the pattern of breakage that is characteristic of the crystal. Flat surfaces formed when a crystal is broken will lie in planes that fit the axes of symmetry as described above.

Hardness. The relative hardness is judged by whether or not one mineral scratches another. The one that does the scratching is harder than the one that is scratched. If each will scratch the other, they are of equal hardness. A ten point scale called *Mohs* scale is usually used for rating hardness. The minerals talc, gypsum, calcite, fluorite, apatite, feldspar, quartz, topaz, corundum, and diamond are the standards for hardness numbers of 1 through 10, respectively.

Streak. Streak is the color of the powder formed by rubbing the mineral on an unglazed procelain plate. Commercial procelain plates are available for testing streak, but the unglazed side of a porcelain tile can also be used.

Color. A given mineral may occur in more than one color, but typically there are only a few colors that are likely to be found in a particular mineral. The color of the specimen and the color of the streak are not necessarily the same.

Luster. The term luster is more or less synonymous with shine. Two major and easily distinguishable types of luster are metallic and nonmetallic. Within the nonmetallic type, however, more subjective terms like silky and greasy are used.

Light Transmission. A thin edge of the mineral may be opaque, translucent, or transparent to light. A more technical term for this property is *diaphaneity*. A related property used in identifying calcite is *double refraction*. Calcite is transparent and will show a double image when one looks through it.

Acid Reaction. A drop of hydrochloric acid can be used as a test for certain minerals containing carbonates. Calcite (calcium carbonate), for example, will bubble rapidly with acid. Dolomite (calcium magnesium carbonate) will react with concentrated or hot acid, but shows little reactivity with cold, dilute acid. Since hydrochloric acid is highly caustic to the skin even when dilute, the acid should be carried in an unbreakable bottle and dispensed cautiously from a dropper that also serves as a cap for the bottle.

Taste. Taste can be used as an aid in identifying some minerals, notably halite (rock salt).

Magnetism. A magnet can be used a test for certain minerals containing iron, notably magnetite.

Fluorescence. Some minerals will fluoresce under ultraviolet light, whereas others will not. A battery powered black light can be used for detecting fluorescence in the field.

Specific Gravity. As mentioned earlier in the discussion of soil properties, the specific gravity of a material is the weight of a given volume of the solid divided by the weight of an equal volume of water. Although not impossible to do in the field, an accurate determination of the specific gravity is more conveniently done in the laboratory. If the mineral is finely divided, a pycnometer can be used to determine the specific gravity as described for soils. If the mineral is in massive form, the specific gravity is determined by weighing first in air and then with the mineral submerged in water. According to the Archimedes principle the weight of an equal volume of water is given by the difference between the two weights. Balances rigged to allow weighing the mineral while it is submerged in water are called specific gravity balances. Gross differences in specific gravity can be judged by the heft of the mineral.

Other Laboratory Tests. Several laboratory methods are available for studying and identifying minerals that would not be practical for field use. Some of the common methods will be mentioned here without going into detail. A blowpipe flame can be used to study flame coloration and fusibility characteristics of certain minerals. Fusing with a flux such as borax will also produce characteristic types of beads or sublimates. The refractive index of transparent minerals provides a diagnostic property that can be determined by immersion in liquids of a known refractive index under a microscope. Likewise the microscopic study of thin sections is a standard petrologic technique. X-ray scattering (diffraction) patterns and radioactivity provide means of examining molecular and atomic structures.

Most introductory geology texts contain keys for identifying minerals based on the above properties. Several such texts are listed in the bibliography of this chapter. There are also a number of popularized field guides that provide descriptions and photographs of commonly occurring minerals.

Rocks. Rocks are solid aggregations of minerals. They are classified according to the conditions under which they form, as reflected in texture and mineral composition. The three main classes of rocks are *igneous, sedimentary,* and *metamorphic.*

Igneous Rocks. Igneous rocks are those that have crystallized when a mass of molten material (magma) cooled. They are characterized by interlocking crystals of minerals which have high melting points. Igneous rocks fall into two major subclasses, being *intrusive* if they formed below the earth's surface or *extrusive* if they solidified after being ejected in volcanic activity. The rapid cooling after ejection gives extrusive rocks such a fine texture that individual crystals are not evident. Patterns of lava flow and escaping gases also leave their mark in extrusive rocks. Intrusive and extrusive rocks are further subdivided on the basis of mineral composition and texture.

Sedimentary Rocks. Sedimentary rocks are those that form by consolidation of particles after settling out of air or water. The particles may be derived from other types of rocks, skeletal remains of organisms, or chemical precipitates from water. Likewise the cementing agent that binds the particles together may be of several different types. These differences provide the basis for classification within sedimentary rocks. Since they settle out of air or water under the influence of gravity, sedimentary rocks typically have a layered structure.

Metamorphic Rocks. Metamorphic rocks are those that have been secondarily modified through forces of heat and pressure. Differences in original materials and the degree of modification form the basis for classifying metamorphic rocks.

Geologic processes are continually working to perform interconversions between rock forms. Weathering and erosion break down all types of rocks and deposit the fragments where they may be recemented into sedimentary rocks if undisturbed for a long enough period of time in the presence of a binding agent. Metamorphic forces alter both igneous and sedimentary rocks as well as those that have been altered previously. If metamorphic forces become sufficiently intense, the rock may melt and become a source of magma for the formation of new igneous rocks. Therefore with rocks, as with soils, one will find all degrees of intergrades between the more typical forms described in diagnostic keys and tables.

Most introductory geology texts and some popularized field guides give diagnostic keys or tables for identifying rocks. Several such references are listed in the bibliography of this chapter. Most of these references are recognizable by their titles.

Lithologic Units and Formations. In the same way that minerals combine to form rocks, rocks combine to form larger structural units. The term *lithologic unit* is often used in describing the geologic structure, but this term is not very specific since "lithologic" roughly translates to "rock." When a lithologic unit can be traced through the geologic structure of a fairly extensive area, it is assigned a name reflecting the area in which it was first described. The lithologic unit thus attains the status of a *formation.* There are parallels between the naming of soil series and geologic formations, except that there is no formal hierarchy of higher level groupings applied to formations.

Landforms and Structural Features. Although the geology of an area usually involves several formations or other lithologic units, one or a few natural processes may have been dominant for a considerable period of time. Geologic processes such as glaciation, erosion, sedimentation, and volcanic activity produce characteristic landforms. An association of related landforms constitutes what we think of as a landscape. Landforms, formations, lithologic units, rocks, and minerals comprise the subject matter of geological surveys.

Of particular interest are planes of weakness such as faults and surfaces of contact between contrasting units. These surfaces are typically warped and contorted rather than flat. Such subsurface planes are expressed on the surface as straight or curved linear features (lineations). *Strike, dip, trend,* and *plunge* are used to describe the spatial orientation of surfaces and linear features.

Strike. The orientation of a two-dimensional plane in three-dimensional space can be described by two axes intersecting at right angles. One of these axes is determined by the *strike,* which is the compass direction of a horizontal line in a slanted plane. For consistency the strike of a planar feature is read from the north end of the compass needle. See Compton (1962) for a discussion of alternative ways of determining the strike and their relative merits.

Dip. Dip fixes the second axis of a planar feature. *Dip* is the angle of downward slant from the horizontal as measured at right angles to the strike line. The dip is usually measured with a clinometer having a scale graduated in degrees.

Trend and Plunge. Trend and plunge are used to describe the orientation of linear features. *Trend* is the compass direction of the lineation. For example the plane of contact between two rock beds intersects the ground surface along a line called the *trace* of the contact plane. If the trace is relatively straight, its compass direction could be determined and recorded as trend. The trend is northerly or southerly according to the direction in which the linear feature slants downward from the horizontal. If the feature is horizontal, then the trend is read from the north end of the needle. The angle at which a linear feature slants downward from the horizontal is called its *plunge.*

4.3.2 Sources of Information on Geology

Detailed geological surveys are among the most difficult environmental surveys to conduct. Opportunities for direct observation of bedrock are limited to outcrops, roadcuts, and mines, unless deep drilling equipment is available. Since drilling operations are slow and expensive, there is a good deal of educated guesswork in linking the few pieces of the three-dimensional jigsaw puzzle which are actually observed. This kind of extrapolation has little meaning unless it is done by someone with a good background in structural geology who has examined all the existing geological literature pertinent to the area. Therefore a detailed geologic mapping project is as much research as it is operational survey.

Conventional geologic mapping projects are fairly expensive operations requiring professional geologists with access to libraries, laboratories, drilling equipment, and seismic apparatus. Therefore operations of this type are usually conducted mostly by federal agencies, state agencies, and companies concerned with mineral exploration. One does not have to be a geologist, however, to obtain useful information on local geology by correlating outcrop information with topography, drainage, soils, vegetation, and land use.

Geologic Maps. The usual sort of geologic map is an attempt to depict the bedrock surface as it would appear with the overburden of soils and other loose materials stripped away. A color or line pattern is chosen for each major type of bedrock. It is used to shade the area over which that type of rock forms the upper layer of bedrock. The bedrock surface pattern is a product of the interplay between geologic structure and topography. This is illustrated in Fig. 4.4 for a simple series of horizontal beds having equal thickness but transected by a variable slope. Most introductory geology texts show a series of block diagrams which help to illustrate this relationship for more complex structural and topographic patterns. However, a bit of simulation provides an even better way to develop a feeling for the relationship between topography, geologic structure, and bedrock exposure pattern.

Use different colors of modeling clay to build blocks with different internal structures that mimic geologic structures. Then carve the upper side of the blocks to simulate typical topographic forms. As the carving progresses,

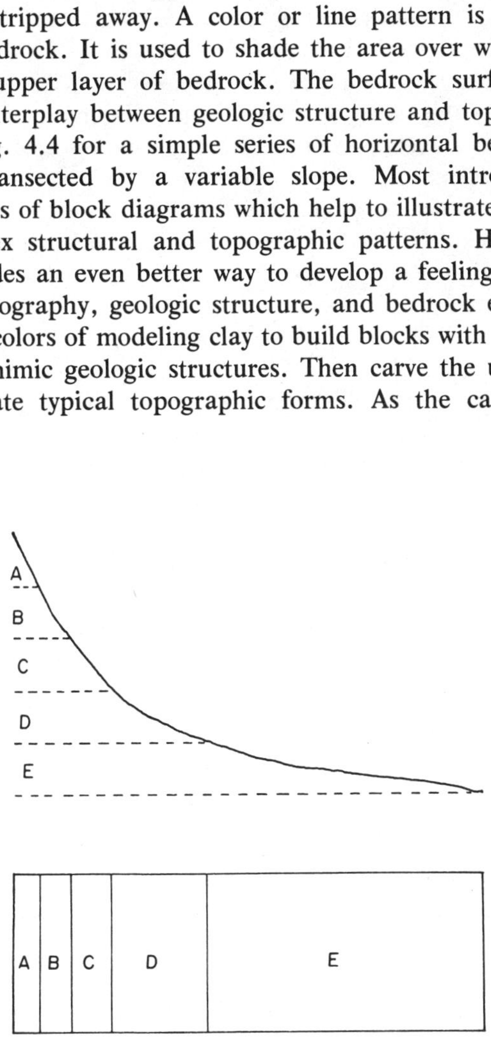

Figure 4.4 Bedrock surface pattern produced by sloping terrain and horizontal beds.

the relationship between structure, topography, and map pattern will soon become apparent.

The topographic factor is handled on geologic maps by superimposing surface contour lines on the bedrock pattern. Likewise the strike, dip, trend, and plunge for structural features are shown with conventional symbols made up of lines and arrow points as explained in the legend of any standard geologic map. A given combination of topography, bedrock exposure pattern, and orientation of structural features pretty well determines the type of subsurface geology to be expected. Block diagrams accompanying the geologic map help the user to understand the subsurface structure which has been worked out with the aid of drill holes and seismic studies.

As with soil maps, the names of formations carry little meaning for the average user. Therefore the utility of a geologic map is largely determined by the interpretive information available. Again like with soils, such information is derived from a combination of laboratory tests and observations of present or past land use. Pertinent information includes rock types, the mechanical strength of rock materials, permeability to water, chemical composition, the type of soils that develop from the rock as parent material, the likely yield of commercial minerals, and so on. Formations that serve as aquifers are especially important.

An aquifer is part of the geologic plumbing system that collects, transports, and stores groundwater. There are three main conditions for the existense of an aquifer. One is to have an interconnecting system of permeable materials. The second is that the permeable materials must be exposed somewhere at the bedrock surface to form the intake. The third is to have impermeable rock types forming a seal around the permeable ones to contain the water. Aquifers are often responsible for cross drainage between watersheds. Therefore the underground of *phreatic* drainage pattern may not correspond with the surface drainage pattern.

So that the maps will be usable in a compact form, brief interpretive information is incorporated in the legend. For additional interpretive information one must consult the detailed survey reports. These are often quite lengthy and technical.

At the federal level geologic maps are prepared and distributed by the U.S. Geological Survey (USGS) in the USDI. Generalized geologic maps at very small scale (1:5,000,000) are available for the entire country, but the availability of larger scale coverage varies both between and within states. The status of USGS mapping for a particular area along with ordering information can be obtained from the National Cartographic Information Center, 507 National Center, Reston, Virginia 22092.

Most states also have agencies responsible for geologic mapping, with the map products being available to the public. Inquiries regarding these maps can be directed to the information office of the respective state capitol for forwarding to the appropriate agency. Geologic maps prepared by private companies in the process of mineral exploration are likely to be proprietary.

Photogeology. When standard geologic maps are not available and there is neither time nor money for a full blown geological survey, airphotos can provide a partial solution. Stereo viewing of airphotos is particularly convenient for the observation of landforms. The various landforms have characteristic shapes and patterns which are often difficult to see from the ground because of poor vantage points. Airphotos provide an excellent vantage point for observing the terrain, and all major topographic variations are clearly revealed by the stereo model. Given an understanding of the shapes and relationships among landforms, one can develop a tentative landform map from careful study of the photos.

Particular attention must be given to lines visible on the imagery which are not due to cultural features. These *lineaments* or *lineations* often correspond with traces of geologic features. Faults, contacts between rock types, old beachlines, and other geological features may be detected in this way. The topography and drainage pattern taken together give clues to the relative hardness, erodibility, and permeability of geologic materials. The species and condition of vegetation also reflect the chemical and physical nature of underlying rocks through the connecting link of soils. In some situations the vegetation that has not been disturbed by man may almost duplicate the pattern of shading shown on a geologic map. Outcrops of bedrock can be located on the photos and visited to verify tentative photo interpretations.

Since several textbooks on the interpretation of airphotos devote substantial portions of their coverage to photogeology, the subject will not be pursued at length here. Strandberg (1967) and Reeves et al. (1975) are both valuable references in this regard. Wanless (1969) has compiled an excellent set of stereograms to aid the study of photogeology. Way's (1973) book on terrain analysis is helpful not only in identifying geological features but also in developing interpretations to guide the use of environmental resources.

Geological Reconnaissance in the Field. Lacking either geologic maps or airphotos, one is forced to do the entire job of geological reconnaissance in the field. Since a map is essential and there are many details that must be plotted in their correct locations, a plane table and alidade are commonly used to develop the map in the field as the survey progresses.

For reconnaissance work of low accuracy a simple traverse board and folding alidade will serve the purpose. For better accuracy a plane table mounted on a tripod with Johnson head should be used in combination with a telescopic alidade. An introduction to the use of a plane table and alidade is given in Sections 3.4.3 and 3.8. For a much more detailed discussion of these instruments as applied in geologic mapping see Compton (1962).

The usual strategy is to establish a control net of known points on the map sheet and then work from one outcrop to another and along traces of geologic features by traversing.

4.4 TOPOGRAPHY, SOILS, AND GEOLOGY IN ENVIRONMENTAL DATA SYSTEMS

As discussed in this chapter, the topography, soils, and geology play either direct or indirect roles in virtually all ecosystem functions. These three environmental components usually change very slowly over time in comparison to other features of ecosystems. Furthermore information on these three aspects of the environment is often available from public agencies. These are all compelling reasons for incorporating data on the topography, soils, and geology into any environmental information system. In fact these features provide a logical starting point for initiating new survey information systems. Existing topographic maps, soil maps, and geological maps will often provide the information needed by most users. In the absence of such maps a good deal of information on these features can be obtained by interpretation of airphotos. The collection of additional information and the verification of existing information can be incorporated into most types of field activity.

The appropriate format for information of this type may be less obvious. If the system is not fully computerized, information on topography, soils, and geology will normally be in map form.

Topography and drainage are conveniently formatted as a set of three overlays which can be superimposed on a base map as needed. One overlay shows topographic contours, the second shows slope classes, and the third shows the drainage network. An overlay showing basic soil classifications such as soil series constitutes the core of the soil information. However, this kind of map does not permit easy extraction of information on the suitability of soils for particular uses. Interpretative soil overlays showing limitations or suitability for common uses will be much more convenient. Unless the information on subsurface geology is unusually detailed, a single geologic overlay map will probably suffice. If the geologic information is sufficiently detailed, however, a series of interpretative overlays can be prepared as with soils.

When the information system is fully computerized, file storage space will be minimized and a greater versatility achieved by storing information in its most basic form and introducing the interpretative analysis during retrieval. Depending on the nature of the data, interpretative analysis can be implemented by mathematical manipulation or table lookup. The choice of display format will depend on the output devices attached to the computer system. system.

4.5 BIBLIOGRAPHY

Allison, L. "Organic Carbon." In C. Black et al., Eds. *Methods of Soil Analysis,* Part 2, Madison, Wis.: American Society of Agronomy, 1965. Pp. 1367-1378.

Allison, L., W. Bollen, and C. Moodie. "Total Carbon." In C. Black et al., Eds. *Methods of Soil Analysis*, Part 2, Madison, Wis.: American Society of Agronomy, 1965. Pp. 1346-1366.

American Association of State Highway Officials. *Standard Specifications for Highway Materials and Methods of Sampling and Testing,* 8th ed. Washington, D.C.: American Association of State Highway Officials, 1961.

American Society of Photogrammetry. *Manual of Remote Sensing,* Vol. 2. Falls Church, Va.: American Society of Photogrammetry, 1975.

Atterberg, A. "Über die physikalische Bodenuntersuchung und über die Plastizität der Tone." *Internationale Mitteilungen für Bodenkunde,* **1** (1911).

Bardsley, C., and J. Lancaster. "Sulfur." In C. Black et al., Eds. *Methods of Soil Analysis,* Part 2. Madison, Wis.: American Society of Agronomy, 1965. Pp. 1102-1116.

Bartelli, L., A. Klingebiel, J. Baird, and M. Heddleson, Eds. *Soil Surveys and Land Use Planning.* Madison, Wis.: Soil Science Society of America and American Society of Agronomy, 1966.

Bauer, K. "Application of Soil Studies in Comprehensive Regional Planning." In L. Bartelli et al., Eds. *Soil Surveys and Land Use Planning.* Madison, Wis.: Soil Science Society of America and American Society of Agronomy, 1966. Chap. 6, Pp. 42-59.

Beasley, R. *Erosion and Sediment Pollution Control.* Ames, Iowa: Iowa State University Press, 1972.

Bertrand, A. "Rate of Water Intake in the Field." In C. Black et al., Eds. *Methods of Soil Analysis,* Part 1. Madison, Wis.: American Society of Agronomy, 1965. Pp. 197-209.

Biship, M. *Subsurface Mapping.* New York: Wiley, 1960.

Black, C., et al., Eds. *Methods of Soil Analysis.* Parts 1, 2. Madison, Wis.: American Society of Agronomy, 1965.

Blake, G. "Particle Density" and "Bulk Density." In C. Black et al., Eds. *Methods of Soil Analysis,* Part 1. Madison, Wis.: American Society of Agronomy, 1965. Pp. 371-390.

Boersma, L. "Field Measurement of Hydraulic Conductivity Below a Water Table" and "Field Measurement of Hydraulic Conductivity Above a Water Table." In C. Black et al., Eds. *Methods of Soil Analysis,* Part 1. Madison, Wis.: American Society of Agronomy, 1965. Pp. 222-252.

Bower, C., and L. Wilcox. "Soluble Salts." In C. Black et al., Eds. *Methods of Soil Analysis,* Part 1. Madison, Wis.: American Society of Agronomy, 1965. Pp. 933-951.

Bowles, J. *Engineering Properties of Soils and Their Measurement.* New York: McGraw-Hill, 1970.

Bremner, J. "Total Nitrogen," "Inorganic Forms of Nitrogen," and "Organic Forms of Nitrogen." In C. Black et al., Eds. *Methods of Soil Analysis,* Part 2. Madison, Wis.: American Society of Agronomy, 1965. Pp. 1149-1255.

Broadbent, F. "Organic Matter." In C. Black et al., Eds. *Methods of Soil Analysis,* Part 2. Madison, Wis.: American Society of Agronomy, 1965. Pp. 1397-1400.

Buckman, H., and N. Brady. *The Nature and Properties of Soils.* New York: Macmillan, 1969.

Buol, S., F. Hole, and R. McCracken. *Soil Genesis and Classification.* Ames, Iowa: Iowa State University Press, 1973.

Burger, D. "Identification of Forest Soils on Aerial Photographs." *Forestry Chronicle,* **33,** 55-60 (1957).

Chapman, H. "Cation-Exchange Capacity" and "Total Exchangeable Bases." In C. Black et al., Eds. *Methods of Soil Analysis,* Part 2. Madison, Wis.: American Society of Agronomy, 1965. Pp. 891-904.

Chow, V., Ed. *Handbook of Applied Hydrology.* New York: McGraw-Hill, 1964.

Cole, J., and C. King. *Quantitative Geography.* London: Wiley, 1968.

Compton, R. *Manual of Field Geology.* New York: Wiley, 1962.

Dapples, E. *Basic Geology for Science and Engineering.* New York: Wiley, 1959.

Davidson, D. "Penetrometer Measurements." In C. Black et el., Eds. *Methods of Soil Analysis,* Part 1. Madison, Wis.: American Society of Agronomy, 1965. Pp. 472-484.

Day, P. "Particle Fractionation and Particle Size Analysis." In C. Black et al., Eds. *Methods of Soil Analysis,* Part 1. Madison, Wis.: American Society of Agronomy, 1965. Pp. 545-567.

Donahue, R. *Soils.* Englewood Cliffs, N.J.: Prentice-Hall, 1965.

Doornkamp, J., and C. King. *Numerical Analysis in Geomorphology.* New York: St. Martin's Press, 1971.

Felt, E. "Compactibility." In C. Black et al., Eds. *Methods of Soil Analysis,* Part 1. Madison, Wis.: American Society of Agronomy, 1965. Pp. 400-412.

Foth, H., and L. Turk. *Fundamentals of Soil Science,* 5th ed. New York: Wiley, 1972.

Gardner, W. "Water Content." In C. Black et al., Eds. *Methods of Soil Analysis,* Part 1. Madison, Wis.: American Society of Agronomy, 1965. Pp. 82-127.

Goodwin, W. "Bearing Capacity." In C. Black et al., Eds. *Methods of Soil Analysis,* Part 1. Madison, Wis.: American Society of Agronomy, 1965. Pp. 485-498.

Greenhood, D. *Mapping.* Chicago: University of Chicago Press, 1964.

Guy, H. *Sediment Problems in Urban Areas.* Washington, D.C.: U.S. Geological Survey Circular 601-E, 1970.

Hamblin, W. K., and J. Howard. *Exercises in Physical Geology.* 4th ed. Minneapolis, Minn.: Burgess, 1964.

Heald, W. "Calcium and Magnesium." In C. Black et al., Eds. *Methods of Soil Analysis,* Part 2. Madison, Wis.: American Society of Agronomy, 1965. Pp. 999-1010.

Hobbs, B., W. Means, and P. Williams. *An Outline of Structural Geology.* New York: Wiley, 1976.

Holtz, W. "Volume Change." In C. Black et al., Eds. *Methods of Soil Analysis,* Part 1. Madison, Wis.: American Society of Agronomy, 1965. Pp. 448-465.

Horton, R. "Erosional Development of Streams." *Geological Society of America Bulletin,* **56,** 281-283 (1945).

Hough, B. *Basic Soils Engineering,* 2nd ed. New York: Ronald, 1969.

Hvorslev, M. *Subsurface Exploration and Sampling of Soils for Civil Engineering Purposes.* Vicksberg, Miss.: U.S. Army Engineer Waterways Experiment Station, CE, 1949.

Keller, E. *Environmental Geology.* Columbus, Ohio: Merrill, 1976.

Klute, A. "Laboratory Measurement of Hydraulic Conductivity of Saturated Soil," "Laboratory Measurement of Hydraulic Conductivity of Unsaturated Soil," "Water Diffusivity," and "Water Capacity." In C. Black et al., Eds. *Methods of Soil Analysis,* Part 1. Madison, Wis.: American Society of Agronomy, 1965. Pp. 210–221, 253–278.

Lambe, T. W. *Soil Testing for Engineers.* New York: Wiley, 1951.

Leopold, L. *Hydrology for Urban Land Planning—A Guidebook on the Hydrological Effects of Urban Land Use.* Washington, D.C.: U.S. Geological Survey Circular 554, 1968.

Linsley, R., Jr., M. Kohler, and J. Paulhus. *Hydrology for Engineers,* 2nd ed. New York: McGraw-Hill, 1975.

Loomis, F. *Field Book of Common Rocks and Minerals.* New York: Putnam's, 1948.

MacFall, R. *Gem Hunter's Guide,* 5th ed. New York: Crowell, 1975.

MacIver, B., and G. Hale. *Laboratory Soils Testing.* Washington, D.C.: Department of the Army, Office of the Chief of Engineers, Engineer Manual EM 1110-2-1906, 1970.

Mader, D. "Sampling Problems Related to Spatial Variation in Soils and Other Aspects of the Physical Environment of Forest Ecosystems." In T. Gunia, Ed. *Proceedings of Symposium on Monitoring Forest Environment Through Successive Sampling.* Syracuse, New York: State University of New York, International Union of Forest Research Organizations, and Society of American Foresters, 1974. Pp. 187–202.

Mason, A. *The World of Rocks and Minerals.* New York: Larousse, 1976.

Mason, B., and L. Berry. *Elements of Mineralogy.* San Francisco: Freeman, 1968.

Michigan Department of State Highways. *Field Manual of Soil Engineering,* 5th ed. Lansing, Mich.: Michigan Department of State Highways, 1970.

Moessner, K., and G. Choate. *Estimating Slope Percent for Land Management from Aerial Photos.* Ogden, Utah: U.S. Department of Agriculture, Forest Service, Intermountain Forest and Range Experiment Station, Research Note INT-26, 1964.

Mokma, D., E. Whiteside, and I. Schneider. *Soil Management Units and Land Use Planning.* East Lansing, Mich.: Michigan State University, Agricultural Experiment Station, Research Report 254, 1974.

Murray, M. *Hunting for Fossils.* New York: Collier, 1967.

Olsen, S., and L. Dean. "Phosphorus." In C. Black et al., Eds. *Methods of Soil Analysis,* Part 2. Madison, Wis.: American Society of Agronomy, 1965. Pp. 1035–1049.

Pearl, R. *Rocks and Minerals.* New York: Barnes & Noble, 1956.

Peech, M. "Exchange Acidity," "Hydrogen-Ion Activity," and "Lime Requirement." In C. Black et al., Eds. *Methods of Soil Analysis,* Part 2. Madison, Wis.: American Society of Agronomy, 1965. Pp. 905–932.

Peele, R., and J. Church, Eds. *Mining Engineer's Handbook,* Vols. 1; 2, 3rd ed. New York: Wiley, 1945.

Perkins, R., and S. Suddarth. *The Estimation of Average Side Slope by Space Vectors.* Lafayette, Ind.: Purdue University, Agricultural Experiment Station Research Bulletin 866, 1970.

Phillipson, J. *Methods of Study in Quantitative Soil Ecology: Population, Production, and Energy Flow.* Oxford: Blackwell, IBP Handbook 18, 1971.

Portland Cement Association. *PCA Soil Primer.* Chicago: Portland Cement Association, 1962.

Pough, F. *A Field Guide to Rocks and Minerals,* 4th ed. Boston: Houghton Mifflin, 1976.

Pratt, P. "Digestion with Hydrofluoric and Perchloric Acids for Total Potassium and Sodium," "Potassium," and "Sodium." In C. Black et al., Eds. *Methods of Soil Analysis,* Part 2. Madison, Wis.: American Society of Agronomy, 1965. Pp. 1019-1034.

Proctor, R. "Design and Construction of Rolled Earth Dams." *Engineering News Record,* Aug. 31, Sept. 7, 21, and 28, 1933.

Quay, J. "Use of Soil Surveys in Subdivision Design." In L. Bartelli et al., Eds. *Soil Surveys and Land Use Planning.* Madison, Wis.: Soil Science Society of America and American Society of Agronomy, 1966, Chap. 8, Pp. 76-86.

Ragan, D. *Structural Geology, An Introduction to Geometrical Techniques,* 2nd ed. New York: Wiley, 1973.

Reeves, R., et al. "Terrain and Minerals: Assessment and Evaluation." In American Society of Photogrammetry. *Manual of Remote Sensing,* Vol. 2. Falls Church, Va.: American Society of Photogrammetry, 1975, Chap. 16, Pp. 1107-1351.

Richards, S. "Soil Suction Measurements with Tensiometers." In C. Black et al., Eds. *Methods of Soil Analysis,* Part 1. Madison, Wis.: American Society of Agronomy, 1965. Pp. 153-163.

Ruhe, R. *Geomorphology.* Atlanta: Houghton Mifflin, 1975.

Sallberg, J. "Shear Strength." In C. Black et al., Eds. *Methods of Soil Analysis,* Part 1. Madison, Wis.: American Society of Agronomy, 1965. Pp. 431-447.

Schneider, I., and E. Whiteside. *Status of Michigan Soil Surveys.* East Lansing, Mich.: Michigan State University, Agricultural Experiment Station Research Report 240, 1974.

Shickluna, J. *Sampling Soils for Fertilizer and Lime Recommendations.* East Lansing, Mich.: Michigan State University, Cooperative Extension Bulletin E-498, 1975.

Skoog, D., and D. West, *Principles of Instrumental Analysis.* New York: Holt, Rinehart & Winston, 1971.

Soil Conservation Service. *Land-Capability Classification.* Washington, D.C.: U.S. Department of Agriculture Handbook 210, 1966.

Soil Conservation Service. *Guide For Interpreting Engineering Uses of Soils.* Washington, D.C.: U.S. Department of Agriculture, 1972.

Soil Survey Staff. *Soil Survey Manual.* Washington, D.C.: U.S. Department of Agriculture Handbook 18, 1951.

Soil Survey Staff. *Identification and Nomenclature of Soil Horizons.* Washington, D.C.: Supplement to U.S. Department of Agriculture Handbook 18, 1962.

Soil Survey Staff. *Soil Taxonomy.* Washington, D.C.: U.S. Department of Agriculture Handbook 436, 1975.

Sowers, G. "Consistency." In C. Black et al., Eds. *Methods of Soil Analysis,* Part 1. Madison, Wis.: American Society of Agronomy, 1965. Pp. 391-399.

Stanton, R. *Ore Petrology.* New York: McGraw-Hill, 1972.

Strahler, A. "Quantitative Geomorphology of Drainage Basins and Channel Networks." In V. Chow, Ed. *Handbook of Applied Hydrology.* New York: McGraw-Hill, 1964. Sec. 4-11.

Strandberg, C. *Aerial Discovery Manual.* New York: Wiley, 1967.

Tilmann, S., and D. Mokma. *Soil Management Groups and Soil Erosion Control.* East Lansing, Mich.: Michigan State University, Agricultural Experiment Station Research Report 310, 1976.

Tilmann, S., and D. Mokma. *Engineering Properties of Soil Management Groups.* East Lansing, Mich.: Michigan State University, Agricultural Experiment Station Research Report 313, 1976.

Travis, M., G. Elsner, W. Iverson, and K. Johnson. *VIEWIT—Computation of Seen Areas, Slope, and Aspect for Land-Use Planning.* Berkeley, Cal.: U.S. Department of Agriculture, Forest Service, Pacific Southwest Forest and Range Experiment Station, General Technical Report PSW-11, 1975.

U.S. Army Corps of Engineers. *The Unified Soil Classification System.* Washington, D.C.: U.S. Government Printing Office, Military Standard 619, 1960.

Wallis, J., and K. Bowden. *A Rapid Method for Getting Area-Elevation Information.* Berkeley, Cal.: U.S. Department of Agriculture, Forest Service, Pacific Southwest Forest and Range Experiment Station, Research Note 208, 1962.

Walsh, L., Ed. *Instrumental Methods for Analysis of Soils and Plant Tissue.* Madison, Wis.: Soil Science Society of America, 1971.

Walsh, L., and J. Beaton. *Soil Testing and Plant Analysis.* Madison, Wis.: Soil Science Society of America, 1973.

Wanless, H. *Aerial Stereo Photographs for Stereoscope Viewing in: Geology, Geography, Conservation, Forestry, Surveying.* Northbrook, Ill.: Hubbard, 1969.

Warncke, D., and L. Robertson. *Understanding the MSU Soil Test Report—Results and Recommendations.* East Lansing, Mich.: Michigan State University, Cooperative Extension Bulletin 937, 1976.

Way, D. *Terrain Analysis.* Stroudsburg, Pa.: Dowden, Hutchinson & Ross, 1973.

Welch, P. *Limnological Methods.* New York: McGraw-Hill, 1948.

Wilde, S., G. Voigt, and J. Iyer. *Soil and Plant Analysis for Tree Culture.* New Delhi: Oxford & IBH Publishing Co., 1972.

Wischmeier, W. "Estimating the Soil-Loss Equation's Cover and Management Factor for Undisturbed Areas." In *Proceedings of 1972 Sediment Prediction Workshop, Oxford, Mississippi.* Washington, D.C.: U.S. Government Printing Office, 1974.

Wischmeier, W., C. Johnson, and B. Cross. "A Soil Erodibility Nomogram for Farmland and Construction Sites." *Journal of Soil and Water Conservation,* **26,**189–193 (1971).

Wischmeier, W., and L. Meyer. *Soil Erodibility on Construction Sites.* Washington, D.C.: Highway Research Board, National Academy of Science Special Report 135, 1973.

Wischmeier, W., and D. Smith. *Predicting Rainfall-Erosion Losses from Cropland East of the Rocky Mountains.* Washington, D.C.: U.S. Department of Agriculture Handbook 282, 1965.

Wisler, C., and E. Brater. *Hydrology.* New York: Wiley, 1959.

CHAPTER 5–THE ENVIRONMENTAL UTILITIES: RADIANT ENERGY, AIR, WATER, AND WEATHER

Radiant energy, air, water, and weather comprise the utilities for habitation Earth. They are considered together in this chapter because of their inextricable linkage in ecosystems.

The primary constituents of air—nitrogen, oxygen, argon, and carbon dioxide—occur abundantly as gases because of their very low boiling points. In company with several less abundant substances these four gases form the atmospheric envelope around the earth, permeate the pore spaces of the soil, and are dissolved in the waters. Because its freezing point lies within the range of ambient temperatures, water occurs in solid, liquid, and gaseous states. The liquid form wets the soil as well as filling the channels and basins. The degree to which the gaseous form mixes with other atmospheric constituents is determined largely by the air temperature. Differential heating from solar radiation stirs the air into winds and causes moisture to evaporate from exposed surfaces. Subsequent cooling produces condensation and precipitation. Thus the interaction of air, water, and radiant energy from the sun generates the continuing cycle of weather patterns. Rapid fluctuations in weather conditions stand in marked contrast to the slow and almost imperceptible changes associated with topography, soils, and geology.

Although the technology does not yet exist for an effective control of either solar radiation or the weather patterns it produces, measurement of these environmental conditions is still crucial. An advance warning of impending severe weather conditions can save many lives as well as huge sums of money. Less spectacular but no less important, weather forecasts facilitate short run planning of outdoor activities in general. Likewise a continuous monitoring of weather conditions is essential for aviation and navigation. In terms of ecosystem analysis, the measurement of solar radiation and its atmospheric effects is essential for determining relationships between natural processes and weather conditions. Likewise a statistical knowledge of the frequency with which extreme weather conditions are likely to be encountered is required for effective land use planning. In addition to direct solar radiation and its derivatives, the environment may also contain ionizing radiation produced by radioactive materials. This high energy radiation must be monitored because of the potential health hazard.

Oxygen is a requirement for respiration of all but a few organisms. It is usually sufficiently abundant in aerial environments, but is often a limiting factor in the soil and in aquatic ecosystems. Because of this, dissolved oxygen levels are often monitored as an index to the status of aquatic ecosystems. Carbon dioxide is an essential ingredient for photosynthesis in plants as well as a byproduct of respiration. Carbon dioxide is seldom in short supply, but the need for its measurement arises in studies of respiratory and photosynthetic activity. The atmosphere also contains a host of other substances—some in gaseous form, some as microscopic droplets of liquid, and some as tiny particles of solid material. Some of these other atmospheric constituents are biologically inert, but many are noxious and affect different types of organisms to different degrees. Monitoring the composition of air is gaining importance as industrial technology increases both the sources and the types of pollutants.

It should scarcely be necessary to elaborate on the importance of water in ecosystem dynamics. Water along with carbon dioxide and sunlight provides the raw material for photosynthesis. Water comprises a large fraction of tissue weight in animals and herbaceous plants. Tremendous quantities are transpired by plant foliage on a daily basis. It serves as the medium for diposal of metabolic wastes in most animals. It is used in a multitude of industrial processes and is the carrier for most human waste disposal systems. It provides a means of travel and transport and is the center of many recreational activities. The high specific heat and heat of fusion of water coupled with its abundance on a global scale provide a moderating system for the climate of the earth. The erosive forces of rain and glaciers play a major role in shaping the surface of the earth. The addition of irrigation water can transform desert areas into highly productive agroecosystems. The list of essential functions goes on and on, so the need for the measurement of quantities and quality of water in the environment should be apparent. Quantities of water are important, but its physical properties and dissolved substances must also be determined to assess the status of an ecosystem.

The variability of these physical factors is often more critical to the proliferation of a species than is the "average" condition. A species may prosper in favorable periods, but suffer severely when the natural variability in some physical factor shifts conditions to the unfavorable end of the range. Without long term monitoring of the physical conditions it is difficult to judge whether the current abundance of a species represents a good adaptation in the long run or reflects instead a chance occurrence of favorable but temporary conditions. Likewise a given species is usually more vulnerable in some stages of its life cycle than in others. The chance occurrence of extreme conditions during the most sensitive period provides the best test of adaptation to the complex of physical factors. The presence of a species in a locality cannot be taken as a priori evidence that the species is well adapted to the site. The occurrence of a species in a given locality is in part an accident of nature, and the present abundance of a species may be due to a chance failure of a better adapted competitor to find its way into the locality or to a chance absence of natural enemies as well as to temporarily favorable physical conditions. Thus long term monitoring of conditions is essential for a full understanding of the dynamics of natural communities, as well as for assessing the probable success of introducing new species.

A distinction must also be made between *microenvironment* and *macroenvironment.* Conditions in small areas such as valley bottoms, exposed slopes, interiors of vegetative stands, and so on may be quite different from those of the surrounding areas. Unless measurements are actually made within the microenvironment, data from surrounding areas are likely to give a very misleading picture of the conditions in the more restricted area. Measuring physical factors in microenvironments often requires miniature sensors or a special design of intake ports for samplers to avoid the influence from surrounding areas.

The measurement of a physical factor can be approached at several levels of accuracy and speed. Devices for making infrequent observations with a low order of accuracy can often be purchased or constructed for a few dollars. These are suitable for use in demonstrations, teaching laboratories, and individual projects in introductory courses; but they would not be sufficiently reliable to serve as a basis for monitoring or planning alterations of ecosystems. Accuracy of the order needed for monitoring or planning modifications of ecosystems comes at considerably higher cost unless one has the special training in physics and chemistry required to fabricate instruments from components and perform the necessary calibrations.

Instruments of this caliber which are operated manually or produce ink tracings on chart strips driven by clockdrives typically cost from one hundred to a few hundred dollars. Costs take another large jump when one moves to the class of instruments which might be described as "autoanalyzers." Such systems operate without direct human supervision, taking measurements either continuously or at preset intervals, and recording the results on a computer readable medium. Costs for systems with this level of sophistication usually run in the thousands of dollars. Our primary concern in this chapter is with the class of instruments in which readings are taken from dials or recorded as ink traces on charts operated by manually wound clockdrives. General descriptions of electronic and "autoanalyzer" systems are intended only to give an appreciation for the capabilities, not to provide operating instructions.

Another matter that should be clarified at the outset is the relationship between rates, averages, and totals on pen trace charts produced by instruments. Most such recordings reflect either current conditions or the rate at which something is taking place. For example, consider a strip chart with a pen trace of fluctuating temperature (°F) against time (hr). In analyzing this record one might want to know the average temperature for the period. In order to find the average temperature for the period one must *integrate* the record by finding the area under the trace for the period of interest.

Since procedures for determining areas are covered in Section 3.7, they will not be discussed here. Suppose that the area under the trace for a 12 hr period is measured to be 30 sq in. When the area under the trace has been determined for the period of interest, the next step is to find the conversion factor between the area and the desired quantity. This can be accomplished by choosing a convenient horizontal line on the chart and measuring the area under this line over a period on the time axis long enough to obtain a reasonably accurate area measurement. Pursuing the temperature chart example, one might choose to measure the area under the 50° line for a length representing 4 hr. This amount of area will represent 50° × 4 hr = 200°hr. Suppose that this reference area measures 8 sq in. Then there are 200°hr per 8 sq in., or 25°hr/sq in. on the chart. Since the total area under the trace for the 12 hr period of interest was 30 sq in., this must represent

$$30 \text{ sq in.} \times 25°\text{hr/sq in.} = 750°\text{hr}$$

Furthermore there were 12 hr in the period, so the average temperature for the period is

$$\frac{750°\text{hr}}{12 \text{ hr}} = 62.5°$$

If the quantity recorded is a rate per unit time, like incident solar radiation, for which it makes sense to determine a total for the period, the total is the figure obtained by multiplying the total area under the trace by the quantity per unit area. The total would be 750°hr in the above example. Likewise the total is equal to the average rate multiplied by the length of the period.

With this as background we will proceed to consider radiant energy in the environment.

5.1 THE ENVIRONMENTAL POWER SOURCE: RADIANT ENERGY

Radiant energy from the sun and gravitational energy are the only known sources that cannot be mined to exhaustion. Fossil fuels represent storage of solar energy from the past, and known reserves will probably not last more than a few centuries. Earth and oceans hold a vast potential for supplying radiant energy through nuclear processes. Despite the environmental hazards associated with nuclear energy, it is only a matter of time before mankind will have to develop this energy source on a large scale. Although nuclear energy potentials are larger by several orders of magnitude than those for chemical energy from fossil fuels, fissionable and fusionable materials are not infinite either.

While the sun will eventually exhaust its nuclear fuels, any remaining inhabitants of Earth had better be prepared to abandon the planet before this happens. Besides its potential for conversion as a direct evergy source, solar energy serves two ecological functions that are even more important. Both of these have been alluded to previously. Through its differential heating effects it serves as the driving force for generating weather patterns. It also serves as the power source for photosynthesis at the base of all food chains. Since man has still not learned to duplicate this process artificially, we are ultimately dependent on the interaction of chloroplasts in green plants with radiant energy from the sun.

Before discussing methods of measurement, it will by helpful to review briefly the nature of radiant energy as it influences ecosystems.

5.1.1 Nature of Radiant Energy

Radiation may consist either of a stream of particles or of electromagnetic waves. Particulate radiation is exemplified by alpha and beta radiation from nuclear sources. Alpha particles are the nuclei of helium atoms, and beta particles are either electrons or their antiparticles called positrons. A positron has the same mass as an electron, but bears a positive charge instead of a negative one. Electromagnetic radiation includes such diverse types of radiant energy as sunlight, X rays, gamma rays, heat rays, and radio waves. While these various forms of electromagnetic radiation all travel at the speed of light, they differ with respect to wavelength. Electromagnetic rays also have corpuscular characteristics, being composed of discrete units called photons. Since these corpuscular units have no rest mass, however, they do not qualify as particles in the usual sense.

Before it reaches the atmosphere, sunlight includes a broad range of wavelengths. The atmosphere, however, absorbs some wavelengths of incoming solar energy to a much greater degree than others. Thus very little of the short wavelength, high energy ultraviolet radiation reaches the surface of the earth. This is a fortunate circumstance since ultraviolet radiation is very deleterious to animal tissue. The ozone in the upper atmosphere is primarily responsible for absorbing in the ultraviolet range, and this is the reason that its possible reduction from fluorinated hydrocarbons in aerosol cans is a cause for concern. Other components of the atmosphere, especially water vapor, selectively absorb radiant energy of other wavelengths. The largest "window" for the passage of solar radiation lies in the visible range of wavelengths which covers the span from about 400 nanometers to 700 nanometers. A micrometer (also called a micron) is one millionth of a meter and is abbreviated μm. A nanometer is one thousandth of a micrometer and is abbreviated nm. This correspondence of the visual range with an atmospheric window is no accident of nature, since natural selection has geared our visual sensors to operate in the range where the most energy is available. Even in the visual range, however, some solar radiation is absorbed by the atmosphere.

Some of the solar radiation which is not absorbed is scattered by the atmosphere. Part of this scattered radiation is directed back out into space, thus contributing to the albedo (or earth-light). The remainder of the scattered radiation bounces around in the atmosphere until it eventually reaches the surface as diffuse skylight. Were it not for this scattering, the sky would be dark instead of light. Also, the atmosphere reradiates absorbed energy in longer wavelengths. Some of this reradiated energy is directed toward the surface. Therefore radiation reaching the surface of the earth has a *direct* component, which comes in on a straight line from the sun passing through the atmosphere without interference, and a *diffuse* component, which comes from various directions after being scattered or absorbed and reradiated in the atmosphere.

There are also several different possibilities for disposition of the radiation that does reach the earth's surface. First, it may be *reflected* back into the atmosphere. This reflected radiation runs the same chance of possible interference by the atmosphere on its outward journey as it did on its inward journey. Thus some of it may go directly out into space as albedo, and some of it may be captured at least temporarily in the atmosphere. Reflected solar radiation constitutes the energy source for conventional aerial photography.

Incident solar radiation that is not reflected by the receiving surface is either *transmitted* or *absorbed.* Energy that is transmitted continues downward until it encounters materials which either reflect it or absorb it. Absorbed energy may simply serve to raise the temperature of the absorbing medium by increasing the kinetic energy of the molecules, or it may serve to power chemical reactions such as photosynthesis. Absorbed energy that produces chemical reactions or changes of state (such as evaporation of water) is stored in the ecosystem until released by the reverse process.

Absorbed energy that produces a temperature increase without chemical reactions or transformations is *reradiated* back into the atmosphere as longer wavelength heat rays when the material cools. Stefan's law states that the intensity of emitted energy (reradiation) is proportional to the fourth power of the absolute temperature ($°K = °C + 273$) of the radiating material. Therefore hot materials radiate energy much more rapidly than cool materials. Wien's displacement law states that the wavelength of maximum emission is inversely proportional to the absolute temperature of the radiating material. Since most environmental materials are relatively cool, they radiate energy at long wavelengths. Water vapor and carbon dioxide in the atmosphere are quite effective in capturing long wavelength heat rays radiated by surface materials, thus producing a "greenhouse effect" which warms the atmosphere. In this manner the continued addition of carbon dioxide to the atmosphere from the combustion of fuels operates to increase the average temperature of the environment. The *net radiation* at any given time and place is the difference between the total incoming radiation and the total outgoing radiation.

Four types of radiation from nuclear sources are also of environmental concern. *Alpha* radiation consists of particles composed of two protons and two neutrons. This is exactly the nuclear structure of the helium atom, so alpha particles become helium atoms upon capture of two electrons. Because of their relatively large mass and double positive charge, alpha particles interact readily with matter and do not penetrate much beyond the surface.

Beta radiation consists of individual electrons or their antiparticles called *positrons.* Electrons (or negatrons) carry a unit negative charge, whereas positrons carry a unit positive charge. Both electrons and positrons have the same mass, which is very small in comparison to the mass of protons and neutrons. Since the positron is an antimatter particle with respect to the negatron, chance collision of a positron with a negatron annihilates them both and transforms their masses into energy. Since beta particles have less mass and charge than alpha particles, the more energetic beta particles can be quite penetrating.

The third type of particulate radiation from nuclear sources consists of a stream of neutrons. Since neutrons carry no net charge, they constitute a very penetrating type of radiation. The interaction with nuclei of atoms having low atomic weight is fairly effective in slowing down (thermalizing) fast neutrons. The tendency of hydrogen nuclei in water to slow down neutrons is used as a means of measuring the soil moisture content. An interaction with heavier nuclei may result in the neutron being captured by the nucleus. Several different types of nuclear reactions may follow neutron capture. If neutrons are not captured by another nucleus, they break up after a short time into a proton, an electron, and a neutrino. Since the neutrino is harmless and almost undetectable, its properties will not be considered here.

Gamma rays constitute the fourth type of nuclear radiation with which we will be concerned. This is a very energetic form of extremely short wavelength electromagnetic radiation. As such it is pure energy and carries neither charge nor rest mass. It is a very penetrating and damaging form of nuclear radiation.

5.1.2 Units of Measure for Radiant Energy

The incidence of radiant energy on a surface is usually expressed in heat equivalents. Thus the rate at which radiant energy is received on a surface is normally expressed in units of *gram calories per square centimeter per minute.* A gram calorie (g-cal or cal) is the amount of heat required to raise the temperature of 1 gram of water by 1 °C. Integrating the rate over time as explained earlier in this chapter gives energy received in gram calories per square centimeter. 1 g-cal/sq cm is also called a *langley* (ly). Thus 1 ly/min is the same as 1 g-cal/sq cm/min. Multiplying langleys or gram calories per square centimeter by the area of the exposed surface then gives the total energy in gram calories. The relationships among other units are: 1 g-cal = 4.186 joules; 1 joule = 10^7 ergs; 1 British thermal unit (Btu) = 1055 joules; and 1 watt = 14.334 cal/min. In ecology, percent of full sun is often used as a relative measure of radiant energy.

Light is radiant energy falling in the visible range of wavelengths. Light energy may be measured in gram calories as above, or by comparison with the international standard *candlepower.* The amount of light radiating through a unit solid angle (*steradian*) from a point source of 1 candlepower is called a *lumen*. In other words, if a candlepower source is placed at the center of a hollow sphere having 1 m, radius, then each square meter of inner surface on the sphere will receive 1 lumen of light. Likewise if a candlepower source is placed at the center of a hollow sphere having 1 ft radius, then each square foot of inner surface area on the sphere will receive 1 lumen of light. 1 lumen/sq m, called a *lux,* is used as the unit of illumination in the metric system. 1 lumen/sq ft, called a *foot-candle,* is used as the unit of illumination in the English system.

None of the above units are suitable for measuring ionizing radiation from nuclear sources. Most instruments for measuring ionizing radiation are built to detect the passage of an alpha particle, a beta particle, or a gamma ray as an individual event. Each such event is counted, with the readout showing either the total count or the count rate. The count per unit time as obtained from the instrument, however, is not the same as the amount of radiation coming from the source.

Some particles or rays may pass through the sensor undetected. The detector has a very short period during which it is refractory (inoperative) associated with each count. *Coincidence* in the arrival of particles or rays may therefore cause the readout to be less than the number of particles or rays passing through the detector. Likewise gamma rays in particular are very penetrating and may pass through the detector without interacting with the materials from which it is constructed.

Furthermore the geometric placement of the detector with respect to the source affects the relationship between counts and the total radiation produced by the source. The source emits radiation in all directions, but the detector is usually exposed to only a portion of this radiation since it does not surround the source completely. It might seem that a correction for this effect could be derived easily by expressing the "window" area of the detector as a fraction of the area of a sphere surrounding the source. However, *backscattering* and *sidescattering* from the walls of the source container may increase the fraction of the total radiation reaching the detector. Conversely self-absorption by the source operates to reduce the amount of radiation reaching the detector.

The most practical way of relating counts from the detector to the actual amount of radiation produced by the source is to compare the count rate from the test source with the count rate obtained from a known source arranged in the same geometry. If the test source is so "hot" that it saturates the detector, absorbing material can be placed between the source and the detector to reduce the intensity of the radiation.

Known sources of radiation are calibrated in *curies.* A curie (Ci) is the amount of radioactive material that produces 3.7×10^{10} nuclear disintegrations per second. Since a curie is a relatively large amount of radioactivity, the millicurie (mCi) = 0.001 Ci and the microcurie (μCi) = 0.000001 Ci are often used as units of radioactivity.

Because nuclear disintegrations are continually taking place, the radioactivity of a standard source continually decreases after calibration. This is expressed by the half-life of the radioactive material. *Half-life* is the time required for half the remaining radioactive atoms to decay. The fact that the half-life is independent of the number of radioactive atoms present implies that the decay rate (disintegrations per unit time) is proportional to the number of radioactive atoms present. The proportionality constant is called the *decay constant*. Formulas involving the decay constant can be used to make adjustments for the decay since calibration if the standard source has a short half-life.

The curie strength of a radioactive source gives the decay rate or the number of disintegrations per unit time. Of more direct interest in environmental problems, however, is the effect that the resulting radiation will have on materials in the ecosystem. This depends on several things such as the ionizing power of the radiation, the nature of the material, and the extent to which the radioactive material is concentrated by chemical processes in sensitive organs of animals. Several units of measure other than the curie are informative with respect to ionizing effects.

The *roentgen* (R) is the amount of X rays or gamma rays that will produce ions totaling 1 electrostatic unit of charge in 1 milliliter of air at standard conditions of temperature (0°C) and pressure (760 mm of mercury). The *rad* (radiation absorbed dose) is the amount of radiation required to deposit 100 ergs of energy per gram of material, whatever the material may be. Since the amount of energy absorbed depends on the type of material, the nature of the material must be stated to make the rad a meaningful unit. The *roentgen equivalent man* (rem) is the quantity of radiation equivalent in the effect on humans to 1 roentgen of X rays or gamma rays. The *roentgen equivalent physical* (rep) is the amount of radiation resulting in the absorption of 93 ergs of energy per gram of tissue (Chase and Rabinowitz, 1967). These roentgen equivalents are used in biological dosage calculations.

5.1.3 Measurement of Solar Radiation

Instruments are available for measuring the several components of radiant energy discussed above. The World Meteorological Organization (WMO) has adopted a terminology for radiation measuring instruments, but this terminology is not always followed too carefully in general practice. Only the more commonly used terms will be given here. The two terms which serve as catchalls for various types of devices that measure radiant energy are *radiometer* and *actinometer*. Other terms have more restricted meanings as described below.

Instruments for Measuring Total Incoming Solar Radiation. Instruments for measuring the total of direct and diffuse solar radiation of all wavelengths falling on a surface are usually called *pyranometers,* but the term *solarimeter* is also used fairly commonly in this context. These instruments may be either rate meters which show incident energy at a point in time, or integrating instruments which show the total amount of energy received over a period of time. Most instruments are of the former type, indicating the rate at which energy is received. These may either be designed for spot readings or equipped with clock-driven recorders for plotting the incidence of energy against time.

Most instruments of the rate reading type are based on the difference between outputs of thermopiles painted black and thermopiles painted white. A *thermopile* is a group of thermocouples. A *thermocouple* is an electrical circuit formed by two wires made of different metals. Manganin and constantan are often used as the metals for constructing thermocouples in pyranometers. When the two junctions of a thermocouple have different temperatures, a current flows in the circuit. The warmer of the two junctions is called the hot junction, and the cooler of the two is called the cold junction. The amount of current flowing is determined by the temperature difference between the two junctions. If the temperature of the cold junction is known, the temperature of the hot junction can be determined from the current. Solar radiation heats the blackened portions of the thermopiles more than the portions painted white. This temperature difference as reflected in thermopile current provides the basis for measuring radiation. The sensitive element is exposed through a hemispherical or bulb-shaped transparent shield.

Factors to be considered in selecting a pyranometer are its sensitivity, response time, cosine response, and stability under changing environmental conditions. Sensitivity refers to the smallest detectable change in incoming radiation. Response time refers to the time required for the new reading to stabilize after a change in intensity of the radiant energy received. Cosine response refers to the accuracy with which slant rays of the sun are recorded. Theoretically the intensity of radiation per unit area of receiving surface decreases as the cosine of the angle between the sun and a line perpendicular to the surface. A good pyranometer should respond to changes in the sun angle according to this cosine law. Of course the better the performance of the instrument in these respects, the higher the price is likely to be.

Since the number of instruments available is expanding rapidly, no attempt will be made to consider specific models. Instruments of this type are included in a survey of instruments for micrometeorology compiled by Monteith (1972) during the International Biological Program (IBP).

A point that bears emphasis is the need for a periodic checking of field instruments against a secondary standard in order to detect any drift in performance.

Integrating pyranometers which indicate the total amount of radiation received may be devised in several ways. One approach is to incorporate integrating circuits in pyranometers as discussed above. Quite a different approach is represented by the Gunn Bellani radiation integrator in which the total radiant energy is measured by the amount of liquid distilled from a blackened receptacle. Similarly the progress of chemical reactions induced by radiation can be used as an index of the radiant energy received. The chemicals are exposed in a tube or impregnated on paper, with the extent of the reaction usually being judged by a color indicator. See Platt and Griffiths (1972) for further information on this latter approach.

Instruments for Measuring Direct Solar Radiation. Instruments for measuring the direct component of solar radiation are called *pyrheliometers.* Since the purpose here is to measure direct solar radiation, the surface of the sensitive element must be oriented so that the sun's rays strike it at a right angle (normal incidence). This can be accomplished in a recording instrument by using a *heliostat* to make the sensitive element follow the sun. The heliostat is a mechanical timing device which maintains a sun synchronous position.

As with pyranometers, pyrheliometers are usually of either the spot reading or the rate recording type. The sensitive element usually consists either of thermopiles or of bimetallic strips. Thermopiles have already been discussed briefly in connection with pyranometers. A bimetallic strip is a laminated strip made of two metals having different coefficients of thermal expansion. This difference in thermal expansion causes a deflection or bending of the bimetallic strip upon heating. The deflection can be magnified mechanically and transmitted to an indicator needle or pen. As with thermopiles, differential heating of black and white elements is used to measure radiant energy. The field of view for the sensor is restricted to a small cone angle so that diffuse radiation is excluded. Most of the considerations discussed above for pyranometers also apply to pyrheliometers, except that cosine response is not involved due to the normal incidence geometry.

Instruments for Measuring Diffuse Radiation. Measurement of diffuse solar radiation alone can be accomplished by equipping a pyranometer with a shield band (occulting ring) which shades the sun's path of travel across the sky. Since this shadow band also cuts out some diffuse radiation, the readings must be corrected to allow for the portion of the sky obscured by the band. A more expensive alternative is to mount a spot shield on a heliostat so that it remains synchronous with the sun's position.

Instruments for Measuring Visible Light. Instruments for measuring the intensity of visible light are called *photometers* or *light meters.* Since "visible" is defined in terms of the human eye, instruments of this type must be sensitive to the same distribution of wavelengths involved in visual response. Such instruments are based on the electrical effects of light energy. Light can be used either to provide a power source for a circuit (photovoltaic effect) or to modify the behavior of circuits charged by other types of power sources (photoconductive and photoemissive effects).

Selenium solar cells which generate a small current when exposed to light are commonly used in constructing light meters. Such meters are rugged and portable, do not require a power source, and respond immediately to changing light intensity. However, they are not very sensitive to dim light, have a nonlinear response, and exhibit some fatigue (Platt and Griffiths, 1972).

The most sensitive type of meter for low light levels is based on changes in the conductivity of semiconductors when exposed to light. This type of meter requires a power source to charge the circuit. Other disadvantages are temperature sensitivity and nonlinear response (Platt and Griffiths, 1972).

Temperature stability and nearly linear response without fatigue can be obtained by using the photoemissive effect in which charged plates of certain metals emit electrons when exposed to light. As with semiconductors, an external power source is required. Such instruments also tend to be more bulky, fragile, and expensive than selenium cells or semiconductor meters.

Sunshine Recorders. Several types of instruments are available for recording periods of bright illumination during the day. These range from the Campbell-Stokes type in which the record is formed by focusing solar rays to burn a trace on a card as the sun moves across the sky, to the Foster sunshine switch in which the difference between the output of shaded and unshaded light meters triggers a switch to make a trace on a clockdriven chart. Almost any type of solar radiation detector can be rigged for activation at some threshold level which provides a definition of "sunshine." This is not necessarily synonymous with cloudless conditions, since many of these instruments will record in the presence of a thin cloud cover as well as in direct sunlight.

Manufacturers' instructions should be followed carefully in operating these instruments since positioning tends to be critical. Likewise it is desirable to observe the behavior of the instrument carefully during initial operation to define the range of conditions over which activation occurs. Platt and Griffiths (1972) provide a more complete discussion of operational principles, proper exposure of instruments, and interpretation of results.

Instruments for Measuring Net Radiation. Net radiation is the difference between total incoming radiation and total outgoing radiation, all wavelengths combined. Instruments for measuring net radiation, called *net radiometers*, are constructed by placing sensitive elements in opposition so that one faces upward and the other faces downward. The difference between the outputs of the opposing elements gives a measure of the net flux (flow) of radiant energy.

Instruments with small elements serve to measure the net radiation at a point, whereas instruments with a series of connected elements exposed in a long horizontal tube give a spatial average of net radiation. Ventilation of the net radiometer is important, and instruments are usually constructed either for complete ventilation or sealed to prevent ventilation. Therefore it is essential to follow manufacturers' instructions for exposure of the instrument. A considerable variety of such instruments is available (see Monteith, 1972).

Measuring the Spectral Distribution of Radiant Energy. Most radiometers are sensitive to a rather broad band (range) of wavelengths. For studying many biological and physical phenomena, however, much narrower bands are of interest. With the simpler types of radiometers, the control of wavelength can be achieved by placing filters over the sensitive element. More sophisticated instruments called spectroradiometers or spectrophotometers are available in which the wavelength is controlled by a dial type selector. The ultimate in this respect is the scanning type of instrument which automatically plots a graph of energy versus wavelength.

5.1.4 Measurement of Nuclear Radiation

Nuclear radiation is assuming ever greater importance in the modern world for several reasons. In view of the extensive publicity given nuclear weaponry, this aspect requires little comment, although less sensational, peaceful uses of atomic energy are fully as important. The use of atomic energy as a power source has also received considerable publicity, both pro and con. Given the increasing demands for energy and the growing scarcity of fossil fuels, there can be little doubt that the production of electricity from atomic energy will assume ever greater proportions.

Power generation through atomic energy requires large concentrations of radioactive materials. Such large concentrations of radioactive materials make environmental contamination from leaks or accidents a very real threat to public safety. Therefore careful monitoring of all operations within such installations and effluents from them is of utmost importance. The constant surveillance and strict safety regulations associated with nuclear installations make large scale environmental contamination from the installation itself extremely unlikely (although not impossible). More difficult to control, and therefore a greater danger, are the transportation of the radioactive materials required by these installations and the disposal of the radioactive wastes that they generate.

Since the radioactive isotopes of an element have the same chemical and physical properties as their stable counterparts, radioactive isotopes can serve as *tracers* for studying chemical, physical, and environmental processes. Radiotracer studies constitute a very important scientific application of nuclear radiation. Likewise nuclear radiation has a host of applications in medicine and industry.

Our focus in this section is on the detection of ionizing radiation. A quantitative measurement of ionizing radiation and the identification of radioactive materials require a much more extensive background than can be provided here. Those interested in the quantitative measurement of radioactivity should consult texts such as Chase and Rabinowitz (1967) or Wang, Willis, and Loveland (1975). Also it should be understood that the handling of radioactive materials is not a matter for the neophyte, and the handling of such materials must be done in accordance with licensing and regulations of the Nuclear Regulatory Commission and state laws.

Gas Ionization Detectors. Thanks to science fiction movies, laymen usually conjure in their minds the clicking of a "Geiger counter" when nuclear radiation is mentioned. In view of this it is natural to consider first the detectors based on the ionization of gases. The Geiger-Muller detector is only one of several in this class of instruments. The sensitive element there is a tube containing charged electrodes with a gas occupying the space between the electrodes. Passage of an alpha particle, a beta particle, or a gamma ray through the tube causes some of the gas molecules to ionize. The negatively and positively charged ions are attracted to the electrodes, causing an electrical current pulse in the circuitry of the detector.

In the simplest of these instruments an electroscope or capacitor is used to provide an initial static potential between oppositely charged elements in the gas filled chamber. The initial static potential may be generated by a charger (either line or battery operated) or even by friction. The detector can be detached from the charger when the charging operation is complete, thus making units of this type highly portable. Ions produced by the direct interaction of radiation with the gas in the chamber cause the static charge to be progressively reduced according to the amount of radiation received. Thus radiation is measured by the amount of the initial charge that has been dissipated. In electroscope models a magnifier permits the user to read the dial at any time, or to observe the rate at which the indicator is moving. In models based entirely on capacitors the detector must be reattached to the charging unit in order to obain a reading. Consequently the latter type can only be used for determining the total dose over a period of time. Pocket size models of either type provide a convenient method of monitoring the amount of radiation to which a person is exposed (dosimeters).

An alternate approach to operating a simple ionization chamber is to use a constant potential across the electrodes rather than a falling potential. In this case the power supply becomes an integral part of the instrument, and radiation is measured by the current that flows when ions are collected on the electrodes as a result of the direct interaction between radiation and the gas filling the chamber. A major difficulty here is that the current pulse produced by radiation induced ionization is very small. Consequently sensitive electronics are required to amplify the weak signal so that it can be registered on readout devices.

Gas amplification can be used as a supplement to electronic amplification in order to reduce the cost of the circuitry. Gas amplification is achieved by increasing the voltage between the electrodes as compared to simple ionization chambers. With higher voltage the ions produced by direct interaction between radiation and the chamber gas move very rapidly toward the electrodes. These rapidly moving ions collide with other gas molecules sufficiently hard to produce secondary ions, which increase the strength of the signal.

Within a certain voltage range the total of primary and secondary ions is proportional to the energy dissipated by the radiation. Instruments operating in this voltage range are therefore called *proportional detectors* or *proportional counters.* Proportional detectors are useful for analyzing the energy characteristics of radiation. Rather expensive electronics are required, however, to maintain the proportionality between ions produced and the output signal.

In many cases interest lies primarily in the number of particles or rays passing through the detector rather than the energy characteristics of the radiation. For these purposes, the cost of the circuitry can be reduced by increasing the voltage to the point where each particle or ray causes essentially complete ionization of the gas in the chamber. This gives a strong signal which can be handled with less expensive circuitry. Ionization chambers operated at these high voltages are called *Geiger-Muller* (*GM*) *counters.* Since each pulse from a GM detector involves essentially complete ionization, the ions must be allowed to recombine into molecules (quenching) before another particle or ray can be detected. Thus GM counters have a short dead time after each pulse during which they are insensitive to radiation. Corrections must be applied for this dead time in order to obtain an accurate count of radiation events. GM counters are the most commonly used type for detecting radiation and measuring its intensity when an analysis of the radiation characteristics is not required.

The type of radiation to be detected also influences the construction of the ionization chambers. As mentioned earlier, alpha particles are not very penetrating and dissipate all of their energy within a short distance of the surface. Ionization chambers are usually constructed with a "window" area through which radiation enters the chamber. The window must be very thin and made of low density material if the chamber is to be used for counting alpha particles. Otherwise the window itself would absorb most of the alpha particles and they would never reach the gas in the chamber. However, the fact that alpha particles dissipate energy rapidly means that they produce a large number of ions per unit length of travel in a substance (high specific ionization). Furthermore alpha particles from a given type of radioactive decay are monoenergetic, which means that they all have the same energy. The high rate of ion production and constant energy characteristics make alpha particles relatively easy to detect once they are inside the ionization chamber.

Beta particles interact with matter less readily than alphas and are not restricted to a single energy level. Since low energy betas are not very penetrating, construction of the ionization chambers is similar for both alphas and low energy betas. Since higher energy betas are more penetrating, heavier windows can be used.

Gamma rays cover an even larger energy range than betas, and are much more penetrating than either alphas or betas. Since the probability of gamma rays interacting with low density materials is rather slight, ionization chambers are relatively inefficient as gamma detectors.

The detection of neutrons is accomplished by means of intermediate nuclear reactions. Thus collision of slow neutrons with boron-10 nuclei in boron trifluoride gas produces alpha particles, and collision of fast neutrons with materials such as paraffin and plastic which contain large amounts of hydrogen produces protons. The secondary alphas or protons resulting from such collisions can be detected by gas ionization.

Scintillation Detectors. In view of the low efficiency of gas ionization for gamma detection, gamma rays are usually monitored through their capacity to produce small flashes of light (called *scintillations*) in certain materials. Sodium iodide crystals are good scintillators and their high density gives a good probability of interaction with gamma rays. Therefore sodium iodide scintillation detectors are commonly used for gamma detection. The small flashes of light are picked up by a light sensitive cathode and magnified by photomultipliers to give a stronger signal which is then transmitted to the circuitry of the counter.

Scintillation can also be used for alphas and betas, but the lower penetrating power requires that the scintillator be in close contact with the radioactive material. This is accomplished by using a liquid scintillation medium and mixing the radioactive sample directly with the scintillation liquid. Liquid scintillation has a great many applications in the biological and physical sciences, but the apparatus tends to be bulky and rather costly.

Semiconductor Detectors. Both gas ionization and scintillation detectors have been in use for quite some time. A newer class of detectors is based on the interaction of radiation with semiconductors such as silicon and germanium containing impurities. The mode of action in these detectors is somewhat analogous to that involved in gas ionization, but radiation produces electrons and "holes" in the crystal lattice of the semiconductor instead of gas phase ions. Without going into further detail we note that semiconductor detectors offer superior sensitivity to radiation, permitting the detection of lower intensities and also possess a greater capability for studying the energy characteristics of the radiation. Although they are still relatively little used for environmental purposes, the importance of semiconductor detectors is certain to increase.

Film Detectors. The simplest and least expensive of all radiation detectors are films similar to x-ray film. When encased in a light tight covering, the exposure of the film by radiation passing through the covering can be used to detect radiation, and the amount of darkening coupled with the length of exposure provides a rough index of radiation intensity. Increasingly heavy coverings can be used to filter out successively higher energy radiation, thus giving information on the nature of the radiation. This approach is widely used in the form of film badges worn by people working in situations where they may be exposed to accidental radiation. Disadvantages of film detectors are their inability to obtain readout prior to developing, their limited ability for quantification, and a very thin sensitive layer. These disadvantages make other detectors preferable in most circumstances.

In closing this brief discussion of nuclear radiation we must again stress the hazards involved in working with radioactive materials and the need for thorough training prior to direct involvement.

5.2 ENVIRONMENTAL AIR CONDITIONING: ATMOSPHERIC CONDITIONS AND AIR QUALITY

Although solar radiation is the basic driving force that generates weather patterns, most of us are more closely attuned to atmospheric conditions of the moment or of the immediate future. Since weather factors have a direct influence on our activity patterns, most people are at least peripherally aware of atmospheric variables such as temperature, humidity, barometric pressure, wind, cloud cover, visibility, and pollution. Likewise there is almost universal acquaintance with the simpler types of thermometers, barometers, and hygrometers. The objective in this section is to extend this casual acquaintance to cover more complex instrumentation and terminology.

5.2.1 Physical Conditions in the Atmosphere

Air temperature is the most familiar physical variable in the atmosphere, so it makes a natural point of departure for considering atmospheric conditions.

Temperature

Temperature Scales. Temperature measures the capacity of materials to transfer heat energy, as determined by molecular kinetic energy. Celsius or centigrade is the preferred scale for expressing temperature measurements. The physical behavior of water provides the basis for this scale, with the freezing point of water being 0°C and the boiling point being 100°C. Most countries have adopted the Celsius scale as a standard, with the United States slowly following suit in conversion from the Fahrenheit scale. However, anyone having the occasion to use past U.S. weather records will continue to need a familiarity with the Fahrenheit scale as well. The freezing and boiling points of water on the Fahrenheit scale lie at 32°F and 212°F, respectively. These points provide the basis for comparison and conversion between the two scales.

The range between freezing and boiling covers 180° on the Fahrenheit scale as opposed to 100° on the Celsius scale. Thus 1 °C is equivalent to 1.8 or 9/5 °F. Conversely 1 °F is equivalent to 5/9 °C. The most straightforward conversion procedure is to begin by expressing the given temperature as a difference from freezing. Celsius temperature is already expressed in this form since its freezing point lies at zero. For the Fahrenheit scale the difference is °F − 32. Next adjust for the difference in the size of a degree. This requires multiplication by 1.8 (or 9/5) in going from Celsius to Fahrenheit, or by 5/9 in going from Fahrenheit to Celsius. The final step is to allow for the difference between the reference point and zero on the scale. Since our reference point is the freezing point of water, we must add 32 °F when going from Celsius to Fahrenheit. Since zero on the Celsius scale coincides with our reference point, no final adjustment is required when going from Fahrenheit to Celsius. Compositing the three steps into a single formula gives:

$$°C = \frac{5}{9} \times (°F - 32)$$

and

$$°F = (1.8 \times °C) + 32 \qquad \text{or} \qquad °F = \left(\frac{9}{5} \times °C\right) + 32$$

However, there is an alternate set of formulas which is easier to remember because the two formulas are more similar. This set is obtained by using as a reference the point at which the two scales give the same reading, which is −40°. The same three-step procedure is used, but this time the temperature is expressed as a difference from −40°. If we let $T°$ stand for the temperature reading on either scale, this difference is

$$T° - (-40°) = T° + 40°$$

This difference is then multiplied by 9/5 in going from Celsius to Fahrenheit, or by 5/9 in going from Fahrenheit to Celsius. To complete the conversion we must adjust for the difference between our reference point and zero degrees on the scale. This consists of subtracting 40°. Thus the conversion formulas become:

$$°C = \frac{5 \times (°F + 40)}{9} - 40$$

and

$$°F = \frac{9 \times (°C + 40)}{5} - 40$$

The only difference between the two formulas in this set is that the multiplier and divisor for the ($T°$ + 40) term are interchanged. A little algebra serves to verify that the two sets of formulas are equivalent since the second set reduces to the first set.

Simple Thermometers. Several types of instruments are available for determining the air temperature. The choice among them will depend on the funds available, the nature of the environment to be monitored, the type of information required, and the need for automatic recording. When readings are to be taken manually and extremely cold conditions will not be encountered, the familiar mercury-in-glass thermometer is suitable and inexpensive. In extremely cold environments alcohol thermometers should be used instead of mercury thermometers. The fineness of graduation on the tube of a simple thermometer is a pretty good indication of the accuracy obtainable with the thermometer. Dial type thermometers activated by a bimetallic strip can also be substituted for simple mercury-in-glass or alcohol-in-glass thermometers. However, the mechanical linkages in dial type thermometers require periodic calibration against other thermometers.

Maximum-Minimum Thermometers. Circumstances often dictate a need to know the maximum and minimum temperatures during the observational period. Specially designed thermometers are available to provide this type of information. One approach to the max–min problem is to use separate thermometers for maximum and minimum observations, as is the case with standard weather instruments.

Thermometers for recording maximum temperature have a narrowing of the tube above the reservoir. The pressure of heat expansion forces the liquid through this constriction up into the tube as the temperature rises, but the liquid does not return to the reservoir as temperature falls. Thus the highest reading is preserved. The liquid is returned to the reservoir by some means such as whirling to create a centrifugal force.

A typical minimum thermometer uses alcohol as a liquid with a light index marker inside the alcohol column resting against the meniscus created by surface tension at the end of the horizontal column. Surface tension draws the marker back as the column contracts with falling temperature, but the marker does not move when the column expands since there is room for the alcohol to flow past it. Thus the end of the marker farthest from the reservoir records the minimum temperature during the period. The thermometer is reset by tilting it so that the marker slides to the end of the column again.

A combination type of maximum–minimum thermometer called the Six's thermometer is also available. The Six's thermometer is a U-shaped tube with reservoir chambers at the upper ends of the arms on the U. The lower portion of the U-shaped tube contains mercury which is relatively heavy in comparison to alcohol. The upper portion of the arms on the U and the left reservoir are filled with alcohol. The right reservoir is empty to receive excess alcohol created by thermal expansion. A spring-loaded marker rides in the alcohol above the mercury in either arm. As the alcohol in the left reservoir expands with rising temperature, it pushes the mercury toward the right side of the tube, and the right marker rides upward on the mercury. As the mercury drops with falling temperature, the right marker remains in place due to the pressure of the spring on the walls of the tube. Thus the lower end of the right marker records the maximum temperature on an upright temperature scale. As the mercury retreats toward the left side of the U, it pushes the left marker upward in the left tube. Therefore the lower end of the left marker records the minimum temperature on an inverted temperature scale. A magnet is used to reset the thermometer. Each marker contains a metal pin, so the magnet can be used to draw it down against the surface of the mercury.

Recording Thermometers. When the instrument cannot be visited daily to take readings or when short term fluctuations of temperature are important, it will be necessary to use a recording thermometer. Moderately priced recording thermometers called thermographs consist of a metallic temperature sensitive element, a chart wrapped around a drum rotated by a mechanical clockspring, and a pen arm to provide the trace. Most of these instruments are designed to be visited once a week for rewinding the clockspring and changing the chart. The sensitivity of such termographs is on the order of 2°F or 1°C. For accuracy better than this one must resort to an instrument which operates electrically, with consequent need for a power source. Three major types of electrical thermometers are available.

The electrical resistance of wires made of pure metals such as platinum or nickel increases quite linearly with temperature, thus providing the basis for one type of electrically activated thermometer. *Wire resistance thermometers* are very accurate and stable, but both the sensitive element and the recorder are quite expensive.

Thermistors also operate on the basis of changing resistance with temperature, but the construction and properties are quite different. A thermistor is a small piece of semiconductor material with wire leads imbedded. Depending on the thermistor, the resistance may either increase or decrease with temperature. In either case the unit change in resistance is quite large, but the amount of change depends on the temperature (nonlinear). Thermistors are quite portable and economical since electrical current requirements are small and the cost of readout equipment is modest compared to that for wire resistance thermometers. However, frequent calibration is needed and the nonlinearity can be a problem.

Thermocouples constitute the third type of electrical temperature sensor. As described previously in connection with pyranometers, thermocouples consist of an electrical circuit formed by two different kinds of wires. The points where the two wires join are called the junctions. When the two junctions have different temperatures, a current flows in the circuit. The current strength depends on the temperature difference between the junctions. If one junction is maintained at a known temperature, the temperature of the other junction can be determined from the current. The thermocouple itself is inexpensive, and the response is essentially linear with temperature. However, thermocouples are not highly sensitive compared to wire resistance sensors, the recorder cost is high because of the need for amplification, and the need for a temperature reference is bothersome.

With electrical sensors the sophistication of the recording system depends only on the funds available. Several sensors can be fed into a multichannel data logger. The output can be obtained on a computer readable medium or fed directly into computers for analysis. It is even possible to use radio transmission to remote receiving sites. Perhaps the ultimate is radio transmission to a satellite with subsequent relay to ground receiving stations which would be out of the radio range for the sensor itself.

Exposure of Thermometers. Thermometers should never be exposed to direct rays of the sun since the reading would be influenced by direct solar heating. Whenever possible thermometers should be housed in a suitable shelter. The shelter should be painted white to reflect the sun's rays and should have louvered sides to permit ventilation. Standard instrument shelters of the type used by the National Weather Service are available commercially. Alternatively serviceable shelters can be constructed along the lines described by Fraser (1961).

Temperature in Microenvironments. Monitoring the temperature in microenvironments usually calls for miniature sensors since conventional thermometers and thermographs would either be affected by conditions outside the microenvironment or alter the microenvironment itself. Wire resistance thermometers, thermistors, and thermocouples can all be miniaturized in varying degrees to meet the needs of monitoring in microenvironments. Thermistors, in particular, can be made very small with extremely fine wire leads and are available in a variety of shapes. A further generalization with respect to the instrumentation for monitoring temperature in microenvironments is difficult, and considerable ingenuity may be required to fabricate suitable sensors and shields.

Humidity. Humidity and temperature are related to such an extent that temperature measurements can be used to determine the humidity. Consequently humidity makes an appropriate sequel to temperature in this consideration of atmospheric conditions.

Ways of Expressing Humidity. The amount of water vapor in a unit volume of air is called *absolute humidity.* Absolute humidity can be expressed either as weight per unit volume or in terms of vapor pressure. The maximum amount of water vapor that a unit volume of air can hold (ignoring the possibility of supersaturation) is controlled by temperature. *Relative humidity* is absolute humidity expressed as a percentage of the maximum for the given temperature. *Saturation deficit* is the difference between maximum and absolute humidity for the given temperature, usually expressed in vapor pressure form. *Dew point* is the temperature at which the given absolute humidity would be maximum or at which the moisture would be on the verge of condensing from the air. *Specific humidity* is the weight of the water vapor per unit weight of moist air, and *mixing ratio* is the weight of the water vapor per unit weight of dry air.

The terms relative humidity and dew point should be familiar from their use in weather reporting. The term saturation deficit is used quite frequently by ecologists and micrometeorologists. Use of the terms absolute humidity, specific humidity, and mixing ratio is restricted mostly to meteorologists. The mathematical relationships among these various expressions of humidity will not be given here, but the formulas are given by Rosenberg (1974). Likewise most meteorology books give the relationships in either tabular or graphical form, called *psychrometric tables* and *psychrometric charts,* respectively.

Humidity Measurement by Psychrometers. Instruments for determining the humidity are collectively called *hygrometers.* Hygrometers can be constructed around several different operating principles. As stated in opening this discussion of humidity, it is possible to determine the humidity from temperature measurements. This is the psychrometric principle, and instruments based on it are called *psychrometers.* A psychrometer consists of two temperature sensors, one of which is dry and the other covered by a moist membrane. Evaporation from the moist membrane causes this sensor to register a lower temperature than the dry sensor. The difference in readings between the two sensors is called the wet-bulb depression. Assuming that the sensors are adequately ventilated, the wet-bulb depression is a measure of the saturation deficit. Given the temperature registered by the dry sensor and the wet-bulb depression, the absolute humidity, relative humidity, dew point, and saturation deficit can be read from psychrometric charts or tables.

The simplest and least expensive type of psychrometer is the *sling psychrometer.* This consists of two mercury-in-glass thermometers mounted in such a way that the whole unit can be whirled rapidly for ventilation. The bulb of one thermometer is covered with moist muslin to provide the wet sensor (or wet-bulb) reading. A series of readings should be taken with whirling between readings to ensure that the wet-bulb reading was stabilized. In some situations there may not be room to whirl a sling psychrometer. Psychrometers that are ventilated by a squeeze bulb are available for use in these circumstances. Psychrometers can be miniaturized and made to drive recorders by using electrical types of temperature sensors with fans for ventilation. There is a limit on the use of psychrometers in microenvironments, however, because the ventilation and evaporation from the wet sensor can alter the microenvironment itself. In these circumstances it may be necessary to accept the lesser accuracy of hygrometers based on hygroscopic elements as described below.

Hygrometers Based on Hygroscopic Elements. Another approach to the construction of hygrometers is to build them around fibers or membranes which absorb moisture from the air and alter their physical properties in the process. A good example is blond human hair which changes length with the moisture content. This is used in constructing hygrographs which operate in the same manner as thermographs. In fact the hygrograph and thermograph are often combined into a single unit called a hygrothermograph or hytherograph. Hygrothermograph charts are divided into two sections with one pen recording on the temperature section and the other recording on the humidity section. Dial type hygrometers are constructed from hygroscopic elements in a similar fashion.

Infrared Gas Analyzer. The absorption of infrared radiation by water vapor provides the basis for a sophisticated approach to measuring the moisture content of air. The infrared gas analyzer operates by generating infrared radiation which is passed through sample and reference chambers. The difference in the strength of radiation leaving the two chambers is measured by a detector and translated into the moisture content of the sample. Infrared gas analyzers are used primarily for laboratory measurements. They are also widely used for determining the carbon dioxide content of the air.

Atmospheric Pressure. Temperature and humidity affect the density of air. Air expands with rising temperature, thus becoming less dense. Water vapor is lighter than dry air, so increasing humidity reduces the air density for a given temperature. Air of low density exerts less pressure than air of high density. However, the air pressure at a given location depends not only on temperature and humidity, but also on the total amount of air in the atmosphere above. Other things being equal, the amount of air above a point on the earth's surface decreases with increasing elevation. The atmospheric pressure situation is also complicated by weather patterns. Weather patterns cause air to pile up in some places and thin out in other places, thus creating high pressure cells and low pressure cells. The atmospheric pressure observed at a given place and time is a composite of all these effects. Because of the relation to weather patterns, air pressure is an important atmospheric variable. The approach of storms tends to be associated with falling atmospheric pressure, and clearing conditions with rising atmospheric pressure.

Expressing Atmospheric Pressure. For scientific purposes atmospheric pressure is expressed in *millibars,* with 1 millibar being 1000 dynes/sq cm. However, other units are also in common use. The oldest of these is to give the height of the mercury column in a mercurial barometer. This height may be stated in inches, centimeters, or millimeters. The nature of the mercurial barometer is explained below. 1 millibar is equivalent to 0.02953 in. of mercury or 0.7501 mm of mercury. The atmospheric pressure may also be stated in pounds per square inch (psi).

The normal practice is to apply corrections for temperature and elevation so that readings taken at different locations and different times will be comparable. The standard temperature is taken as 0°C. Pressure changes by 1/273 of its value at 0°C for each 1°C change in temperature, with the direction of change being the same as that of temperature (assuming constant volume). Given the current pressure p and temperature t°C readings, the equivalent pressure P_0 at 0°C is given by the formula

$$P_0 = \frac{p}{1 + (t/273)}$$

(Donn, 1975).

If two stations differ in elevation by 900 ft (274 m), the pressure at the higher station will be about 1/30 less than that of the lower station (Donn, 1975). This gives an exponential correction formula which is usually handled in table form. Platt and Griffiths (1964) give the approximation of a 3 millibar decrease per 100 ft as being satisfactory for most purposes in environmental monitoring. The mean sea level is used as the standard elevation. The term *station pressure* is used for readings that have not been corrected to mean sea level.

Mercurial Barometers. Instruments for determining the atmospheric pressure are called *barometers.* Two types of barometers, known as *mercurial* and *aneroid,* are in common use. The mercurial barometer consists of a long tube with one open end which is filled with mercury. It is then inverted with its open end below the surface in a small tub of mercury. The level of mercury falls in the tube until the pressure of the air on the mercury in the tub balances the weight of the column in the tube. This normally occurs with a mercury column about 760 mm high. Because of their accuracy and stability, mercurial barometers serve as the standard against which others are compared. However, they are bulky, fragile, manually read, and the readings require more corrections than those of aneroid barometers. A correction must be applied for expansion and contraction of the mercury with temperature. Also, a gravitational correction is required which depends on latitude. Because of these drawbacks aneroid barometers are used more commonly in environmental monitoring.

Aneroid Barometers. The basic structure of the aneroid barometer is a hollow box with air pumped out to form a partial vacuum. Springs inside the box prevent its collapse. Changing air pressure changes the volume of the box, and this change is transmitted to an indicator by mechanical linkages.

Most aneroid barometers automatically compensate for temperature changes. Any systematic error in the instrument can be corrected by a one-time shift in the position of the indicator. The approximate correction for altitude suggested by Platt and Griffiths can be made in the same manner. Recording instruments called *barographs* are constructed by attaching the indicator linkage directly to a pen arm. The portability and convenience of the aneroid makes it the best choice for most applications in environmental monitoring.

Wind. Air moves from areas of high pressure to areas of lower pressure, thus generating winds. However, the rotation of the earth (coreolis force) modifies the original movement along pressure gradients. Likewise one should not lose sight of the role that temperature plays in generating pressure gradients. Heated air expands, becomes less dense, and rises. Cooler, heavier air then flows into the low pressure area left below the rising warm air. Since it is a motion phenomenon, wind involves both direction and speed. The direction is described in terms of azimuths or compass points. The rate of movement is described in terms of either speed or wind run. In the latter approach a counter is activated each time a predetermined amount of movement takes place. For example, the counter might be activated once for each 100 m of air movement.

Wind Direction. The wind direction is determined by either a wind vane or a wind sock in which a fin or streamer orients itself on the downwind side of its pivot. The readout can be either visual, mechanical, or electrical, with visual and electrical being the most common. The electrical readout is accomplished by placing a series of fixed contacts radially around a pivoting contact. The accuracy of the readout is determined by the number of fixed radial contacts. When a contact closes as the shaft pivots, an electrical signal activates a recorder or indicator.

Cup Anemometers. A cup anemometer consists of a set of cups rotating in a horizontal plane about a vertical shaft. Each cup is mounted on the end of an arm, with the arms attached to the collar like spokes of a wheel. The wind catches the open side of a cup causing the collar to rotate. The reverse side of the cup is rounded so that it offers much less wind resistance than the open side. The net result is that the cup assembly rotates in proportion to wind speed. Cup anemometers are equally sensitive to wind blowing from any direction, but very low wind speeds do not create enough force to overcome the friction of the pivot. Low friction pivots are necessary to register light winds accurately, and the starting speed provides one measure of quality for anemometers. Likewise the inertia of the cup assembly will cause some inaccuracy in gusty winds. The shaft must be mounted vertically, and a side-arm support should be used to hold the anemometer away from a tower that is large enough to create appreciable turbulence. The readout may be activated either mechanically or electrically. Cup anemometers are used more commonly than other types.

Propeller Anemometers. Propellers constitute a second type of rotating anemometer. The propeller rotates in a vertical plane on a horizontal shaft. However, propellers must be equipped with a wind vane to keep them facing into the wind. The necessity of adjusting to the wind direction causes some inaccuracy when the wind changes directions rapidly. Otherwise considerations for propeller anemometers are similar to those for cup anemometers.

Heat Dissipating Anemometers. Air motion increases the rate of cooling in materials which are warmer than the air that surrounds them. This principle can be used to overcome the starting speed limitations associated with rotating anemometers. The electrical current necessary to maintain a heated element at a fixed temperature, or the temperature of an element heated by a constant current can be used to measure wind speed. Therefore heat dissipating anemometers are effective for measuring very low wind speeds. Since the heated element can be very small, this type of anemometer is also effective for microenvironments which would not accommodate a rotating anemometer. However, the necessity of providing a power source may be a drawback for some applications. Also a thermometer must be used to adjust for differences in cooling due to air temperature.

Pressure Tube and Pressure Plate Anemometers. The pressure exerted by air movement can be used in several ways to gauge the wind speed. One approach is to use a pitot tube or manometer. In the former the wind pressure lifts a floating ball in the tube. In the latter the wind pressure moves a liquid in a U-shaped tube. Another approach is to use a pressure plate which swings or moves against a spring under wind pressure. Since instruments of this type are unidirectional, the pressure principle is most often used in constructing inexpensive hand held anemometers with relatively low accuracy.

Sonic Anemometers. Changes in the rate of sound wave propagation with air movement provide the basis for the new and sophisticated innovation of sonic anemometers. Since the expense of these instruments is not likely to be justified for general environmental monitoring, we simply note their existence in passing.

Beaufort Wind Scale. When anemometers are not available, the wind speed can be judged roughly by its effect on objects, smoke, and water surfaces. The Beaufort wind scale was devised long ago for making subjective judgments regarding wind speed. This scale consists of 18 rating numbers ranging from 0 for calm air through 17 for winds of intense hurricane force. See Donn (1975) or other meteorology texts for Beaufort rating criteria on land and sea along with the translation of the Beaufort numbers to approximate ranges of wind velocity.

Lapse Rates. All the atmospheric variables discussed in the preceding paragraphs change vertically as well as horizontally. The vertical change in an atmospheric variable is called its *lapse rate.* Lapse rates are of utmost importance for the analysis and forecasting of weather patterns. The vertical profile existing at a given time is compared to a hypothetical standard atmospheric profile in order to draw inferences regarding weather conditions. In order to obtain information on atmospheric profiles, however, one is forced to rely on data collected by the National Weather Service. Most of this information is obtained from radiosonde transmitters carried aloft by balloons which rise at a known rate. This information is supplemented by data gathered from a network of weather satellites. Space does not permit a discussion of methods for an analysis and interpretation of lapse rates, and the interested reader is referred to one of the many meteorology texts available.

Cloud Cover. To the knowledgeable observer clouds provide important clues to approaching weather conditions. This type of insight, however, requires considerable observational practice and much better knowledge of cloud taxonomy than space allows us to impart. All that we can hope to accomplish here is to give a brief introduction which may serve to stimulate further study.

Extent of Cloud Coverage. The amount of cloud coverage is usually reported as tenths of the sky obscured by clouds of all types taken together. This classification by tenths is done visually, but some experience is required to obtain proficiency. In particular, perspective effects are likely to cause the inexperienced observer to overestimate the density of clouds near the horizon. The classification by tenths is supplemented by verbal descriptions and notes on cloud types.

Cloud Taxonomy. There are three basic cloud types, with major variations on these three types giving rise to ten major categories. These ten categories are then subdivided into many different cloud types. Only the ten major categories will be considered here, and these only briefly. The three basic cloud types are cirrus, cumulus, and stratus. *Cirrus* clouds are thin, wispy, high clouds composed of ice crystals. *Cumulus* clouds are isolated, fluffy clouds composed of minute water droplets. *Stratus* clouds form a more or less continuous layer. Like cumulus clouds, stratus clouds are also composed of minute water droplets.

The usual cumulus and stratus clouds are dense, low clouds. Cumulus and stratus clouds forming at higher altitudes appear somewhat thinner and are designated by the prefix *alto* to indicate their altitude. Thus we have the middle altitude categories *altocumulus* and *altostratus.* Low altitude cumulus clouds occurring so close together that they form an almost continuous layer are termed *stratocumulus.* Cirrus clouds that form a nearly continuous layer are termed *cirrostratus,* and bunchy cirrus clouds are termed *cirrocumulus.* All cirrus clouds are high altitude phenomena. The remaining two categories have *nimbus* or *nimbo* in their names, implying that they produce rain. Middle altitude stratus clouds producing rain are termed *nimbostratus.* The other category is the thunderhead with its base at low altitude but extending up a considerable distance vertically. Such thunderheads are called *cumulonimbus* clouds. Each of the major cloud types has an associated symbol for use on weather maps. For these symbols and the many subtypes of clouds within the major categories, the reader is referred to the *International Cloud Atlas* sponsored by the World Meteorological Organization (WMO) or to various publications of the National Weather Service.

Cloud Heights. As the above discussion suggests, the form of a cloud gives some information regarding the altitude at which it occurs. Further information on cloud heights is obtained from radar observations, satellite photos, and reports from aircraft. The height of the first cloud layer present at a given time can also be determined by observation from the ground. One way is to release a pilot balloon that rises at a known rate and time its disappearance into the base of the clouds. At night a strong light called a *ceiling light* can be beamed upward to the base of the clouds. A clinometer is then positioned at a known distance from the light and used to measure the angle to the point where the beam hits the cloud base. A more automated device called a *ceilometer* is operable either day or night. The ceilometer is based on an ultraviolet beam and a photoelectric scanner (Donn, 1975).

5.2.2 Weather Patterns

The intensity of solar heating varies greatly over the surface of the earth and is much greater in equatorial regions than in polar regions but shifting with the seasons. This differential heating sets up patterns of air circulation over the globe, thus giving rise to weather. Generalized descriptions of global circulation patterns can be developed on the basis of theory and extended observation, and these overall patterns are described in meteorology texts. However, variation is the rule rather than the exception, so that weather forecasts for a given locality can look only briefly into the future with much hope of accuracy. *Climate* is weather averaged over time periods long enough to reveal major differences between regions.

Air Masses. Depending on the existing pattern of global circulation, the air over a rather large area may remain more or less stationary for a period of time. When this happens, the air over the region is conditioned by the solar energy and moisture available in that region. With respect to available solar energy, regions may be characterized as equatorial (E), tropical (T), arctic (A), and polar (P), according to latitude and available solar energy. Ocean regions where there is an abundance of moisture are described as maritime (m), and large land areas having less moisture are described as continental (c). An air mass may be described according to the region that has conditioned it as continental polar (cP), maritime tropical (mT), and so on. Changing patterns of circulation will eventually cause the air mass to move out of the source region where it was conditioned, but the effect of the conditioning dominates the weather in and under the air mass for a considerable period of time. Thus a continental polar air mass moving into lower latitudes brings cool, dry weather.

The weather conditions generated within a moving air mass are also affected by the relative temperatures of the air and land surface. A third letter is often added to the air mass designation to indicate this relative temperature situation. The letter w is used to indicate that the air is warmer than the terrain, and the letter k is used to indicate that the air is cooler than the terrain. Warm air moving over cool terrain gives rise to generally stable conditions with little vertical air movement and stratiform clouds. Conversely cooler air moving over warmer terrain tends to generate vertical turbulence as the air in contact with the ground is heated and rises.

Fronts. When an air mass moves under the influence of high and low pressure cells, it displaces the air mass that previously occupied the area. The transition layer along the zone of contact between air masses is typically so narrow that the term *front* is used to describe the interface. If the colder of the two air masses is advancing and the warmer retreating, the interface is called a *cold front.* The term *warm front* describes the reverse situation where the warmer of the two air masses is advancing and the colder retreating. If the interface between the two air masses remains stationary, it is called a *stationary front.*

Frontal movement tends to generate precipitation, but the duration and intensity of the precipitation depends on the type of front. Warm air is lighter than cold air. With a warm front the advancing warm air rides up over the cool air along an interface which slopes gently upward in the direction of frontal movement. Cooling of the warm air with this gradual rise produces slow, steady precipitation which continues for an extended period. With a cold front the interface is more nearly vertical. The warm air in front of the advancing cold air is forced upward rapidly, producing more intense precipitation of shorter duration.

Two other frontal phenomena are also important. One of these develops as a series of traveling waves along a larger front. The leading edge and trailing edge of the wave constitute small fronts. The trailing edge of the wave often overtakes the leading edge forming an *occluded front* in which the main interface between warm and cold air does not extend down to the ground surface but occurs at higher altitude.

The other is a somewhat similar situation on a larger scale in which an upper air mass travels over a much colder and heavier air mass lying near the surface. The leading edge of the upper air mass is called an *upper front.* Although the upper front does not extend down to the surface, it will affect surface weather conditions. Depending on the nature of the upper air interface and its direction of travel, either an upper cold front or an upper warm front can occur. Map symbols used to describe frontal conditions are shown in Figure 5.1.

Weather Maps. The National Weather Service (a division of NOAA) prepares maps showing various aspects of surface weather and upper air conditions.

Surface Weather Maps. Composite weather maps contain a great deal of information. The barometric pressure corrected for temperature and elevation is shown by isobars (lines of equal pressure) using solid lines. In addition to the isobars, high pressure centers are indicated by HIGH or H and low pressure centers by LOW or L. The position and movement of air masses is shown by frontal symbols and code letters describing the nature of the air mass. Dashed or dash-dot lines are used for key isotherms (lines of equal temperature) such as the freezing isotherm. Areas of widespread precipitation are indicated by shading or similar means. Detailed information is given for principal reporting stations through an elaborate station code. A detailed discussion of this station reporting code would require several pages, and will not be given because the code is readily available in publications of the National Weather Service. The general features of the station code are as follows.

Wind is shown by an arrow pointing in the direction that the wind is blowing, and the feathers on the arrow show the wind speed. Clouds and the nature of precipitation are shown by symbols. A variety of other information is shown with numbers differentiated by their positions with respect to a central circle.

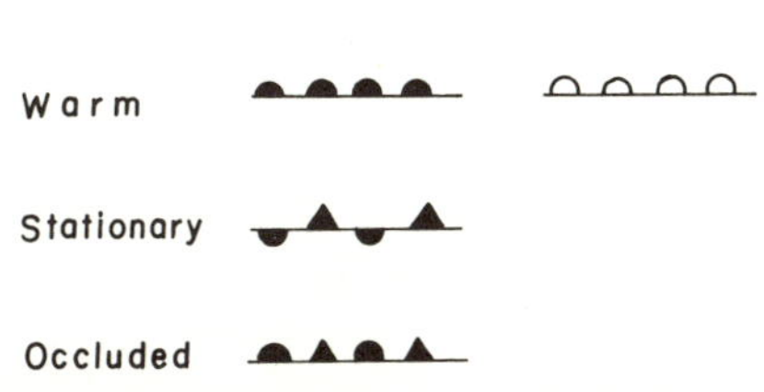

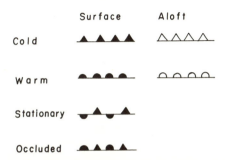

Figure 5.1 Weather map symbols used to describe frontal conditions.

In addition to the composite surface weather map, special maps are also prepared showing such information as max/min temperature and the amount of precipitation.

Upper Air Maps. As with surface conditions, maps are available showing various aspects of the upper atmosphere. The most widely distributed of these is the 500 millibar map. This map contains solid elevation contour lines showing the altitude at which the barometric pressure is 500 millibars. These contour lines define a constant pressure surface. Other information shown for the constant pressure surface includes temperature isotherms (dashed lines) and wind arrows. Similar maps are also prepared for other constant pressure surfaces. Other maps are available showing particular aspects of the upper atmosphere such as the wind speed at various altitudes.

Weather Records. Current weather forecasts and reports are distributed by the National Weather Service over several teletype and facsimile networks. The reporting interval depends on the type of report and the nature of the receiving station. Persons having a need of such information should contact the nearest office of the National Weather Service to determine the local availability. Special reports and forecasts are also prepared to meet the needs of particular user groups such as aviation and agriculture. Information on the availability of these special services can also be obtained from National Weather Service offices.

Needs for past records rather than current data can usually be satisfied from one of several serial publications produced by the NOAA. Haines (1977) has prepared a concise summary of sources for information on weather records and climatic data. His report makes an excellent starting point for locating such information. A NOAA monthly serial entitled "Local Climatological Data" provides detailed records on a local basis. For larger areas of interest the NOAA monthly serial entitled "Climatological Data" will probably satisfy most needs. For information on a national basis the appropriate NOAA publication is "Climatological Data, National Summary." If the needs are not covered by these publications or others mentioned by Haines (1977), inquiries should be directed to the National Climatic Center, Federal Building, Asheville, N.C. 28801.

5.2.3 Air Quality

The primary constituents of the air are nitrogen, oxygen, argon, carbon dioxide, and water vapor with nitrogen accounting for approximately 78% and oxygen about 21%. Nitrogen, oxygen, carbon dioxide, and water vapor play central roles in the functioning of ecosystems, and argon is inert. The quantities of nitrogen, oxygen, carbon dioxide, and argon in the atmosphere are quite stable over time, but the water vapor content fluctuates according to temperature and the surface moisture supply as indicated previously in discussing humidity. Materials other than these five gases make up only a fraction of 1%, but this small fraction is important in ecosystem dynamics because of the noxious effects associated with several of the many substances that may be included. Both natural processes and human activities contribute pollutants to the air, but the major threat comes from the byproducts of human activities.

Air pollutants occur either as gases or as tiny particles. Particulates include both solid fragments and droplets of liquids. The larger particulates are heavy enough to settle under the force of gravity when the turbulence which carried them aloft subsides, whereas the smaller particulates remain suspended under normal air motion. Gases and small particulate aerosols are of greatest concern as direct health hazards since they penetrate deep into the respiratory tract. However, particulates that settle are also important since they create dust, impair visibility, corrode the surfaces on which they settle, and dissolve in surface water.

Common inorganic pollutants of major concern are oxides of nitrogen and sulfur. Common organic pollutants of major concern are nonmethane hydrocarbons. Sulfur oxides are converted to sulfuric acid which is highly caustic. Oxides of nitrogen and nonmethane hydrocarbons are involved in photochemical oxidation. This is a process whereby the ultraviolet energy in sunlight powers chemical reactions between pollutants in the air. The resulting chemical products such as ozone and PAN (peroxyacyl nitrate) are injurious to both plants and animals. Carbon monoxide is a direct respiratory poison. A whole host of other pollutants may present problems on a more local basis depending on the industrial complex.

Several texts and handbooks provide a good introduction to surveys and analyses of air pollutants. The Scientist's Institute for Public Information sponsors a workbook on air pollution authored by Nadler et al. (1970) which is designed to acquaint interested citizens with the nature and scope of the air pollution problem. Warner's (1976) book on the analysis of air pollutants provides a good introduction to the techniques of air sampling and methods of analyzing pollutants. Several books edited by Stern give technical details of methods and procedures; and publications of the National Air Pollution Control Administration of the Environmental Protection Agency describe air quality criteria, analytical methods, and air quality data. Since these references are already available, we will not attempt to duplicate their content here. We shall be content to provide an overview that will help create a context for further reading in these and other references listed in the bibliography at the end of the chapter.

Fortunately an analysis of air pollutants is a more straightforward task than a chemical analysis of soils. Samples must be collected, materials separated, and quantities determined. The quantities determined translate directly to the pollution load. However, pollution levels are quite variable over time, which makes sampling intervals an important consideration. Also the seriousness of emissions at a given rate depends on the weather conditions. With normal mixing of air, pollutants are diluted rapidly and do not reach high concentrations over large areas. However, stable air conditions associated with temperature inversions prevent the vertical movement of air, causing pollution levels to build up rapidly. For the remainder of this brief discussion it will be convenient to consider particulates and gases as separate categories.

Sampling and Analysis of Airborne Particulates. Particulate pollutants large enough to settle out of calm air under their own weight can be sampled very simply and inexpensively. A bucket shaped dustfall collector is placed on top of a post to catch the particulate matter as it settles. Given the area of the bucket's top and the weight of the material collected, the load of settled particulate pollutants is easily expressed in units of weight per unit area per unit time. A small amount of antifreeze may be added to the dustfall collector to help capture the particles and reduce problems with ice and snow. Likewise a ring somewhat larger than the rim of the bucket will serve as a perch for birds so that they do not alight on the rim itself and foul the contents. The material collected can be analyzed chemically and microscopically to determine its composition and origin. *The Particle Atlas* (McCrone, Drafty, and Delly, 1967) is an excellent reference for determining the nature of particulate pollutants from their appearance under a microscope. Local pollution sources can also be sampled at the point of origin and compared microscopically to the contents of dustfall collectors. Microchemical tests and X-ray diffraction provide more sophisticated means of analyzing the contents of dustfall collectors.

Particulate pollutants that are too small to settle under their own weight can be sampled by drawing air through filters with vacuum pumps. In order to obtain quantitative expressions of the pollution load, the volume of air drawn through the filter must be determined by calibrating the pump with air flow meters. Large particulates that would be collected in dustfall collectors can be excluded from the filter sample by causing the air stream to make a sharp bend before reaching the filter. The inertia of the larger particles will prevent them from being carried around the bend and into the filter.

The pollution load of suspended particulates can be expressed most simply on a weight per volume basis such as micrograms per cubic meter of air. However, other ways of expressing the pollution load are also used. For example, the coefficient of haze (COH) is based on optical effects of the pollutants. The percent light transmission of the soiled filter as compared to a clean filter is measured with a densitometer. The optical density of the soiled filter is then calculated as

$$\text{optical density} = \log_{10}\left(\frac{100}{\%\ \text{transmission}}\right)$$

The number of COH units is 100 times the optical density. The number of COH units per 1000 linear feet of air drawn through the filter provides a common index of the pollution load. The linear units of air sampled are given by

$$\text{linear units of air sampled} = \frac{\text{volume sampled}}{\text{area of filter orifice}}$$

Impactors may replace the filter when the purpose is to study the size distribution of the particles. When an air stream is directed against a surface, the inertia and the sudden change of direction will cause particulates to be removed from the stream according to their size and weight and the velocity of the air stream. Particle size separations can be performed by varying the velocity of the air stream as it strikes a series of collecting surfaces.

Sampling and Analysis of Gaseous Pollutants. Most sampling of gaseous air pollutants is done with a variation of the pump and filter approach used for particulates. A container of absorbing solution replaces the filter, and pollutants react with the chemicals in the absorbing solution as air bubbles through it. The absorbing solution must be formulated specifically for the pollutant under study. In order for most of the pollutant to be absorbed into the solution, the air stream must be broken into very fine bubbles. Fritted glass bubblers and impingers are the most common devices for breaking the air stream into fine bubbles. Fritted glass bubblers are familiar to most people because of their use in aerating aquariums. In an impinger a jet of air is directed against the bottom or wall of the container, causing the stream to break up into bubbles. Conventional chemical methods are used to determine the amount of pollutant collected in the absorbing solution. However, the amount of pollutant collected is usually very small, unless the sampling times are long or the pollution loads are heavy. Therefore microchemical analysis is often necessary. The results are expressed either on a volume basis (parts/million) or on a weight per volume basis (micrograms/cubic meter).

Although most sampling of gaseous air pollutants is done in the manner outlined above, other methods are also used. It is possible to use a finely divided solid absorbing medium instead of an absorbing solution. Examples of solid absorbers are Hopcalite for carbon monoxide and Ascarite for carbon dioxide. In some cases low temperatures can be used to induce the condensation of gases into liquids. It is also possible to fill an empty flexible bag with the air to be analyzed (grab sample), and then do a gas phase analysis using methods such as infrared gas analyzers or gas–liquid chromatography.

Air pollution monitoring must be more or less continuous in order to detect high levels of pollution occurring over short periods of time. Likewise sampling must be guided by the location of likely sources of emission. Otherwise localized occurrences of high pollution levels might escape detection. The importance of coupling air quality monitoring with information on weather conditions has already been noted. Characteristic stress symptoms appearing on plants exposed to air pollution can also be used for detection and control operations.

5.3 THE ENVIRONMENTAL PLUMBING SYSTEM: HYDROLOGY

Despite the absence of pipes and valves, water circulates through the environment in an orderly and continually repeated pattern called the *hydrologic cycle.* Solar energy and weather factors discussed earlier in this chapter control the portion of the hydrologic cycle that takes place in the atmosphere. Topography, soils, and geology discussed in the previous chapter are major determinants for the portion of the cycle that takes place on and under the earth's surface. Vegetation and animal life (particularly humans) can alter both the pattern of circulation and the water quality. A brief examination of the hydrologic cycle will help to establish a perspective for monitoring the flow of water through ecosystems.

5.3.1 The Hydrologic Cycle

The hydrologic cycle is powered by the opposing forces of evaporation from solar heating and the downward pull of earth's gravity.

Solar heating evaporates water from moist soil, open water surfaces, and the leaves of green plants. Water vapor is light relative to other components of the air, so it moves upward in the atmosphere. The occurrence and measurement of water vapor as humidity have already been covered in previous sections of this chapter. As air rises it expands and cools. Eventually the rising air cools to the dew point, and water vapor begins to condense into tiny droplets or ice crystals forming clouds. As the droplets or crystals grow in size, the force of gravity eventually pulls them to earth as precipitation (rain, snow, hail, and so on).

If the precipitation falls as snow, it is temporarily stored on the ground surface until melting again makes it subject to the pull of gravity. Rain or melted snow moves downward under the force of gravity along one of two paths. Some of the moisture infiltrates into the soil. Part of this moisture is held on the surface of soil particles. Excess moisture in the soil that cannot be held against the pull of gravity percolates downward as groundwater until it meets geological materials that are less permeable, at which point it may begin to flow laterally underground along the sloping surface of impervious strata. If the rate of precipitation or snow melt exceeds the infiltration rate, the excess flows downslope as surface runoff. The overland flow collects in small streams which join larger streams, which eventually empty into lakes or oceans. Groundwater moving laterally as subsurface flow may also reach streams and rejoin waters that have moved by overland routes.

Snow packs, soil moisture, groundwater, rivers, lakes, and oceans serve as storage reservoirs for water that is available for use in ecosystems. Except for groundwater, some of this stored water moves directly back into the atmosphere by evaporation from exposed surfaces. Soil moisture is absorbed through plant roots and drawn up into the foliage. Large quantities of this water then escape from the foliage into the atmosphere through the process of *transpiration.* Animals drink water from ponds, lakes, rivers, and so on. Part of this water is exhaled directly into the atmosphere and part is excreted as a carrier for metabolic waste products in places where it evaporates rapidly. Man is the only life form that extracts large quantities of water from deep underground storage. This water is eventually returned to the surface where it is subject to evaporation back into the atmosphere. Evaporation and transpiration thus serve as the primary mechanisms by which water is returned to the atmosphere to complete the cycle. The combination of evaporation and transpiration is often called *evapotranspiration.*

The remaining sections of this chapter are devoted to monitoring major phases of the hydrologic cycle and water quality.

5.3.2 Evaporation and Transpiration

We will enter the hydrologic cycle at the point where water is carried up into the atmosphere by evaporation and transpiration. The evaporating power of the air is a composite function of temperature, humidity, and air movement. Although these three factors can be measured quite accurately with instruments already discussed, it is not a simple matter to integrate them mathematically into a composite index of evaporating power. This has led to attempts at finding ways to integrate them physically by measuring evaporation directly. A major difficulty is that evaporation also depends on the nature of the exposed moist surface. Because of this, present approaches tend to be more qualitative than quantitative, but may nevertheless provide useful comparative information.

Evaporation Pans. The National Weather Service uses a standard (Class A) evaporation pan to provide an index of evaporating power. This type of pan is 4 ft in diameter and 10 in. deep. The pan is mounted with its bottom 6 in. above the ground so that air can circulate beneath. The water level in the pan is maintained between 2 and 3 in. below the rim. The change in water level is measured with a hook gauge.

A hook gauge consists of a graduated rod which is bent into a J shape at the lower end, with the free end of the J being sharpened into a point. This hooked rod can be raised and lowered along a support. The support also bears an index mark. The hooked portion of the graduated rod is lowered beneath the water surface and then raised until the point just forms a dimple in the surface but does not break the surface tension. The reading is taken at the index mark on the support. A tubular stilling well around the hook gauge is used to prevent ripples in the surface.

A rain gauge near the pan is used to measure precipitation so that an allowance can be made for water added to the pan from this source. The installation also includes an anemometer and sheltered thermometers. Correction factors or coefficients can be developed which relate evaporation from the pan to evaporation from other surfaces. For example, evaporation from an open lake surface averages about 0.7 times the evaporation from a Class A pan. Other pan designs are also in use, including sunken pans and pans that float on open water surfaces.

Atmometers and Evaporimeters. *Atmometers* or *evaporimeters* are less expensive and more portable than evaporation pans, but they are much less accurate. One atmometer design uses a porous bulb as the evaporating surface. Moisture is supplied to the bulb from a reservoir bottle, with the loss from the reservoir bottle serving as a measure of the evaporating power. The Piche evaporimeter uses a piece of filter paper as the evaporating surface instead of a porous bulb. A set of atmometers or evaporimeters can be calibrated against each other or against evaporation pans.

Lysimeters. Evaporation pans and evaporimeters do not provide a very satisfactory index of the water movement into the atmosphere through transpiration by plants which draw up soil moisture from below the surface of the soil. With herbaceous plants such as grasses and agricultural crops it is feasible to isolate a block of soil in a tank and grow plants in the soil. The water loss can then be measured by changes in the weight of the tank containing soil and plants. The weight of the tank can be determined by placing it on a scale or floating it in water and measuring changes in the water displacement. Installations of this type are called *lysimeters.* This approach tends to be rather expensive.

Indirect Estimates. Instead of using a directly observable index of evaporation and transpiration, one can attempt to construct predictive equations from data on factors which control the process. Such factors include temperature, humidity, wind, nature of the surface, and soil moisture. Several such predictive models are available, but the coefficients of the equations usually have to be adjusted to fit local conditions (see Bruce and Clark, 1966; Linsley, Kohler, and Paulhus, 1975; or Wisler and Brater, 1959).

A second indirect approach is to conduct bookkeeping operations on gains and losses of moisture. If net gains or losses of moisture can be measured directly along with all the other sources of gain and loss, the evapotranspiration can be approximated by the difference between the actual change and changes due to other sources.

Evaporation and transpiration as just discussed are responsible for transforming moisture into the vapor phase and raising it into the atmosphere. Water in the vapor phase constitutes humidity, and humidity determination has already been discussed in Section 5.2. Therefore we shall move on to the next link in the hydrologic cycle which is *precipitation* following the condensation of water vapor in the atmosphere.

5.3.3 Precipitation

Precipitation falls in a variety of forms, with rain and snow being the most common. Most of these forms can be handled reasonably well with some type of rain gauge. Snow presents enough special problems, however, so that it is best considered separately.

Rain Gauges. Major differences between rain gauges arise from the time interval between readings. Simple types of rain gauges are designed for daily readings, but can be read more frequently to give information on a storm-by-storm basis. An automatic recording mechanism is necessary to provide information on variations in the rate of precipitation during a single storm. Gauges with long periods between readings usually require a large storage capacity.

Standard Rain Gauges. The standard rain gauge used by the National Weather Service and most others for routine weather information has a circular orifice 8 in. in diameter. A funnel at the orifice concentrates the precipitation in a collecting tube having a cross sectional area one tenth that of the orifice. This magnifies the depth by a factor of 10, giving more accurate readings. Provisions are made for the collecting tube to overflow into a larger outer container, thus avoiding loss of information during heavy rains.

Rain gauges are mounted with the orifice level, and well away from tall objects that might create turbulence. With strong winds or gusts, however, the gauge itself creates enough turbulence to give inaccurate readings. Various types of shields are available which can be placed around the orifice to minimize such turbulence. *Nipher* and *altar* type shields are most commonly used. The Nipher shield is a solid collar that diverts wind downward from the lip of the orifice, allowing a smooth flow over the gauge. The altar type shield consists of a series of swinging slats hanging on a ring around the orifice of the gauge.

Recording Rain Gauges. There are three common types of recording rain gauges. The *weighing* type collects precipitation in a bucket sitting on scales below the orifice. The scale is linked directly to a pen arm and recording chart. Precipitation of all types is recorded by the change in weight of the reservoir bucket, with the results being expressed in rainfall equivalents.

A second type of recording gauge uses a *float* arm linked to a pen and recording chart to provide the record. Since the float requires a pool of liquid, this kind of gauge does not work for solid forms of precipitation or in subfreezing temperatures unless a heater is provided.

The third type is called a *tipping bucket* gauge. The basic mechanism in this third type is a small divided bucket pivoted under the drain hole of the collecting funnel. Only one part of the little bucket can receive water at a time. When that part of the bucket fills to a certain level, it becomes overbalanced and dumps into a reservoir. Dumping of one part brings the other part into a position where it receives water from the collector, and the process is repeated. Each dump activates an electrical circuit which drives a recorder. Tipping bucket gauges are usable only for liquid forms of precipitation.

Storage Gauges. When rain gauges are situated in remote areas with long periods between readings, provisions must be made to store the water collected and minimize evaporation during storage. Storage requires some sort of tank that can be drained into a calibrated container when the reading is made. A closed tank has less evaporation loss than an open tank, and a surface film of oil or similar material can also be used to control evaporation.

Snow Measurements. The fluffiness of snow creates several problems in its measurement. It is very subject to drifting in wind currents, giving an uneven distribution of snowfall within local areas. Storage containers fill rapidly, and variations in the snow density cause difficulty in converting the snow volume to equivalent rainfall.

Measuring Snow by Rain Gauge. One approach is to remove the concentrating funnel and tube from a standard rain gauge and use the large overflow container to hold snow captured by the orifice. Measurements are taken by melting the snow and pouring it into the measuring tube removed from the rain gauge. Results are expressed as rainfall equivalents. The easiest way to melt the snow is to add a measured amount of warm water and then subtract this same amount from the final measurement. The rain gauge should also be shielded to reduce the effect of turbulence.

Monitoring Snow by Depth. The local variability in snowfall can be assessed by using a cluster of rain gauges as described above. However, this is rather time consuming and expensive. A less expensive approach is to use the snow depth along with conversion factors to rainfall equivalent. Several measurements of snow depth can be made over an area to obtain an average depth. If interest lies only in new snow, clean white boards can be placed on top of old snow to catch and isolate new snow.

A common rule of thumb for the conversion is to take the rainfall equivalent as one tenth the depth of new snow. More accurate conversion factors for new snow can be obtained by measuring both the snow depth and the rainfall equivalent in a rain gauge as described above. Old snow contains more water per unit depth than new snow because of compaction. Conversion factors for the water equivalent of the total snow pack can be obtained by using a tubular sampler to extract a cylindrical sample of known volume which is then weighed on portable scales.

Permanent snow survey routes can be established by setting graduated poles in the ground during the snow free season. If contrasting colors are used along with crossbars of different sizes, depth markers of this type can even be read to the nearest foot with binoculars from low flying aircraft.

Snow Pillows. A more sophisticated approach to monitoring snow packs is based on the use of pressure sensitive pillows to indicate the weight of the overlying snow. Such devices can be connected to remote recorders, thus avoiding the necessity of taking depth measurements and removing samples for weighing. This type of device requires careful calibration and checking against more conventional methods.

Precipitation reaching the ground surface may either infiltrate the soil or move overland to and through drainage channels. Since infiltration rates and soil moisture determination have already been covered in Chapter 4, the next phase of the hydrologic cycle to be considered is runoff and flow through channels en route to lakes and other storage basins.

5.3.4 Runoff and Channel Flow

Runoff from a watershed is usually monitored by measuring the discharge of the main stream as it flows out of the watershed. A graph of the discharge against time is called a *hydrograph*. Much can be learned about the nature of a watershed by studying the shape of its hydrograph following the passage of a storm.

Imagine a rope stretched across the stream above the water surface. The *discharge rate* is the volume of water passing under the rope per unit time. This depends on the shape of the stream's vertical profile along the rope, the water depth, and the nature of the channel downstream. In order to monitor the discharge, one looks for a section of the stream where the channel configuration produces a consistent relationship between water depth and discharge rate. Such a configuration is called a *control section*. Rapids or narrows with steeply sloping banks are likely choices. A stream gauging station is established upstream from the control section to provide a record of the water depth (stage). A *rating curve* or *stage-discharge graph* is then developed to show the relationship between depth and discharge. If a natural control section is not available, a weir can be installed to provide artificial control. A *weir* is a structure that restricts flow of the stream. A common type of weir forces the stream to flow through a V-shaped notch in a concrete barrier.

Depth Gauges and Water Level Recorders. Measuring rods and sounding lines can be used for gauging the depth in situations where the cost of an automatic recorder is not justified. Of course this requires regular visits for taking the readings, and more frequent visits provide better information. Depth rods are often implanted permanently in the streambed so that readings can be taken quickly from the bank without wading or using a boat. Likewise crank and pulley arrangements for sounding lines can be permanently mounted on bridges. Most automatic water level recorders are actuated by a float. The float is operated in a stilling well to remove the effects of waves or ripples and to guard against damage from floating debris.

Current Meters. In order to establish stage–discharge relationships for natural control sections it is necessary to make a series of actual discharge measurements over time at different water depths occurring in the same location. The first step in measuring discharge directly is to map the cross-sectional profile of the streambed at the point along the channel where the discharge is to be measured. The cross-sectional area can be determined from the profile and present water level. The cross-sectional area multiplied by the average current velocity gives the discharge.

The current velocity is determined with a *current meter*. Since air and water are both fluids, it should not be surprising that designs for current meters parallel those for anemometers used to measure wind speed. The most popular current meter (Price type) has rotating cups like a cup anemometer, but propeller models are also common. The operator determines the rotational speed of the meter by counting the clicks transmitted through a set of earphones.

Frictional drag from the banks and eddy currents make it impossible to obtain a satisfactory average current velocity from a single meter reading. In order to obtain a good average value for current velocity, the profile of the stream is divided into a series of equal width vertical strips. Kevern (1973) recommends a single meter reading at 0.4 times depth for shallow strips, and an average of readings at 0.4 and 0.6 times depth for deeper strips. If the meter is slanted, corrections will be required to obtain the horizontal component of current velocity. The average current velocity multiplied by the cross-sectional area of the strip gives a good approximation of the discharge through that slice of the profile. The total discharge is determined by adding the discharges for all strips.

5.3.5 Water Storage

The flow of water into surface or subterranean storage basins completes its movement through the hydrologic cycle. Lakes, ponds, and manmade reservoirs provide surface storage. The quantities of water contained in these surface basins can be determined from the present water level and a hydrographic map of the bottom contours as described in Section 4.1.6. The status of groundwater storage as reflected by the depth of the water table can be monitored by measuring the water level in open wells from which there is no pumping.

The quantities of water and rates of flow through the various phases of the hydrologic cycle are important, but they are not the only concerns in the water balance of ecosystems. Both the physical and the chemical properties of the water affect its biological and human uses. Some of these properties can be determined without drawing water samples, but more often than not it will be necessary to obtain a representative sample of the water for conducting tests.

5.3.6 Water Sampling

Unless there are currents to produce mixing, the physical and chemical properties of water can vary considerably over short distances, both horizontally and vertically. This complicates the problem of obtaining representative samples. There are two strategies for obtaining a representative sample. One is to sample at various horizontal positions and depths, analyze the samples separately, and average the results. The other is to attempt mixing waters from several positions in the process of taking the sample. The former is to be preferred since it gives information on the pattern of variability. However, the latter approach is often taken to reduce the costs of sampling and analysis.

There are also two general mechanisms for operating water samplers. One is to use a pump and hose system to draw water up to the surface and deposit it in sample containers. This approach carries some risks of contamination from the pump or hose. The physical properties may also be altered by the pumping action. The other approach is to lower a container into the water and allow it to fill in place before retrieval. Since it creates less disturbance of the sample, the latter approach is preferable.

Some water samplers are designed to draw a sample from a restricted volume of water and retrieve it intact without mixing. Others are designed to fill gradually, thus giving an integrated sample as the container is raised and lowered. Space does not permit a detailed discussion of the many types of samplers, but they are discussed in several of the references listed at the end of this chapter.

5.3.7 Physical Properties of Water

The routinely monitored physical properties of water include temperature, light penetration, turbidity, and color. Field methods for measuring these properties will be outlined briefly.

Temperature. Since the water temperature usually varies with depth, most readings are taken at specific depths. Standard mercury-in-glass thermometers are of little use for this purpose because the reading changes as the thermometer is being raised to the surface before it can be recorded. The *reversing thermometer* is a modified mercury-in-glass type which avoids this difficulty.

The thermometer is lowered to the desired depth, given time to equilibrate, and then turned upside down before retrieval. This reversal breaks the mercury column, thus preserving the reading. However, the reversing thermometer is not very convenient for plotting depth profiles of temperature because it must be raised to the surface for each reading.

Electrical temperature sensors (resistance type or thermistor) are better for this purpose because the readout device remains at the surface. The sensitive element is lowered to the desired depth, and the reading is transmitted to the recorder at the surface by long wire leads. Thus the entire depth profile can be determined by a single drop of the sensor.

Light Penetration. The Secchi disk provides an inexpensive index of light penetration and water clarity. This is a weighted disk of 10 cm radius painted black and white by quarters. The disk is lowered in the water until it disappears from view, then raised until it reappears. The average of the depths where it disappears and reappears serves as the index.

Since it involves human vision, the Secchi disk has an element of subjectivity. A more objective measure of light penetration can be obtained by comparing readings of paired light meters, one of which is submerged and the other positioned at the surface. The meters must be balanced so that they read the same at the surface, and the vessel containing the submerged meter must be watertight. Results are expressed as a profile of percent light penetration versus depth. Light meters are only sensitive to visible wavelengths of light, but portions of the visible light can be selectively excluded by the use of filters.

Turbidity. Turbidity is a term used to describe the optical interference of particulate matter suspended in water. One part per million of Fuller's earth (finely divided silica) in distilled water is taken as the standard unit of comparison for measuring turbidity (Welch, 1948).

The simplest and quickest method of measuring turbidity in the field is by disappearance of a platinum wire lowered into the water. The USGS has devised an apparatus consisting of a platinum wire, a depth scale graduated in turbidity units, and a positioning guide for holding the wire relative to the eye. See Welch (1948) for a more detailed description of the turbidity rod and its use.

The Jackson candle turbidimeter is another portable and relatively inexpensive device for measuring turbidity. Water is poured into a tube over a standard candle flame until the turbidity of the water obscures the flame. Turbidity is read on a scale corresponding to the depth of water in the tube. The viewing operation must be conducted in darkness.

Both the platinum wire turbidity rod and the Jackson turbidimeter involve subjective visual judgment. Laboratory instruments are available which avoid subjective judgments. The Hellige turbidimeter is a special purpose instrument for measuring the turbidity by the Tyndall light scattering effect of the suspended particles. General purpose photometers can also be used to measure the turbidity by the reduced intensity of a light beam passing through the sample.

Color. The water color is measured by visual or spectrophotometric comparison with platinum-cobalt standards. The standards may be either glass disks for field use of liquid standards for laboratory use. Particulate matter in the samples interferes with color comparisons.

5.3.8 Chemical Properties of Water

The comprehensive reference for a chemical analysis of water is *Standard Methods for the Examination of Water and Waste Water* (APHA-AWWA-WPCF, 1975). Since this authoritative manual sets the standard for analytical procedures, we will confine our coverage to general comments on chemical tests which serve as indexes to the overall status of aquatic components in ecosystems. Several of these tests are available as convenient field kits from commercial suppliers.

pH. As with soils, the pH of water is easily determined and provides an index to several important chemical and biological processes. Color indicators, pH meters, and titration are usable methods of making pH determinations in the field.

Alkalinity. Fresh waters have a buffering system that prevents rapid fluctuations in pH. Important chemical components of the natural buffer system are carbon dioxide, carbonates, bicarbonates, and hydroxides. The total alkalinity expressed as the effective parts per million of calcium carbonate provides an overall index of the capacity to buffer pH changes by neutralizing acids. More detailed information on the composition of the buffer system can also be obtained from the rate of change in pH as acid is added during titration (buffer curve).

Hardness. The concept of hard versus soft water is familiar to most people because of its effect on the use of soaps and detergents. Hardness is a measure of the amounts of certain cations dissolved in the water. Titrating with a soap solution to a foaming end point is a simple and inexpensive but not very accurate method of measuring the total hardness. Better results are obtained by titrating with EDTA.

Conductivity. The electrical conductance of water as measured with a conductivity cell and a Wheatstone bridge gives an index of the concentration for dissolved electrolytes. The results are usually expressed as the specific conductance which has units of micromhos per cubic centimeter of water at a temperature of 25°C.

Dissolved Oxygen and Oxygen Demand. Oxygen has a limited solubility in water, which decreases with increasing temperature. Furthermore it diffuses through water slowly, making replenishment a slow process unless physical mixing takes place to promote aeration. Because of this, dissolved oxygen is frequently a limiting factor for higher forms of aquatic life which require it in fairly large quantities. Thus dissolved oxygen levels provide a good index of the capability of waters to support higher forms of aquatic life.

Dissolved oxygen can be measured in the field either by the Winkler method or electrometrically with a battery operated meter. The azide modification of the Winkler method is commonly used for this purpose.

Another factor in the oxygen balance is the potential consumption by microbes feeding on organic pollutants. The oxygen consumption by microbial metabolism depletes the supply available to higher organisms such as fishes. To evaluate this drain on the oxygen supply, water samples are incubated under conditions favorable to microbial activity and their oxygen consumption is measured. Incubation is done in darkness so that the oxygen production by photosynthesizing algae will not complicate the test. Microbial oxygen consumption is called *biochemical oxygen demand* (BOD).

Dissolved oxygen and BOD are the most important considerations in the oxygen balance. However, some pollutants cannot be readily broken down by microorganisms, but still cause depletion when they are oxidized by chemical reactions occurring in the waters. The combined effect of chemical and biochemical oxidation is called *chemical oxygen demand* (COD).

The combination of oxygen balance, pH, alkalinity, hardness, and conductivity provides a composite index of the conditions in aquatic ecosystems. the reader is referred to *Standard Methods* for more specific types of chemical analysis.

5.4 ROLE OF ATMOSOPHERIC AND HYDROLOGIC DATA IN ENVIRONMENTAL INFORMATION SYSTEMS

Those needing information on current atmospheric and hydrologic conditions would only be hindered by delays involved in data storage and retrieval. They are best served by data routed directly from the monitoring network after summarization. However, statistical information on past conditions is essential to planning for the future. Climatic summaries by regions are available from NOAA, but needs for records on local areas such as watersheds may require independent monitoring.

5.5 BIBLIOGRAPHY

Abrahamson, D. *Environmental Costs of Electric Power*. New York: Scientist's Institute for Public Information, 1970.

Anon. *Snow Survey Safety Guide.* Washington D.C.: U.S. Department of Agriculture Handbook 137, 1974.

Anon. *Automation of the ISCO Water Sampler.* Missoula, Mont.: U.S. Department of Agriculture, Forest Service, Equipment Development Center, Equip Tips 2540, 1975.

APHA-AWWA-WPCF. *Standard Methods for the Examination of Water and Wastewater,* 14th ed. Washington, D.C.: American Public Health Association, American Water Works Association, and Water Pollution Control Federation, 1975.

Berg, G. *Water Pollution.* New York: Scientist's Institute for Public Information, 1970.

Boyd, C. *Evaluation of a Water Analysis Kit.* Auburn, Ala.: Auburn University, Agricultural Experiment Station, Leaflet 92, 1976.

Brenchley, D., D. Turley, and R. Yarmac. *Industrial Source Sampling.* Ann Arbor, Mich.: Ann Arbor Science Publishers, 1973.

Brown, J. *Tables and Conversions for Microclimatology.* St. Paul, Minn.: U.S. Department of Agriculture, Forest Service, North Central Forest Experiment Station, General Technical Report NC-8, 1973.

Bruce, J., and R. Clark. *Introduction to Hydrometeorology.* New York: Pergamon, 1966.

Chase, D., and J. Rabinowitz. *Principles of Radioisotope Methodology.* Minneapolis, Minn.: Burgess, 1967.

Chow, V., Ed., *Handbook of Applied Hydrology.* New York: McGraw-Hill, 1964.

Clark, J., W. Viessman, Jr., and M. Hammer. *Water Supply and Pollution Control,* 2nd ed. New York: International Textbook Co., 1971.

Cunia, T., Ed., *Proceedings of Symposium on Monitoring Forest Environment Through Successive Sampling.* Syracuse, N.Y.: State University of New York, International Union of Forest Research Organizations, and Society of American Foresters, 1974.

Dahlsten, D., et al. *Pesticides.* New York: Scientist's Institute for Public Information, 1970.

Daubenmire, R. *Plants and Environment, A Textbook of Autecology.* New York: Wiley, 1974.

Davis, M. *Air Resource Management Primer.* New York: American Society of Civil Engineers, 1973.

Donn, W. *Meteorology,* 4th ed. New York: McGraw-Hill, 1975.

Doty, R. "A Portable, Automatic Water Sampler." *Water Resources Research* **6**(16), 1787 (1970).

Fosberg, M., W. Marlatt, and L. Krupnak. *Estimating Airflow Patterns Over Complex Terrain.* Fort Collins, Colo.: U.S. Department of Agriculture, Forest Service, Rocky Mountain Forest and Range Experiment Station, Research Paper RM-162, 1976.

Fraser, J. *A Simple Instrument Shelter for Use in Forest Ecology Studies.* Ottawa, Ont.: Canada Department of Forestry, Forest Research Branch, Technical Note 113,1961.

Gay, L. "Temporal Variation in the Forest Environment." In T. Cunia, Ed. *Proceedings of Symposium on Monitoring Forest Environment Through Successive*

Sampling. Syracuse, N.Y.: State University of New York, International Union of Forest Research Organizations, and Society of American Foresters, 1974.

Geiger, R. *The Climate Near the Ground.* Cambridge, Mass.: Harvard University Press, 1966.

Golterman, H., and R. Clymo. *Methods for Chemical Analysis of Fresh Waters.* IBP Handbook 8. Oxford: Blackwell, 1969.

Haines, D. *Where to Find Weather and Climatic Data for Forest Research Studies and Management Planning.* St. Paul, Minn.: U.S. Department of Agriculture, Forest Service, North Central Forest Experiment Station, General Technical Report NC-27, 1977.

Hamilton, E. *Rainfall Sampling on Rugged Terrain.* Washington, D.C.: U.S. Department of Agriculture, Technical Bulletin 1096, 1954.

Herrington, L., and G. Bertolin. "Measurement of the Physical Environment." In T. Cunia, Ed., *Proceedings of Symposium on Monitoring Forest Environment Through Successive Sampling.* Syracuse, N.Y.: State University of New York, International Union of Forest Research Organizations, and Society of American Foresters, 1974.

Hewlett, J., and W. Nutter. *An Outline of Forest Hydrology.* Athens, Ga.: University of Georgia Press, 1969.

Hidore, J. *Workbook of Weather Maps,* 2nd ed. Dubuque, Iowa: Brown, 1971.

Ingebo, P., W. Casner, and G. Godsey. *A Computer Program for Computing Streamflow Volumes.* Fort Collins, Colo.: U.S. Department of Agriculture, Forest Service, Rocky Mountain Forest and Range Experiment Station, Research Note 203, 1971.

Jackson, K. *A Sleeved Pit Gage for Summer Precipitation.* Fort Collins, Colo.: U.S. Department of Agriculture, Forest Service, Rocky Mountain Forest and Range Experiment Station, Research Note RM-256, 1974.

Jackson, M., and J. Newman. "Indices for Expressing Differences in Local Climate Due to Forest Cover and Topographic Differences". *Forest Science,* **13**(1), 60–71 (1967).

Jairell, R. *A Sturdy Probe for Measuring Deep Snowdrifts.* Fort Collins, Colo.: U.S. Department of Agriculture, Forest Service, Rocky Mountain Forest and Range Experiment Station, Research Note RM-301, 1975.

James, L., and R. Lee. *Economics of Water Resources Planning.* New York: McGraw-Hill, 1971.

Jones, J. *Fish and River Pollution.* Reading, Mass.: Butterworth, 1973.

Judson, A., and B. Erickson. *Predicting Avalanche Intensity from Weather Data—A Statistical Analysis.* Fort Collins, Colo.: U. S. Department of Agriculture, Forest Service, Rocky Mountain Forest and Range Experiment Station, Research Paper RM-112, 1973.

Kevern, N. *A Manual of Limnological Methods.* East Lansing, Mich.: Michigan State University, Department of Fisheries and Wildlife, unpublished mimeo., 1973.

Knoerr, K. "Atmospheric Energy and Mass Exchange in the Forest Environment—Models vs. Measurement." In T. Cunia, Ed., *Proceedings of Symposium on Monitoring Forest Environment Through Successive Sampling.* Syracuse, N.Y.: State University of New York, International Union of Forest Research Organizations, and Society of American Foresters, 1974.

Lagler, K. *Freshwater Fishery Biology,* 2nd ed. Dubuque, Iowa: Brown, 1956.

Leaf, C., and J. Kovner. "Sampling Requirements for Areal Water Equivalent Estimates in Subalpine Watersheds." *Water Resources Research,* **8**(3), 713–716 (1972).

Lillesand, T. "Use of Remote Sensing to Monitor Water Quality." In T. Cunia, Ed., *Proceedings of Symposium on Monitoring Forest Environment Through Successive Sampling.* Syracuse, N.Y.: State University of New York, International Union of Forest Research Organizations, and Society of American Foresters, 1974.

Linsley, R., Jr., M. Kohler, and J. Paulhus. *Hydrology for Engineers,* 2nd ed. New York: McGraw-Hill, 1975.

Magill, P., F. Holden, and C. Ackley, Eds. *Air Pollution Handbook.* New York: McGraw-Hill, 1956.

McCrone, W., R. Drafty, and J. Delly. *The Particle Atlas.* Ann Arbor, Mich.: Ann Arbor Science Publishers, 1967.

Minckler, L. *Measuring Light in Uneven-aged Hardwood Stands.* Washington, D.C.: U.S. Department of Agriculture, Forest Service, Central States Forest Experiment Station, Technical Paper 184, 1961.

Monteith, J. *Survey of Instruments for Micrometeorology.* IBP Handbook 22. Oxford: Blackwell, 1972.

Nadler, A., et al. *Air Pollution.* New York: Scientist's Institute for Public Information, 1970.

Platt, R., and J. Griffiths. *Environmental Measurement and Interpretation.* Huntington, N. Y.: Krieger Publishing, 1972.

Rosenberg, N. *Microclimate: The Biological Environment.* New York: Wiley-Interscience, 1974.

Ruttner, F. *Fundamentals of Limnology,* 3rd ed. Toronto, Ont.: University of Toronto Press, 1963.

Schmidt, R., and E. Holub. *Calibrating the Snow Particle Counter for Particle Size and Speed.* Fort Collins, Colo.: U.S. Department of Agriculture, Forest Service, Rocky Mountain Forest and Range Experiment Station, Research Note RM-189, 1971.

Schroeder, M., and C. Buck. *Fire Weather.* Washington, D.C.: U.S. Department of Agriculture, Forest Service, Agricultural Handbook 360, 1970.

Simard, A., and J. Valenzuela. *A Climatological Summary of the Canadian Forest Fire Weather Index.* Ottawa, Ont.: Environment Canada, Forestry Service, Forest Fire Research Institute, Information Report FF-X-34, 1972.

Skoog, D., and D. West. *Principles of Instrumental Analysis.* New York: Holt, Rinehart & Winston, 1971.

Strauss, W., Ed. *Air Pollution Control,* Parts 1;2. New York: Wiley-Interscience, 1971 and 1972.

Todd, D. "Groundwater." In Chow, V., Ed., *Handbook of Applied Hydrology.* New York: McGraw-Hill, 1964.

Vogel, T., and P. Johnson. "Evaluation of an Economical Instrument Shelter for Microclimatological Studies." *Forest Science,* **11**(4), 434–435 (1965).

Voigt, G. "Distribution of Rainfall Under Forest Stands." *Forest Science,* **6**(1), 2–10 (1960).

Wadsworth, R., Ed. *The Measurement of Environmental Factors in Terrestrial Ecology: A Symposium of the British Ecological Society, Reading, 29–31 March 1967.* Oxford: Blackwell, 1968.

Wang, C., D. Willis, and W. Loveland. *Radiotracer Methodology in the Biological,*

Environmental, and Physical Sciences. Englewood Cliffs, N.J.: Prentice-Hall, 1975.

Warner, P. *Analysis of Air Pollutants.* New York: Wiley, 1976.

Welch, P. *Limnological Methods.* New York: McGraw-Hill, 1948.

Wilde, S., G. Voigt, and J. Iyer. *Soil and Plant Analysis for Tree Culture.* New Delhi: Oxford & IBH Publishing Co., 1972.

Wisler, C., and E. Brater. *Hydrology,* 2nd ed. New York: Wiley, 1959.

World Meteorological Organization. *International Cloud Atlas.* Murray Hill Station, New York: Unipub, Inc., 1956.

CHAPTER 6–VEGETATION: A RENEWABLE RESOURCE

All components of ecosystems are dynamic to some degree. Geologic changes are usually slow whereas atmospheric conditions fluctuate rapidly. However, vegetation adds a new dimension of dynamics. That added dimension is life itself. Without becoming embroiled in the metaphysical essence of life, we can recognize the fundamental features of population dynamics in living organisms as being reproduction, growth, and death controlled by genetic patterns and competition. It is of the utmost importance that the nature of these population processes be recognized in ecosystem management. Much of the preservationist philosophy as espoused by laymen appears to be based on the assumption that vegetation will remain in its present state if man will only cease his meddling in "nature." This is simply not the case.

Previously unvegetated surfaces are continually being colonized by vegetation through the process of *succession*. On bare rock, for example, lichens are usually the first plants to establish a foothold because they are very resistant to drought and temperature extremes. The lichen mat provides some protection from moisture and temperature extremes, allowing mosses and hardy herbaceous plants to become established. The herbaceous plants, in turn, moderate the climate still further, allowing shrubby plants to invade. The shrubs then give way to trees. At each successive stage of colonization, the new invaders are taller than the previous plants, cutting down the supply of solar energy and finally choking out the smaller plants. The process eventually stabilizes with the arrival of shade tolerant species which form a closed overstory canopy, yet can grow as seedlings in very dense shade.

There is a corresponding series of increasing dryness which takes place in the colonization of open water. It begins with submerged plants which gradually decrease the water depth as material from dying plants accumulates. This allows rooted plants with floating leaves to follow. These are succeeded by emergents, herbs, shrubs, and finally trees. Such a successional series is called a *sere* with appropriate modifiers. For example, the colonization of water is called a *hydrosere*.

The nature of the final stage or *potential natural vegetation* is determined largely by the climate prevailing in the region and is thus often called the *climatic climax*. Dry areas usually have a grassland or desert shrub climax, whereas more moist areas have a forest climax.

Likewise *secondary succession* takes place when the plant cover is stripped away leaving bare soil. However, secondary succession usually proceeds more rapidly than primary succession since the previous plants leave relatively favorable soil conditions. Kershaw (1973) presents a good discussion of plant succession, as do many other textbooks on plant ecology.

Any vegetational cover that has not yet reached the climax stage can be expected to progress toward the climax unless man or some other influence such as fire or grazing interferes to maintain the subclimax stage. If one wishes to preserve subclimax vegetation, nonintervention is *not* the way to accomplish the objective. More importantly, the phenomenon of plant succession reflects the power of vegetation to modify the physical environment. Manipulating vegetation is an economical and effective means of modifying ecosystems for many purposes.

Time is probably the greatest constraint in managing ecosystems through vegetation. Primary succession is a slow process, often requiring many centuries to reach the climax. Although secondary succession is more rapid than primary, it may take a century or more to replace climax vegetation after an area is denuded. The more advanced the successional stage, the longer its replacement will require. Therefore the destruction of vegetation in advanced stages of succession should not be undertaken casually.

In view of the economic importance of vegetation, the need for survey information on this component of the ecosystem should require little elaboration. Information on the planted area, condition, and probable yield of agricultural crops can forestall famine in underdeveloped countries as well as stabilize food prices and supplies in more developed countries. Range managers require information on forage production and range condition in order to regulate grazing. The forester requires detailed information on forest stands in order to prescribe silvicultural practices and schedule harvests on a sustained yield basis. Food and cover furnished by vegetation are key elements of the wildlife habitat. Vegetation is also a major factor in watershed management and erosion control. Less economic but nonetheless important, vegetation is a dominant element in aesthetics and recreation.

The first order of business in this chapter is to examine the structure in vegetation and note its measurable features.

6.1 STRUCTURAL FEATURES OF VEGETATION

Plants can take on a variety of forms as they grow. Terms like tree, shrub, vine, grass, and herb are commonly used to describe the more typical of these growth habits. Usually a given plant species will be restricted to a single growth form, but this is not always the case. Some species grow as trees under favorable conditions but change to a shrubby form under harsh conditions such as occur near timberline on high mountains. Plant ecologists use the general terms *physiognomy* and *life form* when referring to the growth habit. Rather elaborate classification schemes for describing the life form or physiognomy have been developed by Raunkiaer (1934) and others. Kuchler (1967) summarizes these classification schemes nicely. We will be content here to use laymen's terms such as tree, shrub, and herbaceous in describing growth forms. Whatever terms are used, the fact remains that the most convenient approach to measuring vegetation depends somewhat on the growth habit. It is relatively easy to count individual trees, but practically impossible to identify and count individuals in sod forming grasses.

An obvious, but important, property of terrestrial vegetation is that it grows attached to a surface. Vegetation survey problems in applied ecology are usually stated in terms of area. Interest centers on a population of plants growing on some defined area, rather than on plants wherever they may occur. The fact that plants are stationary greatly simplifies vegetation surveys as opposed to surveys of animal life, since units of area can be used as sampling units and the plants can be relocated relatively easily for measurement on successive occasions. This suitability for area sampling creates almost endless possibilities for survey design. Furthermore many of the measurable properties of vegetation also relate to area in one way or another. Consider a population of plants growing on some relatively small unit of area, and let us examine some of these measurable features.

6.1.1 Number of Individuals and Density

The term *density* is applied to the average number of plant units per unit area. The determination of density involves several considerations. The first requirement is to decide what constitutes a plant unit. This causes little problem with trees since the natural unit is defined by the trunk of the tree. Likewise annual herbs often grow as single and easily recognizable units. With multiple stemmed shrubs there is more difficulty in deciding what the plant unit should be. If the shrubs grow in clumps, either the stem or the clump could be considered as the plant unit. With grasses and perennial plants that grow from root sprouts, however, plant units become very arbitrary and difficult to recognize. Therefore density is not a very objective or useful measure with the latter types of plants.

When a plant unit has been defined, the next step is to decide on area units. Two types of units are involved here. One is the unit on which to base the average. This is usually a fairly simple matter of choosing either a metric or an English base. The usual metric base is the hectare for trees or the square meter for smaller and more numerous plants. Usual area bases in the English system are the acre and the square foot.

The second type of area that needs to be chosed is the unit of area or plot on which to do the actual counting. This is primarily a matter of practicality in the counting operation and variability in the density of plant units. One does not want plots so small that most of the plots are nearly barren. Small plots also have the disadvantage of a large ratio of boundary to plot area, which causes a large number of decisions on whether a plant unit should be included in the plot or not. On the other hand one does not want plots so large that counting the plant units takes an inordinately large amount of time. With such large plots there is a danger of losing track of which units have already been counted. Experience is the best guide in choosing an appropriate plot size for density counts. If experience is lacking, then a little time should be invested in field trials.

6.1.2 Size of Individuals, Biomass, and Volume

Density alone is usually not a very informative measure of the vegetation structure since the individual plant units may be either small or large. In order to give a more complete picture, the density can be supplemented by measures of the average size and the variability in size of the plant units.

The average amount of plant material per unit area is important for both ecology and economics. The amount of plant material as measured by weight is called *biomass.* Biomass has the advantage of being useable for all types of growth forms from grasses to trees. Because of the variability in moisture content, the biomass is usually given in terms of oven dry weight.

The biomass of grasses, herbs, and small shrubs can be determined by the straightforward although tedious method of clipping and weighing. Subsamples can be taken into the laboratory for drying in order to develop conversion factors from fresh weight to dry weight.

The direct determination of biomass in large shrubs and trees is a major undertaking. The usual approach with such large plants is to do special studies in order to develop equations, graphs, or tables for predicting the biomass from easily measured dimensions of plants such as stem diameter and height. These equations, graphs, or tables are then used to estimate the biomass indirectly from the dimensions that serve as independent variables. Regression analysis (see Appendix) is the most efficient method of developing these predictive equations, graphs, or tables.

In forestry the usable volume of wood in trees is used more often than the biomass. As with the biomass, the wood volume is estimated indirectly from tree dimensions via regression analysis.

6.1.3 Area Occupancy and Cover

As mentioned above, the density is rather arbitrary and difficult to determine in grasses and other spreading plants because of problems in deciding what constitutes a plant unit. In such cases the ground area covered by the vertical projection of above ground plant parts provides a more objective measure of area occupancy. The term applied to measures of this type is *cover*. The cover can be measured in either percentage or absolute units of area, and its use is not limited to low growing plants. Both absolute and percentage cover measures are commonly applied to stands of trees. The percentage of area covered by vertical projections of tree crowns provides a measure of canopy closure. The area covered by tree trunks (square meters/hectare or square feet/acre) is popular among foresters as a measure of forest stocking. This is called *basal area*.

The *leaf area index* is another variation on the idea of cover. This is the ratio of the total leaf area to the total ground area. For example, a leaf area index of 3 would mean that the combined areas of the upper surfaces of all plant leaves growing on a plot is three times greater than the area of the plot. Stated another way, lines extended upward from points on the ground surface would pass through an average of three leaves if all leaves were oriented horizontally.

6.1.4 Presence/Absence, Frequency, and Distribution

Still another line of attack on the problem of vegetation occurrence is to state the percentage of plots on which a given kind of plant is found to be present. This type of measure is called *frequency*. The frequency approach is popular among plant ecologists because it is quick and easy to apply in the field. However, there are several problems associated with frequency which make it a relative and somewhat unreliable index rather than an absolute measure.

The most obvious problem is that the likelihood of a plant being present in a plot increases with the plot size. Therefore large plots used in a given stand of vegetation will show a higher frequency than small plots. Frequency, then, is meaningless unless the plot size is stated.

A companion problem arises from the variability in the way plants are distributed over an area. Consider an area marked off into plots and a given number of plants to be set out in the area. If the plants are set out in clumps, there will be more empty plots than if the plants are randomly or regularly distributed over the area. Thus it is possible that a clumped species of plant will show a lower frequency than one which is randomly distributed, even though there are actually more individual plants of the clumped species growing on the area. Many studies have shown that some degree of aggregation is more common than random or regular spacing (Grieg-Smith, 1964; Kershaw, 1973; Mueller-Dombois and Ellenberg, 1974).

6.1.5 Relative Importance and Dominance

The quantitative measures of the vegetation structure discussed thus far have been geared primarily toward describing a particular type of vegetation without reference to other types of vegetation which may also occur on the same area. Of course, comparisons between types of vegetation can be made by the difference between corresponding measures. When the interest centers on the relative importance of different types of vegetation, comparisons are facilitated by shifting the basis for percentages.

The density was defined earlier as a number of plant units per unit area. The cover percent was defined as the percentage of total area covered by plants. Frequency was defined as the percent of all plots in which a given kind of plant was present. All three of these measures can be placed on a relative basis (Curtis, 1959; Mueller-Dombois and Ellenberg, 1974) to facilitate comparisons between different types of plants growing on the same area. The *Relative density* is obtained by dividing the density for a given type of plant by the total density for all types of plants and then multiplying by 100. Likewise, the *relative cover* is obtained by dividing the cover for a given type of plant by the sum of covers for all types and then multiplying by 100. Similarly the *relative frequency* is obtained by dividing the frequency for a given type by the sum of the frequencies of all types and then multiplying by 100. A corresponding relative value could be calculated from the biomass. Any one of these relative measures can serve as a percentage index of importance for one species among all others present on an area.

However, we have seen that the various measures reflect somewhat different dimensions of vegetation structure. Therefore a combination of these relative measures should give a better picture of the importance than any single one. The sum of relative density, relative cover, and relative frequency is often used as a composite measure of importance for one species among all species. This sum has been termed *importance value* by Curtis (1959) and his co-workers. For completeness it should be noted that Curtis used the basal area as a measure of cover in working with tree species.

Dominance is another term that should be mentioned, but used with caution since it means different things to different people. Plant ecologists often use the term as an approximate substitute for cover. Foresters, however, use the term dominance to describe the position of a tree crown relative to the general level of the canopy. Trees with crowns projecting above the general level of the canopy are said to be dominant, and those with crowns forming the general level of the canopy are said to be codominant. Trees with crowns projecting up into the canopy but receiving no direct light from above are said to intermediate. Trees with crowns below the general level of the canopy are said to be suppressed.

6.1.6 Composition and Diversity

In most vegetation surveys the composition of the vegetation is one of the major items of interest. The categories for breaking down the composition may be set up according to several criteria. Most surveys will at least partition the total vegetation according to species or groups of similar species. Other common criteria for partitioning are age classes, size classes, condition classes, and economically important parts of individual plants. Any of the absolute or relative measures of vegetation structure discussed in the preceding sections may be used in quantifying the breakdown unless individual plants are broken down into parts. In the latter case either the biomass or the volume usually serves as the quantitative measure.

In recent years there has been a growing interest in species diversity, and several indexes of species diversity have been developed. Pielou (1975) provides a fairly comprehensive treatment of this subject, but her presentation is at a rather high level of mathematical sophistication. She lists three desirable properties for an index of diversity, and states that the only index having all the desirable properties for sample based data is the Shannon index

$$-\Sigma\ p_i \ln p_i$$

where p_i represents the proportion of the ith species, ln indicates natural logarithm, and the Σ sign indicates summation. However, using the composite sample proportions directly in the formula produces a bias, and the result will also depend on whether the sample is large enough to fairly represent the species composition of the vegetative community. Pielou gives a rather complicated procedure for circumventing the bias and deciding whether or not the sample is sufficiently large. Less mathematically inclined ecologists have approached the problem of the minimal area necessary to fairly represent the species composition by plotting the number of species encountered against the increasing size of the sample and noting where the species–area curve levels off (Cain and Castro, 1959; Grieg-Smith, 1964; Kershaw, 1973; Mueller-Dombois and Ellenberg, 1974). The Shannon index has also been used directly without regard for the bias involved (Benton and Werner, 1972).

6.1.7 Components of Change over Time

The measures discussed thus far have been directed toward a quantitative description of the vegetation at a point in time, or the *standing crop*. Many problems of ecosystem management also require information on the change over time.

One component of change that operates to increase the standing crop is the growth in size of existing plant units. An appropriate term for this component of change is *accretion*. Accretion produces changes in the biomass and cover but does not alter the density. The frequency may or may not be changed by accretion, depending on how the frequency is determined. If plants have a single central stem and a plant is considered present in a plot only when the center of its stem is located in the plot, then the frequency will be unchanged by accretion. If the presence of any part of a plant causes the plant to be counted in the plot, then the frequency is likely to change with accretion. Obviously a component of change can only be assessed by measures that are affected when the change takes place. Accretion is most accurately evaluated by a successive remeasurement of the same plant units.

A second component of change which increases the standing crop and affects all measures of vegetation structure is the creation of new plant units through some reproductive process. An appropriate term for this component of change is *regeneration*. Regeneration cannot be assessed by remeasuring the same plant units, since it is necessary to detect and measure new plant units.

A component of change that operates to decrease the standing crop is the removal of parts of existing plant units through some process such as grazing by animals. *Reduction* might be used as a term for this component of change. If the material removed is put to some "useful" purpose, then such a change can be considered part of the *harvest*. Like accretion, this component of change affects the biomass and cover, does not affect the density, and may or may not affect the frequency depending on how it is determined. However, this component of change is more difficult to assess than accretion for two reasons. Accretion is a relatively uniform phenomenon since all healthy plants can be expected to grow. In contrast, the removal of plant parts is often quite localized, as would be the case with selective grazing or browsing. Larger samples are needed to detect localized changes than uniform changes. The second problem arises from the need to measure a part of the plant which is no longer present. Special procedures such a paired comparisons between protected and unprotected plants may be necessary to circumvent this problem.

A second component of change that operates to reduce the standing crop is the death of plant units. If the plant units are killed by removal for some "useful" purpose, then this component of change is included in the *harvest*. If the dead plant units do not serve any "useful" purpose, then the term is *mortality*. Mortality can be further broken down into *epidemic* and *endemic* types. Epidemic mortality is the extensive death of plant units caused by disease outbreaks and other disasters such as storms or fire. Endemic mortality is the more gradual and less spectacular rate of death normally expected in the absence of such disasters. The death of plant units affects all measures of vegetation structure.

Ingrowth is another component of change, but one that is an artifact of economics rather than biology. This term is used when perennial plants do not become economically useful until they reach some threshold of size. This is the case with boles of trees destined to be cut into lumber. Previously unmerchantable plants growing into merchantable size classes through accretion are called ingrowth.

6.2 SAMPLING STRATEGIES FOR VEGETATION SURVEYS

Section 6.1 has been concerned with an overview of ways in which the vegetation structure may be quantified. Hopefully this overview gives the impression that most vegetation surveys provide descriptions of either standing crop or changes in standing crop through measurements of density, cover, frequency, and biomass broken down by species groups, size classes, and conditions. Further details of the procedure for making these measurements are covered in Sections 6.3, 6.4, 6.5, and 6.6. However, it should be fairly apparent that plants are usually so numerous as to make complete vegetation measurements economically prohibitive, even for small areas. Therefore one must be content with estimates derived from samples. In this section we consider sampling strategies for vegetation surveys.

6.2.1 Stratification

Most areas contain variations in vegetation which are recognizable by inspection of aerial photographs. The number of samples needed to achieve a specified level of precision in estimates will be reduced by dividing the area into subpopulations which appear on the airphotos to contain relatively uniform vegetation. These subpopulations are called *strata* in sampling jargon, and the process of delimiting the strata is called *stratification*. The strata need not be contiguous, and any other prior information available in addition to or in lieu of airphotos can also be used in stratification. Sampling operations are done separately in each stratum, and estimates are also compiled separately by stratum. There estimates for the individual strata are then pooled or combined into a composite estimate for the entire population. Procedures for making composite estimates from separate stratum estimates are discussed in Appendix A.3. Except for very small areas, gains from stratification are usually sufficiently large to make mapping out strata a worthwhile preliminary step in conducting vegetation surveys.

6.2.2 Sampling Units

As mentioned at the beginning of this chapter, most approaches to vegetation sampling take advantage of the ability to associate units of vegetation with units of area or points on a tract of land. The possible variations on this theme are almost endless, but they fall into a few general categories.

Plots or Quadrats. The simplest approach is to use plots of fixed size and shape as sampling units. The term *quadrat* was apparently meant in its original usage to describe square plots, but it has since been broadened to include circular and rectangular shapes. Three general considerations are involved in the choice of the size and shape for simple plots.

The first of these considerations is the practicality in locating plot boundaries and taking measurements. This tends to set a practical upper limit on the plot size. If plots are too large, it will be difficult to locate the boundaries and easy to lose count of which plants have been measured and which have not. The most practical plot size, however, depends on the type of vegetation being measured. For instance, plot sizes can be much larger for mature trees than for tree seedlings.

The second consideration is the edge bias. When a plant unit is situated near the boundary of a plot, one must make a decision whether or not to include the unit in the plot. Some people tend to include more plants in the plot than they should or vice versa, thus producing a bias. With small plots or plots that are long in one dimension, a larger proportion of the plant units occur near an edge, and thus more decisions of this type are required. The fraction of plant units requiring such decisions is reduced by using large, circular plots. Likewise square plots involve fewer such decisions than rectangular plots. When plant units are to be measured or counted as individuals, each individual unit must be associated with a single point on the ground. For example, a tree is included in the plot only if the center of its stem is located in the plot.

The third consideration involves the balance of effort between measuring a few large plots or many small plots. The key factor here is the nature of the sampling error. The greater the difference between estimates from individual plots, the larger the sampling error and the more uncertain the combined estimates from all plots. On the other hand the sampling error decreases in inverse proportion to the square root of the number of plots. Thus an increase in the number of plots reduces the sampling error. If plots are too small, a large fraction of them will be barren or nearly so, thus increasing the variability between plots and the sampling error also. Furthermore small plots mean a larger fraction of time spent traveling between plots. On the other hand the larger the plot size, the fewer the plots that can be measured. The relationship between plot size and sampling error depends on the structure and distribution of the vegetation. If information is available from past studies or field trials, it can be used in choosing an advantageous plot size. See Freese (1962) for further advice on this matter.

Data from field plots provide the raw material for computing estimates. The estimation process often appears more complex than it actually is. The main reason for the apparent complexity is the use of computational shortcuts which conceal the basic logic involved in the estimation. The basic procedure is to make a separate estimate from each plot and then average these estimates. A simple change of area base may be the only operation necessary to obtain an estimate from a plot. As an example consider estimating the density of trees from a plot that is ¼—acre in size. Assume that the density is to be expressed as the number of trees per acre, and that 25 trees are found on the plot. Then

$$\frac{25 \text{ trees}}{1 \text{ plot}} \times \frac{1 \text{ plot}}{0.25 \text{ acre}} = \frac{25 \text{ trees}}{0.25 \text{ acre}} = 100 \text{ trees/acre}$$

The same sort of proportional expansion by area applies to many other types of estimates from single plots. However, two types of estimates do not require such an expansion when derived from plots of a single size. Percentages do not require expansion, and neither do measures of the average size of individual plant units.

Strip Transects. Another time honored approach to area sampling of vegetation is that of strip transects. A strip transect is essentially a series of long and narrow rectangular plots placed end to end, forming a straight strip of uniform width. Like plot size, the strip width is chosen to suit the type of vegetation being studied. Strip transects have both advantages and disadvantages as compared with plots of more compact shape. An obvious disadvantage is the large amount of edge in relation to the area, which introduces a strong possibility of edge bias. This is further aggravated by the difficulty of keeping the strip straight and its width uniform. On the positive side, a strip is more likely to pick up variations in clumped or otherwise nonuniform vegetation, especially if the strips are oriented at right angles to topographic variations that might cause vegetational gradients. Depending on the way strip transects are laid out, there may be less time spent in unproductive travel than would be the case with plots since measurements are taken as one moves along the strip. Since strips are essentially a series of rectangular plots placed end to end, estimation procedures are similar to those for plots. Proportional expansion according to the area is a key step in the estimation with strips as well as plots.

Compound Plots. As discussed earlier, the best choice of plot size depends on the type of vegetation being surveyed. In general, larger and more widely spaced plants call for larger plot sizes. This creates a problem when working with different types of vegetation in the same survey. The solution to this quandary lies in using a compound plot with each kind of vegetation being measured on the type of plot that fits it best. For example, a forester might use a 1/5 acre plot for trees having stem diameters greater than 8 in. (as measured 4.5 ft above ground), a 1/10 acre plot for trees with diameters between 4 and 8 in., a 1/50 acre plot for trees with diameters between 1 and 4 in., and a 1/500 acre plot for young trees less than 1 in. in diameter. These plots might all be circular in shape and situated around a common center. The compound nature of the plots must be taken into account when estimates are made, especially when the estimate combines data from different types of plots. A difficulty most often arises from attempts to add data that are not on the same area base.

As a simple example of the computations, take the problem of estimating the density of trees with diameters greater than 4 in. when trees from 4 to 8 in. are counted on a 1/10 acre plot and trees greater than 8 in. are counted on a 1/5 acre plot. Assume that the 1/10 acre plot contains 15 trees of the 4 to 8 in. class, and that the 1/5 acre plot contains 20 trees greater than 8 in. The first step is to bring the different size classes to the same area base. With an area base of one acre, we have

$$\frac{15 \text{ trees}}{1 \text{ plot}} \times \frac{1 \text{ plot}}{0.1 \text{ acre}} = \frac{15 \text{ trees}}{0.1 \text{ acre}} = 150 \text{ trees/acre}$$

for the 4 to 8 in. class, and

$$\frac{20 \text{ trees}}{1 \text{ plot}} \times \frac{1 \text{ plot}}{0.2 \text{ acre}} = \frac{20 \text{ trees}}{0.2 \text{ acre}} = 100 \text{ trees/acre}$$

for the class with diameters greater than 8 in. Now that these classes are on a common area base of one acre, they can be combined to give

$$150 \text{ trees/acre} + 100 \text{ trees/acre} = 250 \text{ trees/acre}$$

A more subtle pitfall for the estimation lies in the calculation of the average size for individual plant units. Unless the size of the subplot is taken into account, plant units from the large subplots can unduly influence the average. Let us modify the previous example a little bit to illustrate the point.

Suppose that we are interested in estimating the average height of maple trees having diameters greater than 4 in. Suppose further that the 1/10 acre subplot contained two maple trees in the 4 to 8 in. class having heights of 30 and 40 ft; and that the 1/5 acre plot contained 2 maple trees larger than 8 in. diameter having heights of 55 and 60 ft. This compound plot indicates a greater density for small maples than for large maples, so a simple average of the four heights would be considerably larger than the true average. The proper procedure is to determine the density contribution of each measured tree and then use these density contributions as weighting factors in estimating the average. The density contribution for a measured tree on a 1/10 acre plot is

$$\frac{1 \text{ tree}}{1 \text{ plot}} \times \frac{1 \text{ plot}}{0.1 \text{ acre}} = \frac{1 \text{ tree}}{0.1 \text{ acre}} = 10 \text{ trees/acre}$$

and the density contribution of a measured tree on a 1/5 acre plot is

$$\frac{1 \text{ tree}}{1 \text{ plot}} \times \frac{1 \text{ plot}}{0.2 \text{ acre}} = \frac{1 \text{ tree}}{0.2 \text{ acre}} = 5 \text{ trees/acre}$$

The general formula for a weighted average is

$$\frac{w_1X_1 + w_2X_2 + \cdots + w_nX_n}{w_1 + w_2 + \cdots + w_n}$$

where the X's are the n things to be averaged and the w's are the weights associated with the respective X's. With the density contributions as weights we have

$$\frac{(10)(30) + (10)(40) + (5)(55) + (5)(60)}{10 + 10 + 5 + 5} =$$

$$\frac{300 + 400 + 275 + 300}{30} = \frac{1275}{30} = 42.5 \text{ ft}$$

as the weighted average height. The unweighted average would have been 46.25 ft, which is too large because it does not take into account the greater density of small trees.

There are two important points to note about compound plots. The first is that the subcomponents of the compound plot must be placed in the same geometric relationship to each other every time a plot is established. This allows some central or corner point to be used for geographic referencing in the information system. Entirely separate survey operations for the various types of vegetation would not permit a comparison of the vegetation components on a point by point basis. The second point is that the compound plot is an integral unit, and its component subplots cannot be treated as being statistically independent when estimates are made. These two points are really just different ways of saying the same thing.

It should not be difficult to see the extension of the compound plot idea to the compounding of strips. This consists of using different strip widths along the same centerline for measuring different types of vegetation.

Bitterlich or Variable Radius Sampling. The ultimate development of the compound plot idea is to use a different plot size for each different size of vegetation unit. Since vegetation units may take on an infinite number of sizes, this means effectively using an infinite number of different plot sizes. The best example of such a system is the one extensively used by foresters and variously called Bitterlich sampling, variable radius sampling, or point sampling. The essential feature of this system is the use of nested circular plots with the plot radius being a constant multiple of the tree diameter as measured at breast height.

The breast height is taken at 4.5 ft above ground in the English system and at 1.3 m in the metric system. In the eastern United States the plot radius is often made 33 times larger than the tree diameter. If the tree diameter is measured in feet, then the plot radius in feet is simply 33 times the tree diameter. If tree diameter is measured in inches, then plot radius in feet is $33/12 = 2.75$ times the tree diameter. The multiplier for converting the tree diameter to the plot radius is often called the *plot radius factor.* To keep the discussion general, let us use the letter C to mean that the plot radius is C times the tree diameter when both are measured in the same units.

This system would be rather cumbersome to apply in the field if the basic boundary check were necessary for every tree in the vicinity. This basic boundary check is to measure both the diameter of the tree and the distance from the axis of its trunk to the common center point of the nested plots. Multiplying the diameter of the tree by the plot radius factor gives the plot radius for a tree of this particular size. If the distance from tree to center point does not exceed the plot radius, then the tree is within its allowed plot. For example, assume that a tree with diameter of 18 in. is situated with the axis of its bole 40 ft from the plot center. Suppose also that $C = 33$ is being used for this survey. The plot radius factor for diameter in inches and the radius in feet is $33/12 = 2.75$. Therefore the plot radius for an 18 in. tree is $2.75 \times 18 = 49.5$ ft. Since the tree is actually only 40 ft from the plot center, it lies within its plot and should be included in the sample.

Fortunately one can be eliminate most of the boundary checking by standing on the center point and optically projecting a fixed angle from the plot center toward each tree. The angle is chosen so that the tree trunk will appear to fill the angle exactly when the tree stands on its plot boundary (see Figure 6.1). Only those trees that appear to be approximately the same size as the angle need be checked by measurement to decide whether they are in or out of the plot. The appropriate angle α for the chosen value of C can be found be examining Figure 6.1 and noting that

$$\frac{\text{tree radius}}{\text{plot radius}} = \sin\left(\frac{\alpha}{2}\right) = \frac{1}{2C}$$

Thus the appropriate angle is

$$\alpha = 2 \arcsin\left(\frac{1}{2C}\right)$$

With $C = 33$, for example, the angle to be projected is 1.7363°.

A simple device for projecting such an angle can be made by placing a small crosspiece on the end of a stick to form a T shape. Then a peepsight is placed on the other end of the stick. The crosspiece will define an angle when viewed through the peepsight. The proper relationship between crosspiece width and stick length can be determined from the fact that

$$\frac{\text{crosspiece width}}{2 \times \text{stick length}} = \tan\left(\frac{\alpha}{2}\right)$$

Therefore

$$\text{crosspiece width} = 2 \times \text{stick length} \times \tan\left(\frac{\alpha}{2}\right)$$

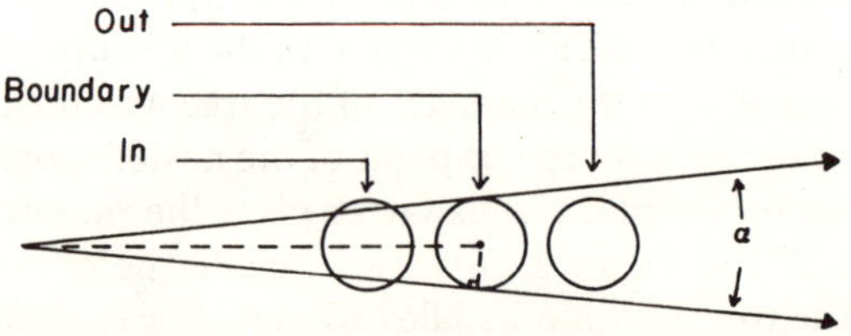

Figure 6.1 Angle projection for determining boundary trees in variable radius sampling. Solid circles represent tree trunks of equal diameter; solid lines represent sides of angle α; dashed lines represent tree radius and plot radius.

Since the sine and tangent functions are nearly equal for small angles, a useable approximation is

$$\text{crosspiece width} = 2 \times \text{stick length} \times \sin\left(\frac{\alpha}{2}\right)$$

$$= 2 \times \text{stick length} \times \frac{1}{2C}$$

$$= \frac{\text{stick length}}{C}$$

so that

$$\text{stick length} = C \times \text{crosspiece width}$$

With $C = 33$, then, we might use a 1 in. crosspiece on a 33 in. stick.

More elaborate devices for projecting angles are also available. Thin prisms are popular angle gauges among foresters. Dr. Walter Bitterlich, who originally conceived this sampling system, markets a sophisticated angle gauge called the Relaskop.

Since the variable radius sampling system is an extension of the compound plot idea, the estimation procedure is also an extension of that for compound plots. The basic operation is to compute the density contribution for each tree in the sample. In order to do this one must know the area of the plot in which the tree was included. The area of a circular plot is π times the square of its radius. In variable radius sampling the plot radius is C times the tree diameter provided that both the tree diameter and the plot radius are measured in the same units. If we denote the tree diameter by D, the plot area becomes

$$\text{plot area} = \pi(C \times D)^2 = \pi C^2 D^2$$

With linear units in feet and an area base of 1 acre the density contribution for a sample tree is

$$\frac{1 \text{ tree}}{1 \text{ plot}} \times \frac{1 \text{ plot}}{\pi C^2 D^2 \text{ sq ft}} \times \frac{43{,}560 \text{ sq ft}}{1 \text{ acre}} = \frac{43{,}560}{\pi C^2 D^2} \text{ trees/acre}$$

For example, take a tree with 18 in. diameter and let $C = 33$. Converting the diameter to feet, we have 18 in. = 1.5 ft. Then its density contribution is

$$\frac{43{,}560}{\pi C^2 D^2} = \frac{43{,}560}{\pi (33)^2 (1.5)^2} = \frac{43{,}560}{7697.6872} = 5.66 \text{ trees/acre}$$

Thus each 18 in. sample tree represents an expectation of 5.66 trees of this size per acre. All components of this formula except D^2 could be condensed into a single value for computational convenience, and the conversion of the tree diameter from inches to feet could also be incorporated. However, such a condensation would tend to mask the logic behind the formula.

To estimate the total density one simply adds up the density contributions for all the sample trees around the common center point. Two sample trees that have the same diameter would also have the same density contribution since they have the same plot size. To compute other types of estimates from a set of sample trees around a common center point, use the density contributions as weights.

Notice that the tree diameter is needed to compute density contributions, and density contributions are needed to compute other estimates. It would appear that one must measure the diameter of every sample tree in order to compute any kind of estimate. This is true, with one important exception. Suppose that we are interested in estimating the cross-sectional area (basal area) of tree trunks on a per acre basis. According to the general procedure outlined above, we should multiply the cross-sectional area of each sample tree by its density contribution and add the results over all sample trees measured around the common center point. If the basal area is to be expressed in square feet per acre, the computational formula for any given tree would be

$$\pi\left(\frac{D}{2}\right)^2 \text{ sq ft/tree} \times \frac{43{,}560}{\pi C^2 D^2} \text{ trees/acre} =$$

$$\left(\frac{\pi D^2}{4} \times \frac{43{,}560}{\pi C^2 D^2}\right) = \frac{43{,}560\pi D^2}{4\pi C^2 D^2} \text{ sq ft/acre}$$

with the results being summed over the sample trees. Note, however, that most of the components in this formula can be canceled, thus reducing it to the constant

$$\frac{10{,}890}{C^2} \text{ sq ft/acre}$$

With $C = 33$ this would further reduce to

$$\frac{10{,}890}{C^2} = \frac{10{,}890}{33^2} = \frac{10{,}890}{1089} = 10 \text{ sq ft/acre}$$

Thus each sample tree, regardless of its size, makes the same contribution to the estimate of basal area per acre. This constant contribution to the basal area per acre is called the *basal area factor.* For the case of $C = 33$ the basal area factor is 10 sq ft/acre. Therefore one need only count the sample trees and multiply by 10 to obtain an estimate of the basal area in square feet per acre.

In summary, the density contribution of a tree with a diameter of D ft is

$$\frac{43{,}560}{\pi C^2 D^2} \text{ trees/acre}$$

and the basal area factor is

$$\frac{10,890}{C^2} \text{ sq ft/acre}$$

The corresponding formulas using meters instead of feet and hectares instead of acres are

$$\frac{10,000}{\pi C^2 D^2} \text{ trees/ha}$$

and

$$\frac{2,500}{C^2} \text{ sq m/ha}$$

It should be clear that the choice of the factor C relating the plot radius to the tree diameter is the crucial decision to be made in applying the variable radius approach.

A final point with respect to the variable radius system is that all measurement and projection of angles is assumed to be done in a horizontal plane. If angles are projected along a slope, too few trees will be included in the sample. The Relaskop incorporates an automatic correction for the slope, but other devices for projecting angles do not. When these other types of gauges are used on sloping ground, trees that appear to lie just outside the plot should be checked by horizontal measurement. When a wedge prism is used to project the angle, an approximate correction for the slope can also be made by holding the lower edge of the prism at a slant which equals the slope angle (Husch, Miller, and Beers, 1972).

Line Transects. The use of parallel lines for measuring an area was explained in Section 3.7.2. The vegetation cover can be estimated by a fairly straightforward application of that principle. One simply runs parallel lines across the tract, and the total length over which these lines transect a particular type of vegetation provides a measure of the area covered by that type of vegetation.

Let the spacing between lines be S. Let T stand for the total length of all lines, and let t stand for the total length over which the lines transect a particular type of vegetation. Then an estimate of the total tract area is

$$S \times T$$

and an estimate of the area covered by the vegetation is

$$S \times t$$

Therefore an estimate of the percent cover for the vegetation is

$$\frac{S \times t}{S \times T} \times 100 = \frac{t}{T} \times 100$$

Likewise the composition of the vegetation on a percent cover basis can be estimated by letting T stand for the total measured length in all types of vegetation and t for the total measured length in a particular type of vegetation.

Less apparent, however, is the fact that lines can be used to implement another extension of the compound plot idea when vegetation units are roughly circular. Assume that rectangular plots of length L are to be situated along a common centerline, and that the plot width is to be equal to the diameter of the vegetation unit. Then there will be as many different plot widths as there are different diameters of plant units. As with variable radius sampling, then, there are potentially an infinite number of different plot sizes. Again the system would be extremely awkward to apply if one actually had to lay out boundaries for a large number of plots. However, it becomes very easy to tell which vegetation units lie inside their respective plots if one notes that the center point of a circular plant unit can lie inside the boundary only if the centerline of the plot transects the plant unit. Thus any vegetation unit transected by the centerline is included in the sample. Since the centerline is sufficient for deciding whether or not a unit is to be included in the sample, there is no need to lay out boundaries at all. Only a line transect is needed to define the common centerline of the several plots for which the width varies according to the diameter of the vegetational units.

In parallel with variable radius sampling one must compute a density contribution for each sample unit and then use these density contributions as weights for making other estimates. The density contribution for a sample unit depends on its diameter D and transect length L according to the formula

$$\text{density contribution} = \frac{1}{L \times D}$$

where L and D are expressed in the same unit of measure, and the area base is the square of the linear unit. For a vegetation unit with a diameter of 1.5 ft transected by a line 100 ft long, the density contribution would be

$$\frac{1 \text{ unit}}{100 \text{ ft} \times 1.5 \text{ ft}} = \frac{1 \text{ unit}}{150 \text{ sq ft}}$$

Shifting the area base from square feet to acres gives

$$\frac{1 \text{ unit}}{150 \text{ sq ft}} \times \frac{43{,}560 \text{ sq ft}}{1 \text{ acre}} = 290.4 \text{ units/acre}$$

Note that the diameter of each unit must be measured in order to compute estimates other than the cover.

The plot width can also be made equal to (diameter + K) by using pairs of parallel lines which are K units apart. In this case a vegetation unit is included in the sample if it is transected by either line or lies entirely between lines. Under this strategy the density contribution for a sample unit would be

$$\text{density contribution} = \frac{1}{L \times (D + K)}$$

Sampling of Points. Yet another strategy in vegetation sampling is to locate a series of individual points and let the plants which occupy these points comprise the sample. This approach has characteristics somewhat like those of sampling with line transects. If the points are distributed systematically over the tract, then one essentially has a dot grid for determining the area as described in Section 3.7.2 Suppose that the points are arranged in rows, with the rows being parallel and equally spaced. Also let the spacing between dots in the rows be fixed. If R is the distance between rows and D is the distance between points in a row, then each point represents an area of $(R \times D)$ with the area units being the square of the linear units used to measure the distance between points and rows. Let T be the total number of points and let t be the number of points that are occupied by a particular type of vegetation. Then an estimate of the area covered by vegetation is given by

$$t \text{ points} \times \frac{(R \times D) \text{ area units}}{1 \text{ point}} = t \times (R \times D) \text{ area units}$$

and an estimate of the total area in the tract is given by

$$T \text{ points} \times \frac{(R \times D) \text{ area units}}{1 \text{ point}} = T \times (R \times D) \text{ area units}$$

An estimate of the percent cover thus takes the form

$$\frac{t \times (R \times D)}{T \times (R \times D)} \times 100 = \frac{t}{T} \times 100\% \text{ cover}$$

Likewise the percentage composition of the vegetation as measured by the cover can be estimated by letting T be the total number of points occupied by all types of vegetation and t the number of points occupied by a particular type of vegetation.

Irrespective of the way points are positioned, this can be viewed as a special case of variable radius sampling for circular units. The special case is that plot diameter and plant diameter are equal. When the plot diameter and the plant diameter are the same, the center of the vegetation unit can lie inside the plot only when the plot center lies inside the circular vegetation unit. Therefore the sampling point is actually the common center of the nested circular plots. Although the angle gauge formulas do not apply, estimation formulas are obtained by setting $C = 0.5$ for the formulas given earlier on variable radius sampling.

Note also that both variable radius sampling and sampling by points can be applied to vegetation units other than trees as long as these units are roughly circular as viewed from above. Bushes, clumps of grasses, and tree crowns are usually close enough to being circular that the procedures are usable.

Satellite or Cluster Plots. When vegetation grows in clumps, a plot of compact shape is likely to fall either entirely inside or entirely outside a clump. In either case only one phase of the vegetation will be sampled and the other missed. One possibility for including more of the variation in each sampling unit is to use strip transects, but this possibility has already been discussed. An alternative approach is to break the area of the plot into several segments and spread out the segments in some fixed geometric pattern around a master plot marker. For example, if each plot is to contain 1/4 acre, this plot area might be spread out into five circular subplots of 1/20 acre each, arranged in a satellite pattern as illustrated in Figure 6.2(a). The U.S. Forest Service has made frequent use of a cluster of 10 variable radius center points arranged as shown in Figure 6.2(b). The distance between any two adjacent center points is 70 ft. The intention with this 10 point cluster is to spread the observations over an area of approximately 1 acre. Satellite systems can be used effectively, but their properties must be clearly understood. Even though an estimate could be developed from each subplot, these estimates are correlated because of the fixed pattern linking the subplots and their proximity to each other. Therefore data from all subplots should be combined and used to produce a composite estimate.

Individual Plants as Sampling Units. As the foregoing discussions indicate, most vegetation sampling systems involve a plot of some sort. However, individual plants may also serve as sampling units somewhere in the process of sample selection. For example, one might use a fairly large plot of fixed size for determining the density by a simple count. As individual units are counted, they could also be assigned numbers. A random subsample of individual plant units on the plot could then be selected for making more difficult measurements. The so called 3-P approach used by foresters in sampling trees is an elaborate system of this nature. The 3-P approach will be outlined in Section 6.3.9.

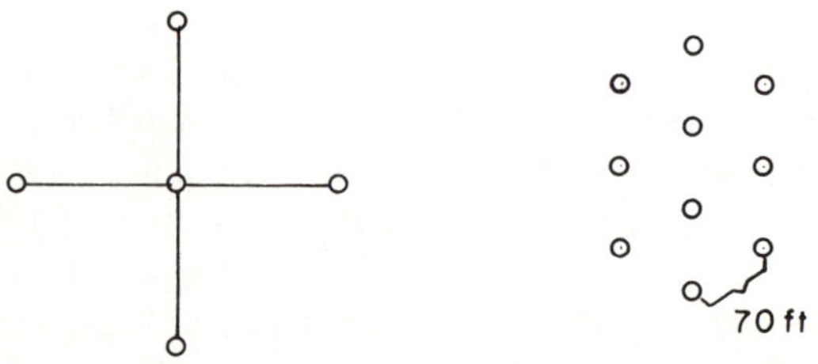

Figure 6.2 (a) Satellite arrangement of circular subplots. (b) Ten-point cluster arrangement of point centers for variable radius sampling used by U.S. Forest Service.

6.2.3 Random versus Systematic Layout of Plots

When a decision has been made on the type of plot to be used, one must still decide on how the plots should be distributed over the area to be surveyed. Any given kind of plot can be distributed in a great many ways. However, there are only two broad categories among the many possible layouts that will produce unbiased estimates of averages per unit area. These two broad categories are *systematic* and *randomized* plot distributions.

As discussed in Section 2.4.3, systematic layouts are those in which plots are distributed in a regular fashion over the area of interest. Furthermore even systematic layouts can contain bias unless one is careful to ensure that every point on the tract has an equal chance of being included in the sample. An important property of systematic layouts is that the position of the first plot determines the positions of subsequent plots. A consequence of this property is that estimates of variability are quite likely to be biased. The regular spacing of systematic layouts does tend to give convenient patterns of travel for field work. This is a major reason why systematic layouts remain popular despite the problems with estimates of variability.

Randomized distributions are those in which each point on the tract has a known chance of being included in the sample, and in which the position of one plot provides no information on the positions of other plots. Randomization makes it possible to obtain unbiased estimates of variability as well as averages per unit area. However, randomized layouts tend to give inconvenient travel patterns for field operations. Methods of achieving systematic and randomized layouts of plots and strips will be covered in the next section.

6.2.4 Plot and Strip Layouts

The most common approach to laying out plots systematically is through line-plot transects. A line-plot transect is a line along which plots are situated at regular intervals. In order to achieve a systematic distribution of plots, parallel line-plot transects are spaced at equal intervals across the tract. Two things must be determined in setting up line-plot transects. One is the spacing between adjacent lines, which we will denote by L. The other is the spacing between adjacent plots on a line, which will be denoted by P. Each plot is associated with a rectangle of size $(L \times P)$ from which it furnishes a sample of vegetation. There are two variations on the design problem in laying out line-plot transects.

One situation is when the area of the tract is not known, but it is desired to establish an average of one plot for every A units of area. In this case the basic relationship is

$$A = L \times P$$

where L and P are measured in the same linear units and the area unit is the square of this linear unit. Two of these three quantities can be chosen for convenience, and the third is determined by solving the equation. For example, suppose that it is desired to establish an average of one plot for every 10 acres in the tract. The first step is to make sure that our units of measure are compatible. Chains and square chains make a convenient set of units for this problem, so we begin by expressing the 10 acres as 100 sq chains. Suppose that we might see a spacing of 10 chains between lines as convenient for field work. We then use the equation above to solve for the required spacing between plots along a line and find

$$A = L \times P$$
$$100 \text{ sq chains} = 10 \text{ ch} \times P$$
$$P = 10 \text{ chains}$$

Thus this plot spacing should be 10 chains. If we had purposely set out to make the line spacing and the plot spacing equal, this could be accomplished by using

$$L = P = \sqrt{A}$$

The second situation is when the area of the tract is known, and it is desired to distribute a specified number of plots systematically over this area. The basic relationship in this case is

$$\frac{A}{n} = L \times P$$

where A is the total area, n is the number of plots, L is the line spacing, and P is the plot spacing. Again the units of measure for A, L, and P must be compatible. As an example, suppose that 100 plots are to be distributed over 300 acres. As before, either L or P is chosen arbitrarily, and the equation is solved for the other. With chains and square chains as units of measure, we might choose to space the lines 5 chains apart. Then the required plot spacing is

$$\frac{3000 \text{ sq ch}}{100} = 5 \text{ chains} \times P$$

$$P = 6 \text{ chains}$$

In order to avoid possible bias, the location of the first plot should be determined by chance. To accomplish this, set up a pair of lines at right angles to each other, oriented so that one line is parallel with the direction that the lines will be run in the field and positioned so that the lines are tangent to the boundary of the tract. Use a random number table to choose a number between zero and L and another between zero and P. Use the tangent lines as axes and the two random numbers as coordinates to locate a potential position for the first plot. If the point lies outside the tract, move parallel to the line axis in steps of P until you reach a point inside the tract. Lines should normally be oriented so as to cross major topographic features of the area.

In order to locate plots completely at random (as opposed to systematically), a map or airphoto of the tract must be available. Set up tangent axes as described in the preceding paragraph. Use a random number table to draw pairs of numbers that span the maximum dimensions of the tract. Use these pairs of random numbers as coordinates on the axes to locate points. If a point falls outside the boundary, simply ignore it and draw a new pair of random numbers. Make a list of plot coordinates and use these to compute the direction and distance of travel from one plot to the next.

The traditional layout of strip transects has been to use parallel and equally spaced transects which run continuously across the tract. The objective in layouts of this kind has usually been to include a specified fraction of the tract within the strips. The basic equation for accomplishing this is

$$F = \frac{W}{S}$$

where F is the area fraction, W the strip width, and S the spacing between centerlines. For example, assume that 1% of the area is to be included in strips 1 m wide. The required spacing between strip centerlines is

$$0.01 = \frac{1\text{ m}}{S}$$

$$S = \frac{1\text{ m}}{0.01} = 100\text{ m}$$

Strips spanning one dimension of the tract can also be located randomly along the other dimension. However, an irregularity in the shape of the tract makes it almost impossible to specify the exact fraction of the tract to be included in randomly spaced strips.

Strips of fixed length can be laid out in much the same manner as plots. However, this is likely to give a large number of partial strips along the boundaries. The area of the strip as determined from its actual length should be used in making estimates from a partial strip. In fact, the actual area should be used in making estimates from boundary plots of any shape. With more compact plot shapes, however, it is not easy to measure the actual area of a boundary plot. This has led to a frequent use of an easy, but not too desirable, expedient of shifting boundary plots. Plots with centers lying inside the tract boundary are shifted so that they lie entirely inside the tract. Plots with centers outside the boundary are not measured at all.

6.2.5 Permanent versus Temporary Sampling Units

When measurements or observations are to be made at more than one point in time (multitemporal surveys), a choice must be made between setting up entirely new plots on each occasion or establishing permanent plots which can be relocated and remeasured.

Permanent plots give better estimates of change than temporary plots. The reason is that temporary plots involve different plant units on each occasion. Two sets of plant units are unlikely to be identical, even if measured on the same occasion. This component of difference, which is due to a variation in space instead of time, inflates the sampling error. In contrast a remeasurement of permanent plots involves only a variation over time and thus gives a smaller sampling error for change.

This gain in sampling error is counterbalanced by the increased cost of establishing permanent markers and additional time in relocating the permanent plots for remeasurement on the next occasion. It is also possible that the method of marking may influence the change between occasions. For example, markings that are too obvious may attract the attention of vandals.

The pros and cons of using permanent plots have received considerable attention in forestry, where extensive networks of permanent plots are the key component of *continuous forest inventory* (CFI) systems.

A compromise solution is called *partial replacement*. The latter approach involves marking part of the plots for remeasurement on the next occasion. The rest of the plots are new for the next occasion. Partial replacement has many desirable properties, but simplicity of the estimation equations is not one of them. For the statistics of partial replacement see Ware and Cunia (1962).

6.2.6 Multistage or Subsampling Systems

It may happen that the vegetation to be surveyed consists of a set of small tracts scattered individually over a wide area. It then becomes a relatively expensive and time consuming proposition to conduct survey operations in each and every individual tract. The strategy known as multistage sampling is often useful for reducing survey costs in these situations.

The sample is selected in several stages, with measurements of the actual vegetation being taken only in the final stage. The earlier stages of sampling serve mainly to localize field operations, but might also involve the mapping of boundaries and the area measurement from airphotos. With a set of scattered tracts the tracts might be used as sampling units in the first stage. A sample of tracts would be selected in which to conduct field sampling operations. Plots of some sort would probably constitute the sampling unit in these field operations.

The basic procedure, then, is to take samples of samples, with sampling units becoming successively smaller at each stage. Estimation proceeds in the reverse order. The smallest units are used to make estimates for the next larger unit in which they lie. In turn these estimates are used to make estimates for the next larger unit, and so on. The hierarchical nature of the process makes *subsampling* appropriate as an alternative term to *multistage sampling*.

Another setting for using the multistage strategy would be the problem of surveying vegetation on a large area about which little is known. One might begin with very small scale aerial imagery from satellites and map out blocks that can be flown with aircraft to obtain larger scale airphotos. A sample of these blocks could next be chosen for the actual flight. The larger scale photo blocks could again be subdivided into blocks for flying at a still larger scale until one finally has airphotos with sufficient detail so that they can be used for laying out ground survey operations in the localized areas which have been flown at this large scale.

Unfortunately the localized distribution of ground survey units resulting from subsampling does not pick up as much of the variability in the population as would an equal number of ground survey units distributed randomly over the entire area to be surveyed. As a consequence, the sampling error for multistage designs tends to be larger than the sampling error for completely random designs with an equal number of plots. Therefore the cost reduction obtained from a multistage design depends on the relationship between travel costs and the costs of installing plots on the ground. Large travel costs and small plot costs tend to favor the use of a multistage approach.

6.2.7 Multiphase Systems

There is often a considerable difference in the cost of making different kinds of measurements on the same set of plant units. In working with trees, for example, we find that the bole diameter at breast height and the length of the bole are relatively easy and inexpensive to measure. In contrast, the volume of the bole requires that diameter measurements be taken at short intervals of length so that the segment volumes can be approximated as frustums of cones or paraboloids which are then added together as an approximation of the total volume. Similarly the fresh weight of grasses and herbs can be determined by weighing in the field; but the dry weight requires slow dessication in an oven prior to weighing. In both cases, however, measuring both easy and difficult variables on the same plant units provides the opportunity for developing relationships by which the difficult quantity can be estimated indirectly from measurements that are less troublesome. A regression equation can be developed to predict the volume from the diameter at breast height and bole length. The ratio of dry weight to fresh weight can be used to estimate the former from the latter.

In order to take advantage of such relationships, one can make the easier type of measurement on a large number of sample units, and then make the difficult measurements on a much smaller subsample in order to derive a relationship for an indirect estimation of the difficult quantity on sampling units where only the easy measurement was taken. Again we have a stepwise or subsampling strategy, but this strategy involves measurements of one kind or another at every step, instead of just at the final step as was the case in multistage sampling. Sampling systems of this nature are called *multiphase sampling.* The computations involved in the total estimation process depend on the nature of the relationship to be used for indirect estimation.

6.2.8 Distance Sampling

Plant ecologists have made some use of a sampling strategy quite different than those so far considered. The idea in this case is to use measurements of the distance between plants or the distance between points and plants to estimate the square root of average area per plant. To appreciate the logic behind this approach, consider a plot of area A containing M individual plant units. The density is measured by M/A and the mean (average) area per plant is A/M. The square root of the mean area per plant provides a measure of average distance D between plants. If one could estimate this same average distance by direct distance measurements in the field, the density could be estimated as

$$\frac{1 \text{ plant}}{D^2 \text{ sq ft}} \times \frac{43{,}560 \text{ sq ft}}{1 \text{ acre}} = P \text{ plants/acre}$$

or

$$\frac{1 \text{ plant}}{D^2 \text{ sq m}} \times \frac{10{,}000 \text{ sq m}}{1 \text{ ha}} = P \text{ plants/ha}$$

Other quantities could then be estimated by multiplying the average size of plants by the density.

Unfortunately the relationship between the distance as measured in the field and the square root of the average area per plant has proven rather elusive. Various ways of measuring the distance in the field have been investigated in the hope of finding one or a combination of several that bears a consistent relationship to the square root of the mean area (Mueller-Dombois and Ellenberg, 1974). Field trials have been made using the distance between neighboring pairs of plants, between a plant and its second nearest neighbor, between a point on the ground and the nearest plant, and several variations of the nth nearest neighbor distance. There would be no problem if plants were randomly distributed over an area, but this is seldom the case, and departures from random distribution take a variety of forms.

One of the more promising approaches is the *point-centered quarter method* (Cottom and Curtis, 1956). To apply this method one establishes points in the field in much the same manner as for centerpoints of circular plots. The area in the vicinity of each point is divided into quarters or quadrants by lines at right angles to each other. The distance from the centerpoint to the nearest plant in each of the four quadrants around a point is measured. The average of these four distances has given a good approximation to the square root of the mean area for several nonrandom plant patterns. As a note of caution, however, the distance methods have been applied mostly in studies of plant ecology where errors in density did not carry serious economic consequences. Field trials should be carried out to compare any proposed distance method against plot methods before using it in economically oriented surveys.

The more important approaches to vegetation sampling have now been covered in general terms, and we can proceed to consider adaptations for particular types of vegetation.

6.3 SURVEYING FOREST TREES

Whenever forest trees of potentially commercial size occur in substantial numbers, information on this component of vegetation should be collected in a format that is usable for forestry purposes. Since the requirements for information to support forestry operations are more stringent than those for other uses of forest trees, this information will also satisfy most other potential users of the data base. It should be noted that foresters often use the term "cruising" as a synonym for surveying when they speak of making measurements on standing trees.

6.3.1 Stand and Stock Tables

The information most frequently required in forestry operations can be summarized in the form of two tables.

A *stand table* shows the average number of trees per acre broken down by diameter classes and species groups. The diameter is measured outside the bark at 4.5 ft above ground. The diameter measured in this position is called *diameter breast high* (dbh). Dbh in either 1 in. or 2 in. classes forms the rows of the stand table. Species or groups of similar species form the columns of the stand table. Thus each cell in the body of the tables shows an average number of trees per acre for a particular diameter class and species. The last column on the right shows the totals by diameter class, and the last row on the bottom shows the totals by species. The format for a stand table (without data) is shown in Table 6.1.

The format for a *stock table* is exactly the same as that for a stand table, but the merchantable volume of wood per acre replaces the number of trees per acre as entries in the table. Therefore Table 6.1 serves as a model for both stand and stock table formats. Similar types of tables can be constructed for metric data with an area base of 1 ha and diameters measured in centimeters. Our next concern is with methods for collecting data necessary to compile stand and stock tables.

Table 6.1 General format for forest stand and stock tables. Stand table entry is number of trees per acre; stock table entry is volume per acre

Dbh class	Species A	Species B	Species C	Dbh total
Class 1				
Class 2				
Class 3				
.				
.				
.				
Class *n*	———	———	———	———
Species total				Grand total

6.3.2 Tree Diameters

Several different diameters can be defined in reference to a tree bole, and there may be economic or scientific interest in measuring more than one of these. Diameters at some points along the lower bole can be reached for measurement from a standing position. Out-of-reach diameters are much more difficult to measure. Diameters encompassing both wood and bark are much easier to measure than diameters inside the bark. Overbark diameter measurements at breast height (4.5 ft) are quick, easy, inexpensive, relatively accurate, and closely correlated with the tree volume or biomass. Small wonder, then, that diameter breast height (dbh) is the most frequently measured variable in forestry.

Dbh. The dbh can be measured to the nearest 0.1 in. with either diameter tape or calipers. Less accurate dbh measurements are usually made with a Biltmore stick.

Diameter Tape. The diameter tape is simply a measuring tape with a hook on one end for implanting in the bark and graduations scaled to read in units of diameter instead of circumference when the tape is wrapped around the tree. Care must be taken to hold the tape level when wrapping it around the tree, or the reading will be larger than it should. Likewise the tape gives readings that are too large for trees which are not circular in cross section. The desired measurement for such noncircular trees is the diameter of a circle having a cross-sectional area equal to that of the tree. Since the circumference of a noncircular tree is always greater than that of a circular tree having an equal area, the diameter tape gives a positive bias. However, the diameter tape is light, compact, and equally good for both large and small trees. Furthermore the diameter tape readings are highly reproducible. Therefore it is usually the instrument of choice for dbh measurements.

Calipers. Accurate dbh measurements for small and moderate sized trees can also be obtained with a simple caliper. The caliper consists of a pair of parallel arms, one arm fixed in position and the other sliding along a scale. The arms are placed in contact with opposite sides of the tree, and the tree diameter is read from the scale as indicated by the movable arm. Since there is often some play in the movable arm, care must be exercised to ensure that it is perpendicular to the scale. The scale must also be perpendicular to the axis of the tree trunk. The accuracy of caliper readings for noncircular trees depends on the way the measurement is taken. Common practice is to average a reading of the largest dimension and a reading at right angles to the largest dimension. Calipers are awkward for large trees because the caliper scale must be at least as long as the diameter of the tree.

Biltmore Stick. The Biltmore stick is a rather primitive device for making quick but approximate dbh measurements. It consists of a scale marked on a stick. The stick is held against the tree with outstretched arm. The zero mark on the stick is aligned with the line of sight to the left side of the tree, and the dbh is read where the line of sight to the right side of the tree crosses the stick.

The geometry of a Biltmore stick is diagrammed in Figure 6.3. The right portion of this diagram shows how the stick is held in relation to the tree, and the left portion is a mirror image of the right showing similar triangles by hatch lines. In addition to the notation of Figure 6.3, let G be the distance from the zero mark on the stick to the point where the D inch graduation is placed. Thus $G = 2g$ and $D = 2r$. From the similar triangles,

$$\frac{g}{r} = \frac{R}{S}$$

so

$$\frac{G}{D} = \frac{R}{S}$$

and

$$G = \frac{DR}{S}$$

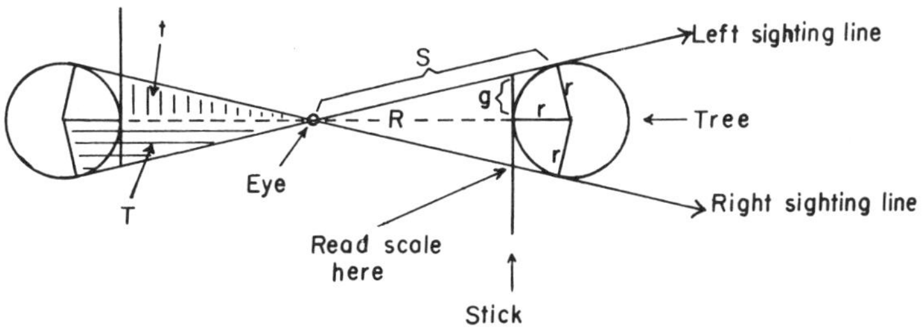

Figure 6.3 Diagram of geometry for a Biltmore stick. R is reach; r is tree radius. Other notation is explained in text.

Squaring both sides gives

$$G^2 = \frac{D^2R^2}{S^2}$$

From the Pythagorean theorem,

$$S^2 = (r + R)^2 - r^2 = r^2 + 2rR + R^2 - r^2 = 2rR + R^2 = R(D + R)$$

Therefore

$$G^2 = \frac{D^2R^2}{R(D + R)}$$

Canceling an R gives

$$G^2 = \frac{D^2R}{D + R}$$

and taking square roots, we obtain

$$G = \sqrt{\frac{D^2R}{D + R}}$$

R in this formula is the distance between the eye and the stick when it is being held against the tree. This is commonly called the "reach."

To construct a Biltmore stick from materials at hand, first determine your reach. Next solve the formula for each graduation to be placed on the stick. Then measure and mark the graduations. Suppose that your reach is 25 in. Then the graduation for a 10 in. tree diameter would be placed

$$G = \sqrt{\frac{D^2R}{D + R}} = \sqrt{\frac{10^2 \times 25}{10 + 25}} = \sqrt{\frac{2500}{35}} = \sqrt{71.43} = 8.45 \text{ in.}$$

from the zero mark.

Reasonable accuracy with a Biltmore stick depends on maintaining the geometric conditions in Figure 6.3. The most important conditions are that the reach be correct, that the stick be perpendicular to the line of sight to the center of the tree and that the head not be moved during the sighting operation. The stick must also be perpendicular to the axis of the trunk. Because of the difficulty in satisfying all these conditions simultaneously, the best that one can hope to obtain is the measurement of the diameter to the nearest inch.

Other homemade instruments can also be fashioned. For small trees and few diameter classes one can cut a series of nested notches in a board with notch widths corresponding to diameter classes. Another approach is to use a single V notch and place calibrations according to the way circles fit in the V notch.

Upper Stem Diameters. It would obviously be possible to measure out-of-reach diameters by climbing the tree, but this is slow and carries a high risk of personal injury. Therefore climbing would ordinarily be restricted to research studies. A variety of optical devices for measuring upper stem diameters while standing on the ground are available, although they are relatively expensive. The generic name for such devices is *dendrometer*. Among the more widely used models are the spiegel Relaskop, wide scale Relaskop, Telerelaskop, Wheeler optical caliper, McClure optical caliper, and the Barr and Stroud dendrometer. Operating procedures for these various dendrometers are beyond our scope, and the interested reader should consult catalogs of suppliers or manufacturers' instructions.

As discussed by Grosenbaugh (1963) these dendrometers are of three general types. *Optical calipers* project parallel lines of sight, do not require distance measurements, and give direct readings. *Optical forks* project angles and require that the distance from the sighting station to the base of the tree be determined. The several Relaskops operate on this second principle. Short base range finders comprise the third category and are typified by the Barr and Stroud dendrometer. They utilize right triangles, require considerable calculation or table lookup, and are expensive because of their precision optics.

Diameter Inside Bark. Inside bark diameters (dib) are determined by first measuring the diameter outside bark (dob) and then deducting the bark thickness. The bark thickness is measured with a depth probe which can be driven through the bark but lodges against the denser wood. The Swedish bark gauge is the standard instrument for this purpose. It consists of a depth probe which is driven into the bark with the palm and a flange to define the outer surface of the bark.

6.3.3 Tree Height

The tree height need not be measured for assembling a stand table, but the height is closely correlated with the volume and therefore is used either directly or indirectly in assembling stock tables. As with the diameter, the height can be defined in more than one way. The two most common types of height are *total* and *merchantable* height.

If the tree is growing on level ground, the total height is a straightforward measurement from the tip of the tree to the ground level. If the ground is sloping, one uses a level plane through the base of the tree as representing the ground. This procedure is followed even for leaning trees, so the measured height may be less (but never greater) than the actual length of the slanting bole. The total height is measured in either feet or meters.

The merchantable height is measured from the upper limit of usable material to the top of an imaginary stump. The stump height is usually taken as 1 ft.

There are several ways of defining the upper limit of merchantability. One way is to specify a minimum diameter with smaller material assumed to be unmerchantable. Both dib and dob have been used for this purpose. A second way is to use the point at which the diameter falls to some specified fraction of the dbh. Still a third way is to limit the merchantable height at the point where large branches, forks, or other major irregularities occur. The third is also used in combination with the first two. It is common practice in the United States to measure the merchantable height in terms of "logs" and half logs. A log is taken to be 16.3 ft, with the fraction of 0.3 being a "trim allowance" for squaring up the log after cutting.

The usual procedure for making relatively accurate height measurements is to use a hand held clinometer as described in Section 3.6.5. A *Merritt hypsometer* is often used for quick but rough estimates of the merchantable height to accompany the dbh as estimated with a Biltmore stick. Like the Biltmore stick, the Merritt hypsometer is just a scale marked on a stick. The Biltmore and Merrit scales are often placed on opposite sides of the same stick to form a so called "cruiser stick." A simple inch scale and a T-type angle gauge for variable radius sampling are often added to the cruiser stick.

The person using the Merritt hypsometer stands at a specified distance from the tree and holds the Merritt stick vertically at arm's length. The zero mark is aligned with the top of the stump, and the height is read in logs and half logs where the line of sight to the upper limit of merchantability crosses the stick. This geometry is diagrammed in Figure 6.4.

The right side of Figure 6.4 shows how the stick is positioned during use. The left side is a mirror image showing similar triangles which form the basis of the measurement. The portion of the stick above eye level h_1 and the reach R form a right triangle which is similar to the right triangle formed by the upper part of the tree H_1 and the base distance B. Therefore

$$\frac{h_1}{R} = \frac{H_1}{B}$$

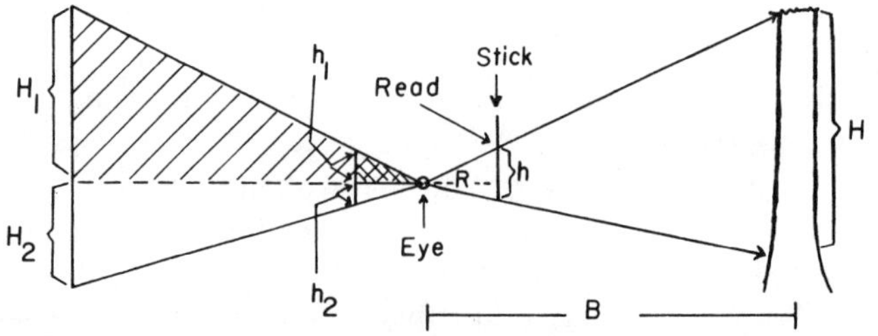

Figure 6.4 Diagram of geometry for Merritt hypsometer. R is reach; H is height of tree; B is base distance. Other notation is explained in text.

and

$$H_1 = B \times \frac{h_1}{R}$$

There is a second pair of similar triangles below the horizontal for which

$$\frac{h_2}{R} = \frac{H_2}{B}$$

and

$$H_2 = B \times \frac{h_2}{R}$$

Since $H = H_1 + H_2$, we have

$$H = H_1 + H_2 = \left(B \times \frac{h_1}{R}\right) + \left(B \times \frac{h_2}{R}\right) = B\left(\frac{h_1 + h_2}{R}\right) = \frac{Bh}{R}$$

or

$$h = H \times \frac{R}{B}$$

Thus R/B provides a constant of proportionality relating the length intercepted on the stick to the height of the tree, and any convenient stick can be graduated accordingly. For example, a reach of 25 in. and a base distance of 1 chain gives a correspondence between 1 log on the tree and 6.17 in. on the stick.

The key elements of the geometry are correct reach, correct base distance, vertical stick, and head stationary. The difficulty in maintaining this geometry limits the expected accuracy to ½ log.

For very rough work the merchantable heights of trees are sometimes determined by ocular estimate. Ocular estimates are very likely to produce bias unless checked frequently by actual measurement.

6.3.4 Tree Form

Given that tree boles are roughly circular in cross section and approximately straight, the three most useful predictors of the merchantable volume should be the dbh, merchantable height, and taper. Measurements of the first two have been discussed, and we turn now to taper or form.

The problem of describing the tree form has been approached at several levels of sophistication. The more sophisticated approaches have involved attempts to find some standard geometric solid such as a conoid, paraboloid, or neiloid which approximates the form of the bole. Such studies have shown that the lower bole is shaped like a neiloid, the upper bole like a paraboloid, and the tip like a conoid (Husch, Miller, and Beers, 1972). Given two diameter measurements separated by a known distance, an equation for the approximating solid can be derived by analytical geometry. However, this knowledge is of little practical utility in the field since a single approximating solid does not fit well over the entire length of the bole.

One solution has been to express the tree volume as a fraction of the volume of some selected solid such as a cylinder having the same dbh and height. This would be called a *cylindrical form factor.* Form factors are also rather impractical for field use because of the need for a series of diameter measurements at regular intervals along the bole to give a direct estimate of the volume.

The most practical solution has been to use the ratio of an upper diameter to a lower diameter as an index of the tree form. Such indexes are called *form quotients*. The form quotient that has found most favor in the United States is called *Girard form class.* The Girard form class uses the dbh as the lower diameter, and the dib at 17.3 ft above ground (top of first log) as the upper diameter. The Girard form class is the ratio of these upper and lower diameters expressed as a percentage. Different indexes of the form quotient type are used in other countries. The dbh usually serves as the lower diameter, with the difference appearing in the choice of the upper diameter.

6.3.5 Tree Volume and Weight

Tradition in the market for primary forest products makes the determination of the tree volume more complicated than is really necessary. Logs destined to be sawed into lumber have been sold in terms of *board feet,* and those meant for other uses have been sold mostly in terms of *cords.*

Board Foot Measure. The board foot is a unit of measure for rough sawn lumber consisting of a piece 12 in. square and 1 in. thick or the equivalent. Somewhat less material than this actually remains after the lumber is finished. However, a rather large amount of speculation is required to use the board foot as a unit of measure for the volume of standing trees. The amount of lumber obtained from a standing tree depends on how the tree is cut into logs, what sawmilling equipment is used to saw the logs, the sawing pattern used, and the type of defects encountered. This speculation takes the form of a *log rule* for predicting the rough lumber yield of a log given its dib at the small end and its length. Although a great variety of log rules have been used in the past, the current practice is limited to one of three rules. These are the International ¼ in. rule, the Scribner rule, and the Doyle rule. Log rules are usually set up in tabular form, and are often printed on the face of a scaling (measuring) stick so that log volumes can be read directly as a log is being measured. Husch, Miller and Beers (1972) give a good summary of the logic behind these three log rules.

The International ¼ in. rule is the most accurate of the three. This rule treats a log as a series of 4 ft cylinders with the first having a diameter equal to the small end dib of the log. The diameter is increased by ½ in. for each successive section to make allowance for log taper. The board foot volume V of a 4 ft section is given by the formula

$$V = 0.2D^2 - 0.64D$$

where D is the diameter (inches) of the cylindrical section.

The Scribner rule was originally developed by drawing diagrams of boards in circles representing the ends of logs. Bruce (1925) found that the formula

$$V = 0.79D^2 - 2D - 4$$

with D being the small end dib (inches) gives a reasonable approximation to Scribner volumes for 16 ft logs. Since the Scribner rule does not make any taper allowance, shorter logs can be handled by reducing the volume from this formula in proportion to length. Scribner log volumes are usually rounded to the nearest 10 board feet, and the final zero is dropped in setting up the table. This is called a "decimal" tabulation. The letter C is used to distinguish the current version of the Scribner rule from other versions that have been proposed in the past. Thus the full name for this rule is usually given as "Scribner decimal C." The Scribner rule underestimates the actual lumber yield, especially for large logs. Such an excess of lumber yield over the log rule estimate is called *overrun*.

The Doyle log rule estimates the board foot volume as

$$V = L \times (0.0625D^2 - 0.5D + 1)$$

with D being the small end dib (inches) and L the length (feet). The Doyle rule gives a large overrun bias for small logs.

Overruns produced by the Scribner and Doyle rules are taken into account by price differentials in marketing logs. The application of log rules for estimating volumes of cut logs is detailed in the *National Forest Log Scaling Handbook* (U.S. Forest Service, 1969).

Cord Measure. A *cord* is an orderly stack of wood occupying 128 cu ft of total space including bark and voids between pieces. The term *face cord* is sometimes used for a stack having an end area of 32 sq ft including gaps between the ends of pieces. The amount of actual wood in a stack of given size depends on such things as size of piece, straightness, and care in stacking. Therefore, the application of the cord measure to standing trees involves uncertainties similar to those for the board foot measure.

Cubic and Weight Measures. Since the board foot and cord measure involve considerable guesswork as well as assumptions regarding the use of the wood, it would seem preferable to measure the volume of standing trees in simple cubic units and then apply conversion factors as necessary for particular product units. Fortunately there is a trend in this direction. In addition to cubic feet and cubic meters, the *cunit* has found some popularity. A cunit is simply 100 cu ft of wood.

The ease with which truckloads of logs can be weighed has encouraged the use of weight as an alternative to volume. Interconversions between weight and volume require a knowledge of the wood density (specific gravity) and moisture content.

Indirect Estimation of Volume and Weight. Both volume and weight are difficult to determine directly in standing trees. The volume must be approximated by measuring the diameter at regular intervals up the bole, calculating the approximate volume of each section, and then summing the sectional volumes. The weight determination also requires estimates of specific gravity and moisture content. Such detailed measurements would be economically prohibitive for large numbers of trees. This can be avoided by developing regression equations (see Appendix) for predicting the volume from less costly measurements of dbh, height, and form. Dbh, height, form indexes, the products of these three, and their logarithms usually serve as independent variables for the regression equation. If compilations are to be done manually rather than by computer, the regression equation is translated into tabular form.

Tables with dbh as the only entry are called *local volume tables* because their use is limited to forests in the area from which data for the regression were collected. Tables with dbh and height as entries are called *standard volume tables* or *double entry volume tables.* Tables having dbh, height, and form class as entries are called *form class volume tables* or *triple entry volume tables.*

Data for the regression equation may be collected either in a special research project or by subsampling in a larger survey operation.

6.3.6 Basal Area and Crown Closure as Indexes of Stocking and Composition

The basal area and crown closure deserve special mention because of the ease with which they can be estimated. As explained in Section 6.2.2, the area occupied by tree boles can be estimated by counting trees with an angle gauge in variable radius sampling and then multiplying the tree count by the basal area factor for the angle gauge. The species composition can be assessed by keeping the count separately by species or species groups. This is an extremely quick way of making estimates in the field. Guidelines for management can be constructed through special studies in which desirable stocking is related to the basal area. The basal area thus becomes a very convenient index for controlling forest management operations.

Ground reconnaissance is necessary for estimating the basal area, but the percentage of ground covered by tree crowns (crown closure) can be estimated from airphotos with a dot grid or special crown closure comparator. Although not as good as the basal area, the crown closure can be used as an aerial index for making preliminary management decisions. The crown diameter can also be measured on large and medium scale airphotos. A comparative ground and photo analysis may reveal a possibility for estimating the basal area indirectly from a combination of crown closure and average crown diameter. This relationship will differ between species and may also vary according to growing conditions.

6.3.7 Site Quality

Foresters use the term *site quality* to describe the tree growth potential of an area as determined by the combination of physical, chemical, biological, and climatic factors influencing the forest vegetation. Since the site quality is an expression of an area's potential for producing wood, it enters into all phases of timber management. Good sites can be managed more intensively than poor sites because they give a larger and earlier return on the money invested. Poor sites are usually more sensitive to mistreatment than good sites because poor sites often have thin topsoil and/or steep slopes. Harvesting operations must be conducted with care on such sites, and it is often difficult to obtain reproduction after harvesting. Furthermore the less thrifty condition of trees on poor sites usually increases their susceptibility to insects and disease. Several techniques have been developed for assessing the site quality, but none of these approaches is entirely satisfactory. Some considerations for choosing a measure of site quality are as follows.

The first requirement of a measuring technique is that it be both accurate and consistent. Therefore any feature to be measured as a reflection of the site quality must be closely correlated with the capacity of the area to produce forest growth. Differences in the indicator must imply differences in the growth potential. Conversely the indicator of site quality must be sensitive enough to variation in the growth potential to discriminate between sites that have a real difference in this respect. Furthermore the indicator must function well under a wide range of conditions. An indicator that works only under certain conditions cannot be adopted as a standard measure. An ideal measure of site quality should be simple, inexpensive, and objective. It seems to be some sort of law of nature that the accuracy of measurement increases with complexity of the technique. However, field personnel will view simplicity as a plus.

A good measure of the site quality must be somewhat independent of the stand structure. Most stands for which the site quality is to be determined will not be fully stocked. Therefore a good indicator of the site quality must be relatively insensitive to changes in the level of stocking or stand density.

Different species are adapted to different environmental conditions and have genetically controlled differences in growth rates. Furthermore different races or varieties of the same species may grow at quite different rates on the same area. Thus a given value of a site indicator cannot mean the same thing for all species. However, it is highly desirable to have a measure of site quality which can be translated from one tree species to another. This would allow the determination of an area's suitability for a species that is not part of the present flora. This is very important in situations where type conversion is contemplated.

Also it is desirable that the age structure have a minimal influence on the applicability of the technique. For instance, the age structure becomes a problem in the application of techniques that require the measurement of dominant and codominant trees of substantially the same age. In uneven-aged stands the crown position is controlled jointly by age, growth rate, and competitive ability. Therefore trees of the same size may have quite different ages in an all-aged stand.

Nonforested areas present one of the most difficult situations for achieving a good determination of site quality. However, the site quality of nonforested areas must be evaluated in order to make the correct choice of species for reforestation or afforestation of denuded areas. In such situations techniques for assessing the site quality solely on the basis of tree characteristics have an obviously limited utility.

There are basically two approaches to the assessment of site quality. On the one hand our interest lies in the growth of forest vegetation. Therefore it seems logical to use the presence, age, size, and other characteristics of forest flora as indicators of site quality. On the other hand environmental factors such as nutritional status, physical factors, and climatic factors are the underlying causes of variations in site quality. Thus it is equally logical to approach the assessment of site quality through measurement of environmental factors. The vegetational approach will be considered first.

A considerable amount of European work and some American effort have been directed toward the assessment of site quality from the presence and relative abundance of understory vegetation. This approach places a premium on the botanical savy of the forester.

The simplest use of indicator plants is to identify each site class by two or three key species of understory plants which are most characteristic of the site class and best differentiate it from other site classes. In this system the site classes are usually designated by the scientific names of the key species. However, the key species approach is quite vulnerable to chance variations in plant distribution and local disturbances which may cause the key species to be absent from the site class that they would otherwise characterize.

This problem can be largely circumvented by noting the presence and relative abundance of a whole series (or spectrum) of indicator plants which are listed in the order of site quality on a prepared form. The point along the list at which the flora of the area centers then serves as an indicator of the site quality. In relation to the use of indicator plants it should be noted that understory vegetation is strongly influenced by the presence and character of the overstory. This complicates the use and interpretation of indicator plants.

When overstory trees are present, the trees themselves provide the most direct source of information on the site quality. Height as a function of age forms the basis for the most commonly used measure of site quality in the United States. The primary advantage of height as a measure of site quality is that the total height of dominant and codominant trees bears little relation to the stand density except at the extremes of the density range. Dominant trees are those with crowns projecting above the level of the general canopy, and codominant trees are those with crowns forming the general upper level of the canopy.

The *site index* is the average height of the dominant and codominant trees at some predetermined index age. The index age is usually 50 years, but it may be as few as 35 for short rotation species and as many as 100 for long rotation species. Since stands are usually not at index age, it is necessary to prepare a set of site index curves for each species. A site index curve shows the relationship between height and age for a given site quality. The range of site qualities is usually divided into three to five classes, and a curve is prepared for each class. Thus a set of site index curves contains a curve for each site class, with the entire set of curves drawn on the same graph. The height forms the vertical axis of the graph, and age forms the horizontal axis. The site quality is determined with the aid of a set of site index curves by taking a sample of dominant and codominant trees. On each tree the total height is measured and the total age is determined either from records or by using an increment borer. An increment borer has a hollow bit that can be twisted into a tree with a detachable handle. When the bit has been driven into the center of the tree, the core of wood in the hollow bit is removed by inserting an extractor, giving the handle a reverse turn, and then sliding out the extractor containing the core. Annual rings on the core are then counted to determine the age of the tree. The average height and average age for the sample trees are calculated. These two averages determine a point when plotted on a set of site index curves. Interpolation between the curves lying above and below the plotted point gives the site index.

A set of site index curves may be prepared in two ways. One method is to take height measurements and borings for age on standing trees covering the full range of ages and site qualities. A guiding curve is then prepared from the entire data set. The shape of the index curves for all site classes is controlled by the shape of the guiding curve. The difficulty with this method is that site index curves are sometimes polymorphic, that is, the curves should have different shapes for the different site classes. Since the present method gives curves which all have the shape of the guiding curve, it will lead to errors for cases in which curves are polymorphic.

A more satisfactory, but tedious, method of preparing site index curves is through stem analysis. In this approach each sample tree is dissected to determine the age at regular intervals of height up the bole. The presence of polymorphism can be detected by plotting separate growth curves for single trees or subgroups of trees. Trees of different ages should be included for each site in order to smooth out the effects of local variations in the growing conditions and unfavorable growing seasons.

The site index works relatively well for even-aged stands. An even-aged stand is one in which the trees are all of approximately the same age. However, the site index is by no means an ideal or universal measure of the site quality. The increment borings necessary to determine the age are relatively difficult and time consuming. Furthermore it is necessary to add a rather subjective allowance for the number of years required to reach breast height where borings are taken. There is some question as to the desirability of including codominant trees, and some sets of site index curves are based only on dominant trees. The average age in relation to dominance and codominance may cause difficulty in uneven-aged and all-aged stands. The site index for one species cannot be directly translated to another species. The site index is not applicable in nonforested areas. Finally increment borings cannot be used in tropical areas where trees do not form annual rings.

The second general approach is to measure the site quality through the environmental factors that are the underlying causes of variation. All the nutritional, physical, and climatic factors of the environment exert an influence on forest growth. However, many of these environmental variables are difficult to measure on a local scale. Microclimatic factors require expensive instrumentation and are highly variable over a small area. Climatic data from nearby weather stations, on the other hand, apply to such a large area as to be virtually useless for discerning local variations in site quality. Another limitation on the use of environmental variables is the high degree of intercorrelation among the variables. This complicates the formulation of a mathematical model to integrate the various factors into a single index of site quality. The best strategy in attempting to measure the site quality through environmental variables is to isolate and work with the limiting factors.

The environmental approach has been most successful where the available soil moisture is the limiting factor to forest growth. The availability of soil moisture is largely determined by the physical properties of the soil. Where the available soil moisture is a limiting factor, the soil properties furnish a reasonably good basis for assessing the site quality. The depth of soil horizons, the soil texture, and the slope will usually be among the more important factors. Site classification on the basis of soil may be done with varying degrees of sophistication. On the one hand, broad site classes may be set up with soil and topographic specifications given in descriptive form. On the other hand, regression models may be formulated with soil characteristics as independent variables. Soil-site has the advantage of being independent of vegetative cover. However, a single soil-site system is not universally applicable because a different area may have a different complex of limiting factors.

6.3.8 Growth Projections

Since forest trees are long-lived, the measurement of growth is important in timber management planning. The problem of measuring the plant growth has already been discussed in general terms. However, the forester is faced not only with the problem of measuring past growth, but also the more uncertain venture of predicting future growth over periods ranging from 5 to 20 years. Our primary concern here is with this problem of prediction. The most advantageous way of making such predictions depends on the stand structure and also the amount of detail needed in the predictions.

The fewest problems are encountered when dealing with stands having a simple structure. By simple structure we mean that the stand is relatively even-aged and contains a single species. One can exploit this simplicity by basing predictions of future performance on past experience with stands very similar to the one being considered. This past experience is recorded in the form of *yield tables* which describe the expected development of stands over time. Age and site quality serve as classification variables for determining which of the tabled stands is most like an actual stand.

Age and site quality are the only classification variables for *normal yield tables* which present numerical descriptions of fully stocked stands. Values shown in normal yield tables must be reduced in accordance with the stocking percentage of the actual stand as measured by an index such as the basal area. *Empirical yield tables* are similar to normal yield tables except that there is no attempt to seek out "fully stocked" stands in collecting the original data from which the yield tables are derived. Adjustments must still be made for the difference in stocking between the tabled stand and the actual stand, but the tabled stand is not purported to represent the nebulous concept of "full stocking." *Variable density yield tables* contain an additional classification variable such as the basal area to represent different levels of stocking.

There are some drawbacks associated with yield tables. One drawback is the large amount of data collection required to construct yield tables in the first place. Another is that the yield and structure of forest stands are altered through management. If the management regime for tabled and actual stands is greatly different, development of the actual stand over time cannot be expected to parallel the sequence shown in the table. The yield table approach soon breaks down with an increasing complexity of the stand structure. The greater the number of age classes and species present in the stand, the more difficult it becomes to prepare yield tables for the many possible combinations of structural variables. Likewise, the greater the number of subjective adjustments required for the difference between tabled and actual stands, the more uncertain the growth projections become.

When yield tables are not available, growth projections must be based on past performance of the stand being studied as determined from survey samples. Data on diameter growth can be obtained either by remeasurements over time of by measurement of increment cores. Data on the height growth can be obtained either by sectioning stems (stem analysis) of by remeasurement over time. Endemic mortality must also be taken into account.

When a breakdown by size class is not required in the growth projections, the two-way or total stand projection method can be used (Spurr, 1952). In this method it is assumed that the relationship between basal area per acre or per hectare, average height of trees, and average volume per unit area will not change during the projection period. The nature of this relationship depends on the average form of trees comprising the stand. The assumption that the tree form will not change greatly over a period of 5 to 10 years is usually quite reasonable. This average form relationship can be evaluated from survey data. Given estimates of mortality and growth in both the average basal area per acre and the average height, projections of future basal area per acre and future average height can be developed. The present stand form relationship can then be used to estimate future volume.

Stand table projection methods are appropriate when a breakdown by dbh classes is required. Like total stand projection, stand table projection involves the assumption that form relationships do not change during the projection period. Form relationships are incorporated into a local volume table that gives tree volume as a function of dbh. The procedural steps in stand table projection are as follows.

Inventory data are used to assemble a stand table showing the present distribution of trees by dbh class and species. This present stand table is converted to a present stock table via the local volume table. Growth data from increment cores are used to determine dbh growth rates by diameter class and species. Each cell in the present stand table is then examined to determine which class the trees would occupy after growing during the length of the projection period. The result of this operation is a projected stand table showing how the trees are expected to be distributed by diameter class and species at the end of the projection period. The local volume table is then used to convert the future stand table to a future stock table. The difference between present and future stock tables gives the projected growth. There are several variations in the treatment of growth data and in the way the present stand table is translated into a future stand table. See Husch, Miller and Beers, (1972), Avery (1975), or Spurr (1952) for details of this and other growth projection methods.

6.3.9 The 3-P Approach to Forest Sampling

This section contains a brief outline of a sampling strategy which has become increasingly popular among foresters since its introduction by Grosenbaugh (1965). Since it has been used primarily in the timber inventory, it is covered here rather than in Section 6.2. Also it is not a plot oriented system like those covered in the earlier section. The name 3-P stands for "probability proportional to prediction." The essential features of this system are as follows.

Each tree or other elementary unit in the population is treated as an individual. Each of these individuals is visited and a prediction is made of the value that would be obtained by measuring the individual. The predicted value is recorded, and a random number is drawn to see whether an actual measurment should be taken. The probability that the individual will be selected for actual measurement is proportional to the size of the ocular estimate. Thus the actual measurement effort is concentrated on the more important units. On the subsample of individuals actually measured, the ratio of measured to predicted value is calculated. These ratios are averaged, and the total of the estimated values for all individuals is multiplied by the average ratio to obtain the final estimated total for the population. Thus the average ratio of actual to predicted values is used as a correction factor for the total obtained by ocular estimate.

The more consistent the estimates, the smaller will be the sampling error of the final estimate. The mechanics of designing and carrying out 3-P timber inventories are covered by Husch, Miller and Beers, (1972), Space (1974), and Wiant (1976). Schreuder, Sedransk, and Ware (1968) present a good critique of 3-P sampling. Lund (1975) has compiled an annotated bibliography on 3-P sampling. Actually 3-P sampling is a general system with its potential utility extending beyond timber inventory. However, care must be exercised in defining sampling units since each one must be visited. 3-P sampling can also be used as a second stage of sampling for surveying plant units within plots located in more conventional ways.

6.4 SURVEYING SHRUBS AND TREE SAPLINGS

There are some rather obvious differences between trees and shrubs, and these differences must be taken into account in surveys. However, the material already covered in this chapter should permit us to deal with these considerations rather briefly. Since saplings of tree species are similar in many respects to shrubs, many of these same considerations apply.

6.4.1 Density

Shrubs and saplings are typically more dense than trees, and often grow in clumps. Plot sizes and widths of strip transects for shrubs and saplings must be smaller than those for trees in order to avoid counting large numbers of individuals on a single plot or strip. The greater density also makes it more difficult to run out a tape for radius measurements on circular plots. Therefore square plots with boundaries marked by string are preferable to circular plots. Many shrubs have a bushy growth form with multiple stems. This may create difficulties in recognizing individual plant units for counting and also causes problems in deciding whether a plant unit lies inside or outside the plot boundary. These problems may be serious enough to make one abandon density in favor of cover estimates.

6.4.2 Cover

Basal area estimates are of little value for shrubs and saplings because it would take a great many of the small stems to give an appreciable basal area.

However, the area covered by foliage is often a convenient variable to use in surveying this type of vegetation. Because of the small stature, there is little problem with parallax in ocularly projecting foliage down to the ground. Visual estimates of the percent cover on small plots are often sufficiently accurate for many purposes. In order to avoid bias, however, the ocular estimates should be checked periodically by marking off a grid of smaller squares over the larger plot and making separate estimates for each square in the grid. Comparing the conbined estimate from grid squares to the estimate made for the plot as a whole usually reveals any tendency toward bias.

Line transects provide a convenient and objective method of determining cover. It is relatively easy to judge whether the line is running through or under a shrub. The field record consists of the line length and the portion of this length which is covered by shrubs. The covered fraction of the line length multiplied by 100 gives an estimate of the percent cover.

The angle gauge method can be readily adapted for determining the cover of shrubs which grow as separate and roughly circular bushes or clumps. The procedure parallels that for angle gauge sampling of trees, except that circular clumps constitute sighting units and a larger angle is used.

6.4.3 Biomass

Volume is little used in surveying shrubs, but biomass (dry weight) finds frequent application. In contrast to trees, shrubs are small enough that cutting and weighing is quite feasible although time consuming. The amount of clipping and weighing can be reduced by developing regression equations for indirect estimation. The independent variables for such regression equations are density, cover, average height, and average stem diameter. Data for developing the regression equations can be obtained by either subsampling or conducting special research studies. A graduated pole usually works quite well for height determination, except that it may be difficult to maneuver in very dense brush. A small caliper works well for taking diameter measurements. In this case diameter measurements are taken near the ground rather than at breast height. A convenient height for measuring the diameter of many species is 1 ft above ground. The biomass can be broken down into components as necessary.

6.4.4 Browsing

Harvest in the form of animal browsing is often of interest and is one of the more difficult things to measure on shrubby species. Wild animals usually confine their browsing to the more succulent twigs. Cropped ends of twigs thus remain as evidence of browsing activity, and it is often possible to identify the type of animal by the way twigs are severed.

A relatively straightforward index to browsing activity is obtained by counting both cropped and uncropped twigs to determine the percentage of available twigs that have been utilized. If there is difficulty in distinguishing new browsing from older browsing, one may have to premark a large number of twigs and then resurvey these marked twigs to determine the percentage that have been cropped. Combining the taper of uncropped twigs with the diameter of cropped twigs may allow a rough estimate of the amount of material removed by browsing.

In order to assess the total effect of browsing more quantitatively, it is often necessary to make paired comparisons of fenced and unfenced plots in close proximity to each other. Plots protected from animals by fencing or caging are called *exclosures*.

6.5 SURVEYING HERBACEOUS VEGETATION AND SEEDLINGS OF WOODY PLANTS

Further adjustments of the survey technique are required in moving from shrubs and saplings to herbaceous vegetation and woody seedlings. The greater abundance of the latter dictates very small plots, typically on the order of 1 sq m or 0.001 acre. Collapsible frames strung with rope or wires can be used effectively in laying out such small plots. The small stature of the plants often permits ready access from above, making it possible to take vertical photographs and even chart the positions of plants in the plot with a pantograph. Point observations can be made objectively by lowering pins along guides on a frame standing over the vegetation to record "hits" on the vegetation below. Biomass determinations can be made directly by clipping, sorting, and weighing with subsamples for measuring the moisture content in the laboratory. Some of these possibilities are discussed below for the more common parameters of vegetation.

6.5.1 Density

The density determination for seedlings of woody plants is a fairly straightforward matter of counting the seedlings in small plots. However, density is a useful measure for herbaceous vegetation only when plant units are readily recognizable as individuals. This is not the case for plants that grow from underground stems called rhizomes or for those that have spreading and vinelike growth habits above ground. In the latter case cover is a more useful measure.

6.5.2 Cover

Cover is a convenient measure for working with many types of herbaceous vegetation, especially those for which density is not appropriate. Two types of cover can be recognized. The first is the area covered by foliage. The second is the area occupied by the basal parts of the plants as revealed by clipping just above ground level. The latter is useful for plants that grow as clumps or bunches.

The quickest way of assessing the cover is by visual estimates in small plots. It is often helpful to subdivide the plot into a grid of smaller squares by wires running across the frame. If such a grid is used, one must bear in mind the fact that the grid squares are not independently located samples and should be used only for improving the composite estimate for the plot as a whole.

Evaluating the cover by points is another widely used method of sampling herbaceous vegetation. In this method a frame is placed over the vegetation and a pin is lowered along guides on the frame to avoid consciously influencing the place where the pin strikes plant foliage or ground surface. The fraction of pin placements striking vegetation or a particular type of vegetation provides a means of estimating the percent cover.

Some versions of this procedure involve several pin positions on the same frame. In this case estimates must take account of the fact that several pin positions on the same frame are not equivalent to an equal number of independently located pin positions. The use of multiple-pin frames is somewhat parallel to gridding of plots.

Goodall (1952) and others have called attention to the need for sharpening the end of the pin so that it is actually a point rather than a small circle. A sighting device with double crosshairs is an alternative way of implementing this kind of sampling (Mueller-Dombois and Ellenberg, 1974).

There have been suggestions that the pin or sighting device should be inclined at an angle to increase the likelihood of a hit on vegetation. However, this requires that correction factors be developed for obtaining vertical estimates from the slant counts unless the interest lies only in the relative composition of the vegetation. However, there is better reason to consider inclined pins when one is interested in estimating the leaf area index of drooping foliage.

6.5.3 Frequency

Frequency has been a popular measure for herbaceous plants because it is so easy to determine the presence or absence in small plots. Problems associated with the effect of the plot size on frequency data have already been discussed in Section 6.1.4. Plots for determining frequency are often very small.

One common procedure is to toss a small hoop over the vegetation, with the hoop serving to define the plot boundary. The toss supposedly avoids conscious bias in the placement of the hoop. However, a haphazard toss should not be confused with randomization.

Frequency has even been used with a loop less than 1 in. in diameter forming the plot. With such small plots the frequency data have characteristics similar to the data obtained from points defined by pins. In fact, loop counts approach pin counts as the size of the loop decreases. However, loops will always give higher average counts than pins. Therefore the two types of data cannot be considered equivalent.

6.5.4 Biomass

The most accurate and straightforward method of sampling for biomass is clipping, sorting, and weighing with subsamples for moisture content. However, this can become an expensive operation, especially for large areas that require many plots. Economics may dictate that less expensive methods be used in order to distribute more plots over a larger area. There are several possibilities for achieving this.

One possibility is to use ocular estimates on most of the plots with some plots being measured by both ocular estimates and clipping. The ratio of clip estimates to ocular estimates on plots measured both ways provides a correction factor for plots measured only by ocular estimates. Formally this is a type of multiphase sampling called double sampling with ratio estimation.

A second approach is to develop regression equations for estimating the biomass from measurements such as cover and average height. Data for developing regressions can be gathered either by subsampling or in special studies. A modification of pin sampling also offers another variable which may be useful for estimating the biomass indirectly. Normally one records only the first hit as the pin is lowered. One can continue lowering the pin and record any additional hits on plant parts below. The average number of hits per pin position, called *cover repetition*, should bear some relationship to the biomass and therefore be a potential independent variable for regressions.

6.5.5 Growth and Grazing

In the absence of grazing, growth can be estimated by the difference in samples taken on successive occasions. However, grazing complicates the issue since one must contend with simultaneous growth and removal. There is no great problem in estimating net growth, but one would often like to have separate estimates for total growth and removal by grazing.

If grazing is light enough to leave some plants of each species ungrazed in each plot, it may be possible to estimate total growth from the ungrazed plants and removal from the grazed plants. This involves the implicit assumption that ungrazed and grazed plants grow at the same rate, which may or may not be reasonable, depending on the circumstances.

If grazing pressure is heavy enough that all plants are affected, it may be necessary to use paired plots with one plot of each pair being protected by an exclosure.

Another possibility is to do studies in which grazing is simulated by clipping.

However, it is difficult to assure that simulated grazing has the same effect as natural grazing.

6.5.6 Phenology

The term *phenology* refers to the temporal sequence of plant development over the growing season. A word of caution is in order to point out that the character of herbaceous vegetation does vary considerably over the growing season. Herbaceous species start growth at different times, attain their maximum size at different times, flower at different times, and reach senescence at different times. Thus vegetative cover on the same area can have strikingly different appearances at different times in the growing season. In particular, species that are in flower will appear to be more important than they actually are according to objective measures such as density, cover, and biomass. If one is interested in the total vegetation complex it may be necessary to do surveys at several times during the growing season in order to achieve a fair representation for all species. If one in interested in particular species, surveys should be conducted when the plants of interest are easiest to detect and to measure. A knowledge of phenology is especially important for interpreting herbaceous vegetation from aerial photos.

6.6 RANGE SURVEYS

The range manager has a need for several types of information, and space permits only a brief outline of these needs. However, the needs can be largely satisfied by an appropriate combination of the methods covered in this chapter and in Chapter 4.

Species composition is important because livestock show preferences for some species and avoid others. Certain species of plants are even poisonous to cattle, and such poisonous plants may or may not be completely avoided in grazing. The amount of forage available and a breakdown of this forage by palatability is necessary for determining the number of animal units that can be grazed on the range and the duration of such grazing. An *animal unit* is a cow and a calf or the equivalent capacity for consuming forage in other types of livestock. The trend in the range condition is also of critical importance since a decline in condition means that grazing pressure is excessive even though forage remains for livestock.

Range managers often evaluate the range condition in terms of a theoretical vegetative cover (supposedly climax) which might be expected if the range were to be left ungrazed for a sufficiently long period of time to allow the vegetation to return to its original or natural state. When such natural vegetation is placed under grazing pressure, some species of plants will decrease in abundance either because they are sensitive to grazing or because livestock show a strong preference for them. These species are called *decreasers*. Other species present in natural vegetation will increase under grazing pressure. Such species are either avoided by livestock or are tolerant to grazing pressure. In either case the decline in competition from decreasers shifts the balance of growing conditions in their favor. These species are called *increasers*. The decline in competitive pressure will also allow rough plants to become established which otherwise would not be present. These are called *invaders*. Therefore a shift in the vegetative composition toward increasers and invaders signals a decline in the range condition.

The theoretical model for the natural ungrazed vegetation also includes expected proportions of decreasers and increasers. The percentage of natural vegetation remaining provides an index of the range condition. This percentage is composed of all decreasers, and increasers up to the percentage that they would comprise in the natural vegetation. Invaders are not included since they would not be present to any significant extent in natural vegetation. Suppose that decreasers currently comprise 5% of the vegetation, increasers 75%, and invaders 20%. Assume further that increasers comprise 50% of the natural vegetation. Then the present vegetation as a percent of the ideal is $5\% + 50\% = 55\%$. This example is much simplified since the increaser species would actually be judged individually against their respective natural percentage occurrences rather than collectively.

The U.S. Forest Service makes extensive use of a three step procedure for evaluating rangelands. The first step consists of frequency determination by lowering a small loop along a transect line. The second step involves photographs. The third step is a score card rating of other factors such as erosion.

Francis (1978) gives a comparative summary of range inventory procedures used by federal agencies. Stoddart, Smith, and Box (1975) devote a chapter to general considerations in range inventory.

6.7 SURVEYING WOODY DEBRIS

Thus far in this chapter we have been concerned with surveying living plants. There is also reason to measure dead material, especially when evaluating hazard for wildfire control. Leaf litter and duff on the forest floor can be sampled quite readily by collecting the material from small plots. However, this approach does not work so well for the larger material.

Line and planar intersect sampling techniques have been developed to meet the need for surveying larger materials (DeVries, 1973; Roussopoulos and Johnson, 1973; Brown, 1974; Brown and Roussopoulos, 1974; Van Wagner and Wilson, 1976). These methods are based on the assumption that the pieces are scattered randomly over the forest floor. Line and planar intersect techniques are conceptually the same, with planar intersect samples being just short versions of line intersect samples. The term *planar* is actually more descriptive.

Field procedures consist of establishing a series of lines. The direction of each line is chosen randomly to avoid bias. Depending on the line length, the line can be defined by either a bar or a measuring tape. The line defines an imaginary vertical plane. The diameter of any piece crossing this plane is measured and recorded.

The contribution of a recorded piece to the estimate of weight W in tons per acre is

$$W = \frac{11.65D^2S}{L}$$

where D is the diameter of the piece in inches, L the total horizontal length of all lines in feet, and S the specific gravity of the material. If a substantial number of pieces are not oriented horizontally, a correction should be applied as described by Brown and Roussopoulos (1974).

6.8 ASSOCIATION ANALYSIS

Plant ecologists have devoted a good deal of effort toward investigating natural groupings of species or lack thereof in plant communities, and this chapter would not be complete without an attempt to establish a perspective in this matter. There are two major schools of thought in this regard, and there has been considerable controversy between them.

One view is that the physical factors of the environment are the primary determinants of where a given species does or does not occur, and that each species responds more or less independently to environmental gradients. Corresponding to any environmental gradient there should be a continuum of change in the abundance of each species depending on its sensitivity to the factor that changes along the gradient. The overall pattern of plant distribution is a result of superimposing the continua for the several species. This has come to be known as the continuum concept.

The other view is that plant species show a greater degree of association than could be explained by superimposing independent responses to simple gradients in the physical environment. The plant cover on an area is thus seen to be more than just the sum of the species that comprise its parts. There is thought to be a sociological order of interaction among species which would permit a hierarchical grouping of plant communities. The term *phytosociology* embodies much of the connotation underlying this school of thought, although continuum studies also fall under this same general heading.

Ecologists from each school have mostly confined themselves to the methods that are compatible with their respective viewpoints. Those from the continuum school have sought to find orderings in the vegetation that would reflect the underlying environmental gradients. These ordering techniques are called *ordination* methods. The ordination approach is exemplified by the work of Curtis and his associates in Wisconsin (see Curtis, 1959). Ecologists from the other school have concentrated on classifying vegetation into community hierarchies. The classification approach is exemplified by the work of Braun-Blanquet and his associates in Europe (see Mueller-Dombois and Ellenberg, 1974).

As with most controversies, the final answer probably lies between the two extremes. The composition of vegetation is undoubtedly controlled by environmental conditions. Where a marked gradient does exist, vegetation will be distributed in accordance with the continuum concept. On the other hand environmental factors operate as complexes and may not form simple gradients. Furthermore it is well documented that plants alter their environment at least on a microenvironment scale. Thus the species composition on a given area will continue shifting as in succession until a group of species with compatible requirements in terms of total environment finally becomes established. This is a form of competitive interaction between species superimposed on the macroenvironmental control.

The investigative strategies of both schools have value, except that the search for an all encompassing hierarchy seems somewhat futile. The ordination of vegetative cover reveals changes in environmental factors that may be of practical value for ecosystem management. An example is the ordination of understory vegetation as indicators of the site quality for forest trees mentioned in Section 6.3.7. Likewise, natural groupings of vegetation reflect complexes of environmental factors, whether or not they can be arranged in a neat hierarchy. However, there seems to be little point in reviewing the classical techniques of ordination and classification since the developing field of multivariate statistics has made a more powerful battery of techniques available.

Cluster analysis, factor analysis, and canonical analysis can be applied to either ordination or classification. These multivariate techniques are mathematically sophisticated and must be carried out by computer. The mathematical sophistication places them beyond the scope of this book. See Williams and Lambert (1959, 1960, 1961) for an introduction which is now out of date. Sokal (1974) provides an overview framed in terms of all types of organisms. Unfortunately the mathematics have also discouraged many ecologists from using these methods to best advantage.

6.9 ROLE OF VEGETATION DATA IN ENVIRONMENTAL INFORMATION SYSTEMS

The economic as well as the ecological importance of vegetation and the possibilities for using vegetation to alter the environment make it an essential component of information systems for ecosystem management. Maps showing the vegetative cover in rather broad categories may be available from several sources. However, relatively short term changes in vegetation make it unlikely that the necessary data can be lifted directly from existing data banks. Some amount of field work using appropriate combinations of methods described in this chapter is certain to be required. These same changes make it necessary to update the information at regular intervals. This need for updating makes a network of permanent sample plots quite attractive. An effective use of air-photos and other forms of remote sensing can help to reduce the costs of field data collection.

6.10 BIBLIOGRAPHY

Avery, T. E. *Forester's Guide to Aerial Photo Interpretation*. Washington, D.C.: U.S. Department of Agriculture, Agricultural Handbook 308, 1969.

Avery, T. E. *Natural Resource Measurements,* 2nd ed. New York: McGraw-Hill, 1975.

Bell, H. *Rangeland Management for Livestock Production.* Norman, Okla.: University of Oklahoma Press, 1973.

Benton, A., and W. Werner, Jr. *Manual of Field Biology and Ecology,* 5th ed. Minneapolis, Minn.: Burgess, 1972.

Brown, D. *Methods of Surveying and Measuring Vegetation.* Bucks, England: Commonwealth Agricultural Bureaux, Commonwealth Bureau of Pastures and Field Crops, Bulletin 42, 1954.

Brown, J. *Handbook for Inventorying Downed Woody Material.* Ogden, Utah: U.S. Department of Agriculture, Forest Service, Intermountain Forest and Range Experiment Station, General Technical Report INT-16, 1974.

Brown, J., and P. Roussopoulos. "Estimating Biases in the Planar Intersect Method for Estimating Volumes of Small Fuels." *Forest Science*, **20,** 350–356 (1974).

Bruce, D. "A Formula for the Scribner Rule," Journal of Forestry **23,** 432–433 (1925).

Cain, S., and G. Castro. *Manual of Vegetation Analysis.* New York: Harper, 1959.

Chapman, S., Ed. *Methods in Plant Ecology.* New York: Wiley, 1976.

Cottom, G., and J. Curtis. "The Use of Distance Measures in Phytosociological Sampling." *Ecology* **37,** 451–460 (1956).

Curtis, J. *The Vegetation of Wisconsin: An Ordination of Plant Communities.* Madison, Wis.: The University of Wisconsin Press, 1959.

DeVries, P. *A General Theory on Line Intersect Sampling with Application to Logging Residue Inventory.* Wageningen, Netherlands: Mededelingen Landbouwhogeschool Wageningen, Nederland, 73-11, 1973.

Dierschke, H. *On the Recording and Presentation of Phenological Phenomena in Plant Communities.* Translated from German by R. Wesell and S. Talbot. Hague, Netherlands: Dr. W. Junk N. V., 1972.

Driscoll, R. *Color Aerial Photography, A New View for Range Management.* Fort Collins, Colo.: U.S. Department of Agriculture, Forest Service, Rocky Mountain Forest and Range Experiment Station, Research Paper RM-67, 1971.

Eberhardt, L. "Transect Method for Population Studies." *Journal of Wildlife Management,* **42,** 1–31 (1978).

Francis, R. "Current Rangeland Inventory Methods—Compatibility Toward an Ecological Base?" In H. Lund et al., Eds. *Integrated Inventories of Renewable Natural Resources: Proceedings of the Workshop, Jan. 8–12, Tucson, Arizona.* Fort Collins, Colo.: U.S. Department of Agriculture, Forest Service, Rocky Mountain Forest and Range Experiment Station, General Technical Report RM-55, 1978.

Freese, F. *Elementary Forest Sampling.* Washington, D.C.: U.S. Department of Agriculture, Agricultural Handbook 232, 1962.

Gates, F. *Field Manual of Plant Ecology.* New York: McGraw-Hill, 1949.

Giles, R., Jr., Ed. *Wildlife Management Techniques.* Washington, D.C.: The Wildlife Society, 1971.

Goodall, D. "Some Considerations in the Use of Point Quadrats for the Analysis of Vegetation." *Australian Journal of Scientific Research,* Sere B, **5,** 1–41 (1952).

Goodall, D. "Statistical Plant Ecology." *Annual Review of Ecology and Systematics,* **1,** 99–124 (1970).

Grieg-Smith, P. *Quantitative Plant Ecology.* London: Butterworths, 1964.

Grosenbaugh, L. "Optical Dendrometers for Out-of-Reach Diameters: A Conspectus and Some New Theory." *Forest Science,* Monograph 4 (1963).

Grosenbaugh, L. *Three-pee Sampling Theory and Program "THRP" for Computer Generation of Selection Criteria.* Berkeley, Cal.: U.S. Department of Agriculture, Forest Service, Pacific Southwest Forest and Range Experiment Station, Paper PSW-21, 1965.

Harlow, W., and E. Harrar. *Textbook of Dendrology,* 5th ed. New York: McGraw-Hill, 1969.

Harrington, H., and L. Durrell. *How to Identify Plants.* Chicago: Swallow Press, 1957.

Heady, H. *Rangeland Management.* New York: McGraw-Hill, 1975.

Hotchkiss, N. *Common Marsh, Underwater, and Floating-leaved Plants of the United States and Canada.* New York: Dover, 1972.

Husch, B. *Planning a Forest Inventory.* Rome, Italy: United Nations, FAO Forestry and Forest Products Studies 17, 1971.

Husch, B., C. Miller, and T. Beers. *Forest Mensuration,* 2nd ed. New York: Ronald, 1972.

Kershaw, K. *Quantitative and Dynamic Plant Ecology,* 2nd ed. New York: Elsevier, 1973.

Kuchler, A. *Vegetation Mapping.* New York: Ronald, 1967.

Loetsch, F., and K. Haller, *Forest Inventory,* vol. 1. Munich, Germany: BLV Verlagsgesellschaft, 1964.

Loetsch, F., F. Zohrer, and K. Haller. *Forest Inventory,* Vol. 2. Munich, Germany: BLV Verlagsgesellschaft, 1973.

Lund, H. *3P Sampling: An Annotated Bibliography.* Upper Darby, Pa.: U.S. Department of Agriculture, Forest Service, State and Private Forestry, 1975.

Milner, C., and R. Hughes. *Methods for the Measurement of the Primary Production of Grassland.* IBP Handbook 6. Oxford: Blackwell, 1968.

Mueller-Dombois, D., and H. Ellenberg. *Aims and Methods of Vegetation Ecology.* New York: Wiley, 1974.

Peterson, R., and M. McKenney. *A Field Guide to the Wildflowers.* Boston: Houghton Mifflin, 1968.

Petrides, G. *A Field Guide to the Trees and Shrubs.* Boston: Houghton Mifflin, 1958.

Pielou, E. C. *Ecological Diversity.* New York: Wiley, 1975.

Pielou, E. C. *Mathematical Ecology,* 2nd ed. New York: Wiley, 1977.

Poole, R. *An Introduction to Quantitative Ecology.* New York: McGraw-Hill, 1974.

Raunkiaer, C. *The Life Forms of Plants and Statistical Plant Geography.* Oxford: Clarendon, 1934.

Raunkiaer, C. *Plant Life Forms.* Oxford: Clarendon, 1937.

Roussopoulos, P., and V. Johnson. *Estimating Slash Fuel Loading for Several Lakes States Tree Species.* St. Paul, Minn.: U.S. Department of Agriculture, Forest Service, North Central Forest Experiment Station, Research Paper NC-88, 1973.

Schreuder, H., J. Sedransk, and K. Ware. "3-P Sampling and Some Alternatives, I." *Forest Science,* **14,** 429–453 (1968).

Shimwell, D. *The Description and Classification of Vegetation.* Seattle, Wash.: University of Washington Press, 1972.

Sokal, R. "Classification: Purposes, Principles, Progress, Prospects." *Science,* **185,** 1115–1123 (1974).

Space, J. *3-P Forest Inventory: Design, Procedures, Data Processing.* Atlanta, Georgia: U.S. Department of Agriculture, Forest Service, State and Private Forestry—Southeastern Area, 1974.

Spurr, S. *Forest Inventory.* New York: Ronald, 1952.

Stoddart, L., A. Smith, and T. Box. *Range Management*, 3rd ed. New York: McGraw-Hill, 1975.

Symonds, G., and A. Merwin. *The Shrub Identification Book.* New York: Morrow, 1963.

U.S. Forest Service. *Techniques and Methods of Measuring Understory Vegetation. Proceedings of a Symposium at Tifton, Georgia, October, 1958.* Washington, D.C.: U.S. Department of Agriculture, Forest Service, 1958.

U.S. Forest Service. *Range Research Methods. Proceedings of a Symposium, Denver, Colo., May 1962.* Washington, D.C.: U.S. Department of Agriculture, Forest Service, Misc. Publication 940, 1963.

U.S. Forest Service. *National Forest Log Scaling Handbook.* Washington, D.C.: U.S. Department of Agriculture, Forest Service, Handbook FSH 2409.11, 1969.

U.S. Forest Service. *Range and Wildlife Habitat Evaluation, A Research Symposium.* Washington, D.C.: U.S. Department of Agriculture, Forest Service, Misc. Publication 1147, 1970.

U.S. Forest Service and National Aeronautics and Space Administration. *Photointerpretation Guide for Forest Resource Inventories.* Houston, Tex: National Aeronautics and Space Administration, Lyndon B. Johnson Space Center, 1975.

Van Wagner, C., and A. Wilson. "Diameter Measurement in the Line Transect Method." *Forest Science,* **22,** 230–232 (1976).

Viertel, A. *Trees, Shrubs, and Vines.* Syracuse, N.Y.: Syracuse University Press, 1970.

Ware, K., and T. Cunia. "Continuous Forest Inventory with Partial Replacement." *Forest Science,* Monograph 3 (1962).

Wiant, H., Jr. *Elementary 3-P Sampling.* Morgantown, W. Va.: West Virginia Agricultural and Forest Experiment Station, Bulletin 650T, 1976.

Williams, W., and J. Lambert. "Multivariate Methods in Plant Ecology: I. Association Analysis in Plant Communities." *Journal of Ecology,* **47,** 83–101 (1959).

Williams, W., and J. Lambert. "Multivariate Methods in Plant Ecology: II. The Use of Electronic Digital Computer for Association Analysis." *Journal of Ecology,* **48,** 689–710 (1960).

Williams, W., and J. Lambert. "Multivariate Methods in Plant Ecology: III. Inverse Association Analysis." *Journal of Ecology,* **49,** 717–730 (1961).

CHAPTER 7—HUMAN INFLUENCES

The necessity for "managing" ecosystems arises not because of any deficiencies in "nature"; but because of human needs, wants, and philosophies. If the human race were somehow just dissolved out of existence, the Grand Designer would continue to paint natural landscapes in material form. The ecosystem cycle of production, consumption, and decomposition is not predicated on the existence of *Homo sapiens.* Other species can perform the simpler actions associated with man's role in nature quite effectively. However, it is also obvious that ecosystems in the absence of technological man are quite different than those under his influence. *Homo sapiens* is an extremely powerful species in terms of its ability to effect alterations in its surroundings. Still, this power does not *yet* extend to the creation of life support systems that involve no other living organisms. Man is not yet an entity unto himself; and it remains a matter of speculation as to whether he will ever become such. Meanwhile man must remain an integral part of *human ecosystems* involving other biological organisms.

Since man must share human ecosystems with other life forms, there is little point in debating whether these ecosystems are "unnatural." However, *responsibility* accompanies man's *ability* to effect control over these ecosystems. Furthermore man must also define the extent to which this responsibility extends beyond the minimal requirement that a human ecosystem serve as a life support system for *Homo sapiens.* Since man is a social species, formulation of this definition of social responsibility requires that a collective opinion be assembled. This, in turn, requires that managers have survey techniques at their disposal for investigating the attitudes, preferences, perceptions, and activity patterns of the people whom they serve. These methods constitute the subject matter of this short concluding chapter.

7.1 LAND USE AND OWNERSHIP

Information on current land use and ownership provides a baseline for land use planning and the formulation of policy for managing ecosystem resources.

7.1.1 Ownership

Direct information on ownership comes from legal records which are public information. The utility of complete ownership information is limited by problems in keeping the information current. In view of the rapid turnover in actual ownership, needs for ownership information must be analyzed critically. In many circumstances one can make do with less than full ownership information. The critical need may be for a breakdown into relatively few categories such as public, corporate, and individual ownerships. Since many of the land transactions take place within the individual ownership category, a file of this type becomes out of date much less rapidly.

7.1.2 Land Use

Detailed information on land use is not easy to obtain since considerable field work is necessary to make fine distinctions. Furthermore it takes much thought and testing to arrive at a detailed classification which contains no overlap between categories and serves a variety of purposes. The *Standard Land Use Coding Manual* (U.S. Urban Renewal Administration, Housing and Home Finance Agency, and Bureau of Public Roads, 1965) provides a point of departure for those who view detailed information on land use as an absolute necessity.

However, detailed land use information is not really necessary for most applications in land use planning and policy formulation. When general information will suffice, land cover can often be used as a surrogate for land use. The ease with which the land cover can be detected by remote sensing has made this approach popular in modern land use mapping efforts. Anderson, Hardy, and Roach (1972) proposed a hierarchical classification system in which first level categories could be assessed from data collected by remote sensors aboard satellites. The categories comprising this first level are (1) urban and built-up land, (2) agricultural land, (3) rangeland, (4) forest land, (5) water, (6) nonforested wetland, (7) barren land, (8) tundra, and (9) permanent snow and ice fields. Categories at the second level are structured so that most of the information can be obtained from aerial photographs taken from conventional aircraft. The third and lower levels were left unspecified for later development according to regional and local needs.

This proposed classification system was widely distributed by the USGS to agencies concerned with land use mapping for their review and comment. Responses to the proposed system led to a revision as described by Anderson, Hardy, Roach, and Witmer (1976). Minor modifications and expansions of the original or revised system have been adopted by many states and other agencies involved in land use mapping.

The many advantages of a uniformity in classification make it advisable to adopt this system with such modifications as seem absolutely necessary.

Regional differences and differences in the purpose for collecting the information make it unreasonable to expect much uniformity at the lower levels of the classification hierarchy. A series of articles contained in the October, 1978, issue of the *Journal of Forestry* provide an excellent review of problems and progress in land classification.

Despite our endorsement of this approach as a source of inexpensive and uniform information on land use/cover, we wish to stress once again the viewpoint that it is preferable to structure information systems so that each component of an ecosystem is documented separately along with the spatial and temporal indexing necessary to formulate land categorizations in several alternative forms by superimposing information on components.

7.2 PERSONAL AND ORGANIZATIONAL INFORMATION

Land use information gathered as outlined above serves many purposes in general land use planning and policy formulation. However, the more detailed phases of project planning and environmental impact assessment will often require information on activity patterns, attitudes, preferences, economic characteristics, and demographic characteristics of the human population that is affected by or forms the clientele group for the proposed undertaking. Some information of this nature is available in census records, but much of it must be gleaned by survey research techniques developed by social scientists.

The main weapons in the arsenal for survey research are questionnaires, interviews, and observations made by trained observers. Public hearings can be regarded as an unstructured type of group interview. Cost and nonresponse are the two main difficulties that plague efforts to obtain this kind of information. The time available for conducting the survey is often an additional consideration. We cannot hope to cover the many details of survey research techniques here, but we can provide an overview of common strategies and point out advantages and pitfalls. For greater depth the reader should consult references such as Babbie (1973), Kerlinger (1973), and Rosenberg (1968).

7.2.1 Questionnaires

Questionnaires can be used to obtain all kinds of information about both individuals and organizations. The outstanding advantages of questionnaires are their low cost and the ability to reach large numbers of potential respondents simultaneously. The outstanding disadvantage of questionnaires is the ease with which they can be ignored by potential respondents.

Because questionnaires are so easy to ignore, the designer of the questionnaire must use every available device to motivate potential respondents to complete and return the questionnaires. The first concern is to catch the interest of the potential respondent so that the person will examine the questionnaire more closely instead of discarding it immediately. Bright colors, sketches, and catchy phrases printed in bold letters may all serve to pique the person's interest enough that he or she will want to scan the questionnaire.

When potential respondents begin to scan the questionnaire, it is essential to convince them that it serves an important purpose. This can be achieved by a carefully worded cover letter or opening paragraph. If the designer succeeds in convincing the potential respondent that the questionnaire is important, the next thing is to convey an impression that completing the questionnaire will not be an onerous task. The key to this is simplicity of design and clarity of instructions. Finally the potential respondent must not feel threatened with invasion of privacy by questions asked. Sensitive material is more likely to be furnished if respondents are convinced that they will remain anonymous and the information will not be used to their detriment.

Suppose that our potential respondent has decided to become an actual respondent and proceeds to complete the questionnaire. We have cleared one hurdle only to encounter others. Does the wording of the questionnaire clearly identify the information we seek? Have we used leading questions that tend to prejudice the response? Are the allowable responses to questions so restrictive that respondents cannot adequately express their opinions? Is there so much latitude in formulating responses that it becomes easy to stray from the main issue? Does the respondent feel threatened by the wording of some questions, causing him or her to become evasive? All these questions must be addressed as the questionnaire is being formulated.

Survey researchers are in the habit of referring to a questionnaire as an "instrument." This terminology reflects the fact that the same care must be taken in constructing a questionnaire as in any other kind of measuring device. Each question must be clearly and concisely worded. Jargon that might be unfamiliar to the respondent should be carefully avoided. Except for agree/disagree questions, the wording should be neutral enough that the response will not be affected by the way the question is stated.

Tabulating responses is easiest with forced choice questions. However, it may be desirable to include an "other—please explain" category to detect possible responses that were overlooked. The answer can always be shifted to one of the existing categories if it appears to belong there on the basis of the explanation for choosing "other." Completely open-ended questions offer insight into the respondent's thinking, but are expensive to tabulate because the responses must be grouped into categories manually or by computerized content analysis.

Questions calling for graded responses (strongly disagree/disagree/undecided/agree/strongly agree) are often used because they offer the possibility of constructing ordinal indexes in numerical form. The response to such a question might be coded 1 to 5 according to the degree of agreement. These numbers can be added over a series of questions relating to the same subject, thus giving a composite index that approaches measurement strength. This is known as *Likert scaling*.

Whether or not Likert scaling is used, multiple questions dealing with the same topic give insight into the consistency of the responses. Likewise it is advisable to include some questions that can be cross checked against external sources of information. This provides an indication of the accuracy of the information provided by the respondent. Questions dealing with sensitive material should be relegated to the end of the questionnaire so that the suspicions they arouse will not color responses to other questions.

Despite great care taken in formulating the questionnaire, it is unlikely that the person designing the instrument can anticipate exactly how it will be perceived by respondents. Therefore a pretest on a sample of respondents prior to full distribution is a must.

When the respondent has completed the questionnaire, the next concern is that it be returned promptly. Self-addressed envelopes or self-mailing questionnaire formats with postage provided serve as an incentive to return mail questionnaires promptly. Also it may help to place mailing instructions as the final item on the questionnaire. If the questionnaire is distributed to users of a park or similar facility with the expectation that they will complete it before leaving, then prominently marked drop boxes should be conveniently positioned at exits.

Even with well structured questionnaires which are convenient to return one must anticipate a high rate of nonresponse, often in excess of 50%. If the only concern were to obtain a specified number of completed questionnaires, it would be possible to compensate for low response by distributing additional questionnaires. However, the real problem lies in the possibility that answers provided by respondents may be different from those which nonrespondents would give if they could be pursuaded to return the questionnaire. The only way that such differences can be detected is by doing follow-up work on a sample of the nonrespondents.

This, in turn, requires the ability to divide the original list of questionnaire recipients into respondent and nonrespondent categories. This can easily be done by numbering the questionnaires or otherwise coding them in some way that is not obvious to the respondent. However, this raises a question of ethics if the respondents have been led to believe that they will remain anonymous. One possibility is to include a separate identification card with each questionnaire along with instructions to return the card separately so that the identification cannot be associated with the responses on the questionnaire. It is possible to send a second mailing to the entire distribution list with instructions to ignore if already completed. However, this is expensive and tends to annoy earlier respondents. Furthermore, it does not allow follow-up by other means than mail. It may be worthwhile to make second and third mailings in the hope of increasing returns, but those who do not respond to a third mailing usually have to be approached by some other means than a self-administered questionnaire.

7.2.2 Interviews

There are less problems with nonresponse to interviews than with self-administered questionnaires, so interviews can be used either as a prime tool for survey research or as a follow-up to nonrespondents with questionnaires. Interviews may be conducted either by telephone or in person. Interviews conducted in person are more expensive than telephone interviews, but both are considerably more costly than questionnaires.

Interviews have the advantage that on can probe deeper into the reasons behind answers than would be possible with questionnaires. However, the interviewer introduces another source of variability in the responses. With a questionnaire all respondents are approached in the same manner. In contrast, no two interviewers behave or react in exactly the same way. Furthermore a given interviewer may behave somewhat differently depending on how he or she feels and events that preceded the interview. Therefore it is essential that interviewers be trained for uniformity of approach. Since the chances of interviewer bias are greater with unstructured than with structured interviews, it is usually best to provide some sort of schedule to be followed in the interview process along with guidelines for permissible departures from the schedule.

The ultimate in structure is to have the interviewer read a questionnaire and record responses in much the same form as they would be recorded with a self-administered questionnaire. Although this minimizes interviewer bias, it does not take advantage of the ability to probe deeper into interesting questions. Telephone interviews limit the interaction between interviewer and respondent, and thus are intermediate between questionnaires and personal interviews.

Responses are not assured even with interviews, because some people will refuse to be interviewed and others will not be at home. Furthermore the logistical problems in supporting a staff of interviewers are considerable.

7.2.3 Trained Observers and Remote Sensing

When one is interested primarily in gathering information on local activity patterns, it may be possible to make use of trained observers to categorize and record the activities taking place. One mode of operation for an observer is to stand unobtrusively in the background and record activities from a distance. Another approach is for the observer to be an apparent participant in the activities. The latter approach is less likely to arouse suspicion among the participants, but it is difficult to do the recording, and care must be exercised that the observer does not alter the activity pattern. Ethical questions may also arise in using a participating observer.

In some situations that would be appropriate for a passive observer it may be possible to replace the observer by a remote sensing system. For example, the U.S. Forest Service has developed an efficient and unobtrusive trail traffic counter which utilizes a beam of infrared light (DeLand, 1976). The unit is lightweight, easy to conceal, not sensitive to weather conditions, and operates two to three months on a set of batteries. The recreational usage of beaches and lakes can be monitored effectively by taking photographs with a 35 mm camera from a light plane. Concealed cameras and time lapse photography is still another possibility.

7.3 SCENIC QUALITY OF LANDSCAPES

Recent controversies over forest clearcutting and other land management methods that alter the visual character of the landscape have generated interest in measuring scenic beauty. This task is neither easy nor straightforward, and the many problems are still being addressed by researchers. Although the U.S. Forest Service (1973, 1974) has an operational system for categorizing and managing visual resources, it is probably too early to go into great detail regarding systems that are likely to undergo considerable evolution. Annotated bibliographies have been prepared by federal forest research agencies in both the United States and Canada (Arthur and Boster, 1976; Murtha and Greco, 1975).

Scenic beauty is a composite of the physical characteristics of the landscape, the vantage point, and the perception of the observer. Proposed systems for assessing scenic beauty vary in emphasis on these three components. The operational Forest Service system emphasizes physical characteristics of the terrain. A system proposed by Daniel and Boster (1976) is based on the observer response to color slides that sample various aspects of the landscape being evaluated. The latter report is interesting reading not only for the proposed system, but also for the concise summary of the problems involved in measuring scenic beauty. Scenic beauty is only one of the many environmental intangibles that will assume increasing importance in the future. Quantification of these intangibles so that they can be managed objectively is one of the more challenging areas of research in survey methods.

7.4 IN CONCLUSION—ENVIRONMENTAL IMPACT ANALYSIS

In this book we have been concerned with methods for collecting data on the various components of ecosystems and with assembling these data into well structured information systems to support ecosystem management and environmental impact analysis. However, it is no small step to go from the data base to a comprehensive environmental impact statement which takes account of the manifold interactions that characterize all ecosystems. Systematic approaches to environmental impact analysis must be used to ensure that all likely effects are considered. Several such strategies have been devised.

Most of these involve some sort of matrix or checklist which arrays different types of ecosystem alterations against possible effects (Schlesinger and Daetz, 1973; Warner and Preston, 1974; Whitlatch, 1976). However, the possibilities are so numerous and interlinked that such charts quickly become unwidely. Computers provide an effective means for cutting quickly through complexity, and the logical extension is to create information systems containing potential effects. Codes for different ecosystem alterations can serve as keys so that only the relevant portion of the impact bank is retrieved and displayed to the analysis team as impact assessment proceeds. Thor, Elsner, Travis, and O'Loughlin (1978) describe a system of this type which they call IMPACT. Further development of systems like this holds great promise for a space age solution to the down-to-earth problem of environmental impact assessment.

7.5 BIBLIOGRAPHY

Amidon, E., and G. Elsner. *Delineating Landscape View Areas.* Berkeley, Cal.: U.S. Department of Agriculture, Forest Service, Pacific Southwest Forest and Range Experiment Station, Research Note PSW-180, 1968.

Anderson, J., E. Hardy, and J. Roach. *A Land-Use Classification System for Use with Remote-Sensor Data.* Washington, D.C.: U.S. Department of the Interior, Geological Survey, Circular 671, 1972.

Anderson, J., E. Hardy, J. Roach, and R. Witmer. *A Land Use and Land Cover Classification System for Use with Remote Sensor Data.* Washington, D.C.: U.S. Department of the Interior, Geological Survey, Professional Paper 964, 1976.

Arthur, L., and R. Boster. *Measuring Scenic Beauty: A Selected Annotated Bibliography.* Fort Collins, Colo.: U.S. Department of Agriculture, Forest Service, Rocky Mountain Forest and Range Experiment Station, General Technical Report RM-25, 1976.

Babbie, E. *Survey Research Methods.* Belmont, Cal.: Wadsworth, 1973.

Bailey, R., R. Pfister, and J. Henderson. "Nature of Land and Resource Classification—A Review." *Journal of Forestry,* **76** (10), 650–655 (1978).

Cowardin, L. "Wetland Classification in the United States." *Journal of Forestry,* **76** (10), 666–668 (1978).

Crapo, D., and M. Chubb. *Recreation Area Day-Use Investigation Techniques: Part I—A Study of Survey Methodology.* East Lansing, Mich.: Recreation Research and Planning Unit, Department of Park and Recreation Resources, Michigan State University, Technical Report 6, 1969.

Daniel, T., and R. Boster. *Measuring Landscape Esthetics: The Scenic Beauty Estimation Method.* Fort Collins, Colo.: U.S. Department of Agriculture, Forest Service, Rocky Mountain Forest and Range Experiment Station, Research Paper RM-167, 1976.

DeLand, L. *Development of the Forest Service Trail Traffic Counter.* Missoula, Mont.: U.S. Department of Agriculture, Forest Service, Equipment Development Center, Report 7700-10, 1976.

Driscoll, R., D. Betters, and H. Parker. "Land Classification Through Remote Sensing—Techniques and Tools." *Journal of Forestry,* **76** (10), 656–661 (1978).

Frayer, W., L. Davis, and P. Risser. "Uses of Land Classification." *Journal of Forestry,* **76** (10), 647–649 (1978).

Gordon, R. *Interviewing: Strategy, Techniques, and Tactics.* Homewood, Ill.: Dorsey, 1969.

Hirsch, A., C. Cushwa, K. Flach, and W. Frayer. "Land Classification—Where Do We Go from Here?" *Journal of Forestry,* **76** (10), 672–673 (1978).

Kahn, R., and C. Cannell. *The Dynamics of Interviewing.* New York: Wiley, 1967.

Kerlinger, F. *Foundations of Behavioral Research,* 2nd ed. New York: Holt, Rinehart, and Winston, 1973.

Kish, L. *Survey Sampling.* New York: Wiley, 1965.

Lacate, D., and M. Romaine. "Canada's Land Capability Inventory Program." *Journal of Forestry,* **76** (10), 669–671 (1978).

Leopold, L., F. Clarke, B. Hanshaw, and J. Balsley. *A Procedure for Evaluating Environmental Impact.* Washington, D.C.: U.S. Department of the Interior, Geological Survey, Circular 645, 1971.

Murtha, P., and M. Greco. *Appraisal of Forest Aesthetic Values: An Annotated Bibliography.* Ottawa, Onto: Canada Forestry Service, Department of the Environment, Forest Management Institute, Information Report FMR-X-79, 1975.

Nelson, D., G. Harris, and T. Hamilton. "Land and Resource Classification—Who Cares?" *Journal of Forestry,* **76** (10), 644-646 (1978).

Oppenheim, A. *Question Design and Attribute Measurement.* New York: Basic Books, 1965.

Payne, S. *The Art of Asking Questions.* Princeton, N.J.: Princeton University Press, 1965.

Rosen, S. *Manual for Environmental Impact Evaluation.* Englewood Cliffs, N.J.: Prentice-Hall, 1976.

Rosenberg, M. *The Logic of Survey Analysis.* New York: Basic Books, 1968.

Schlesinger, B., and D. Daetz. "A Conceptual Framework for Applying Environmental Assessment Matrix Techniques." *Journal of Environmental Science,* **16,** 11-16 (1978).

Smith, S., R. Nuxoll, and F. Galloway. *Survey Research for Community Recreation Services.* East Lansing, Mich.: Michigan State University, Agricultural Experiment Station, Research Report 291, 1976.

Stankey, G., and D. Lime. *Recreational Carrying Capacity: An Annotated Bibliography.* Ogden, Utah: U.S. Department of Agriculture, Forest Service, Intermountain Forest and Range Experiment Station, General Technical Report INT-3, 1973.

Survey Research Center. *Interviewer's Manual.* Ann Arbor, Mich.: Institute for Social Research, 1968.

Thor, E., G. Elsner, M. Travis, and K. O'Loughlin. "Forest Environmental Impact Analysis—A New Approach." *Journal of Forestry,* **76** (11), 723-725 (1978).

U.S. Forest Service. *National Forest Landscape Management,* Vol. 1. Washington, D.C.: U.S. Department of Agriculture, Handbook 434, 1973.

U.S. Forest Service. *National Forest Landscape Management,* Vol. 2. Washington, D.C.: U.S. Department of Agriculture, Handbook 462, 1974.

U.S. Urban Renewal Administration, Housing and Home Finance Agency, and Bureau of Public Roads. *Standard Land Use Coding Manual, A Standard System for Identifying and Coding Land Use Activities.* Washington, D.C.: U.S. Government Printing Office, 1965.

Warner, M., and E. Preston. *A Review of Environmental Impact Assessment Methodologies.* Washington, D.C.: U.S. Environmental Protection Agency, Office of Research and Development, Socioeconomic Environmental Studies ser. EPA-60015-74-002, 1974.

Whitlatch, E., Jr. "Systematic Approaches to Environmental Impact Assessment: An Evaluation." *Water Resources Bulletin,* **12,** 123-137 (1976).

Witmer, R. "U.S. Geological Survey Land-Use and Land-Cover Classification System." *Journal of Forestry,* **76** (10), 661-666 (1978).

APPENDIX–DESCRIPTIVE STATISTICS AND SAMPLING

The purpose of this appendix is to provide a single reference on descriptive statistics and sampling which can be consulted from any point in the text or independently of the text. The presentation is deliberately concise. Examples serve only to illustrate computational procedures, not to represent realistic applications. Examples are based mostly on the area and shoreline data for the hypothetical Mirage Lakes introduced in Section 2.6. These data are reproduced in Table A.1.

Statistical parameters of populations are treated in Section A.1. Section A.2 contains an introduction to probability as a background for the estimation of population parameters by sampling. Direct estimation of these parameters from simple random samples is covered in Section A.3. Indirect estimation by regression is introduced in Section A.4. Nomograms for graphically evaluating regression equations are described in Section A.5. Reduction of the sampling error by stratification is discussed in Section A.6. Reduction of the sampling costs by multistage sampling is discussed in Section A.7. Double sampling strategies are introduced in Section A.8, and sampling with unequal probabilities is introduced in Section A.9. The final section of the appendix is concerned with chi-square procedures for analyzing cross tabulations.

Table A.1 Surface areas and shoreline lengths for Mirage Lakes

Name of lake	Area (ha)	Shoreline length (km)
South Mirage Lake	210	5.7
Long Mirage Lake	202	6.8
Muddy Mirage Lake	48	2.9
Round Mirage Lake	45	2.6
Mirage Bay Lake	84	5.2
Middle Mirage Lake	77	5.0
Misty Mirage Lake	47	3.1
North Mirage Lake	123	5.3
Little Mirage Lake	42	3.0
Big Mirage Lake	240	5.9

A.1 STATISTICAL PARAMETERS OF POPULATIONS

For present purposes the term *population* will describe a data set consisting of all possible measurements on one or more variable quantities. The variable quantities will be called *variables* or *variates*. If there is only one variate, we have a univariate population. If there are two variates, we have a bivariate population. If there are several variates, we have a multivariate population. Thus Table A.1 contains a bivariate population of surface area and shoreline length measurements on a set of 10 hypothetical Mirage Lakes. The *observational unit* for this population is an individual lake. In general there will be one data value for each variate for each observational unit. The matched set of variate values that come from the same observational unit will be called an *observation*.

The larger the population, the more difficult it becomes to comprehend by examining the observations individually. The problem is reduced by developing summary figures which distill most of the information into relatively few numbers. Such summary figures are called *parameters*. Two categories of parameters are routinely calculated. The first type of parameter characterizes the typical or average value for a variate in the population. Parameters of this kind are called measures of *central tendency*.

A.1.1 Mean

The most frequently used measure of central tendency is the *arithmetic mean*. The formula for the arithmetic mean is

$$\text{arithmetic mean} = \frac{\Sigma X}{N}$$

where Σ indicates summation, X indicates any particular value of a variate, and N represents the number of observations in the population. Thus the numerator is obtained by summation over all values of the variate, and the denominator is the number of observations in the population. If there are several variates in the population, other letters such as Y and Z are commonly used in place of X to distinguish the variates. The Greek letter μ is usually used to denote the arithmetic mean of a variate. Let X stand for shoreline length in the lake population of Table A.1 and Y for surface area. Then the mean shoreline length is

$$\mu_X = \frac{\Sigma X}{N} = \frac{5.7 + 6.8 + \cdots + 5.9}{10} = \frac{45.5}{10} = 4.55 \text{ km}$$

and the mean surface area is

$$\mu_Y = \frac{\Sigma Y}{N} = \frac{210 + 202 + \cdots + 240}{10} = \frac{1118}{10} = 111.8 \text{ ha}$$

The arithmetic mean is the optimum measure of central tendency in the sense that the sum of the squared differences between the individual values and the mean is as small as possible, that is,

$$\Sigma(X - \mu)^2 = \Sigma(X^2 - 2\mu X + \mu^2) = \Sigma X^2 - 2\mu\Sigma X + N\mu^2$$

is smaller than for any other value that one might substitute for μ. To prove this we let C be an arbitrary increment to μ, either positive or negative. Then

$$\begin{aligned}
\Sigma[X - (\mu + C)]^2 &= \Sigma[X^2 - 2(\mu + C)X + (\mu + C)^2] \\
&= \Sigma X^2 - 2(\mu + C)\Sigma X + N(\mu + C)^2 \\
&= \Sigma X^2 - 2\mu\Sigma X - 2C\Sigma X + N(\mu^2 + 2\mu C + C^2) \\
&= \Sigma X^2 - 2\mu\Sigma X - 2C\Sigma X + N\mu^2 + 2N\mu C + NC^2 \\
&= [\Sigma X^2 - 2\mu\Sigma X + N\mu^2] + [2N\mu C - 2C\Sigma X] + NC^2 \\
&= \Sigma(X - \mu)^2 + [2CN\mu - 2CN\mu] + NC^2 \\
&= \Sigma(X - \mu)^2 + 0 + NC^2 \\
&= \Sigma(X - \mu)^2 + NC^2
\end{aligned}$$

Since C^2 is always positive, any value other than the mean will always give a larger sum of squared deviations.

The arithmetic mean also has the property that the sum of the deviations from it is zero. This is proved simply as follows:

$$\begin{aligned}
\Sigma(X - \mu) &= \Sigma X - \Sigma\mu = \Sigma X - N\mu \\
&= \Sigma X - \frac{N\Sigma X}{N} = \Sigma X - \Sigma X = 0
\end{aligned}$$

A.1.2 Variance

The second type of parameter characterizes the variability among the values of a variate in the population. Parameters of this type are called *measures of dispersion.* The usual measure of dispersion is the average squared difference from the mean, or *variance.* Thus the definition formula for the variance is

$$\text{variance} = \frac{\Sigma(X - \mu)^2}{N}$$

However, this formula is inconvenient for computation because of all the subtraction required.

A more convenient formulation for computation is obtained by expanding the square in the numerator as follows:

$$
\begin{aligned}
\Sigma(X - \mu)^2 &= \Sigma(X^2 - 2\mu X + \mu^2) \\
&= \Sigma X^2 - 2\mu\Sigma X + N\mu^2 \\
&= \Sigma X^2 - 2\frac{\Sigma X}{N}\Sigma X + N\left(\frac{\Sigma X}{N}\right)^2 \\
&= \Sigma X^2 - 2\frac{(\Sigma X)^2}{N} + N\frac{(\Sigma X)^2}{N^2} \\
&= \Sigma X^2 - 2\frac{(\Sigma X)^2}{N} + \frac{(\Sigma X)^2}{N} \\
&= \Sigma X^2 - \frac{(\Sigma X)^2}{N}
\end{aligned}
$$

Replacing the numerator of the variance by this expression gives

$$\text{variance} = \frac{\Sigma X^2 - (\Sigma X)^2/N}{N}$$

as the computing formula for the variance, where ΣX^2 is the sum of the squared values and $(\Sigma X)^2$ is the square of the sum. The variance is usually denoted by σ^2.

The variance for the shoreline length in Table A.1 is

$$\sigma_X{}^2 = \frac{\Sigma X^2 - (\Sigma X)^2/N}{N} = \frac{227.45 - (45.5)^2/10}{10} = 2.04 \text{ sq km}$$

and the variance of the surface area is

$$\sigma_Y{}^2 = \frac{\Sigma Y^2 - (\Sigma Y)^2/N}{N} = \frac{178{,}920 - (1118)^2/10}{10} = 5392.7 \text{ sq ha}$$

A.1.3 Standard Deviation

Note that the unit of measure attached to a variance is the square of the original unit, that is, square hectares for the surface area. This is somewhat inconvenient for interpretation, so it is common to use the square root of the variance as a measure of variability having the original units of measure. This is called the *standard deviation* and is denoted by σ. The standard deviations for shoreline length and surface area are 1.43 km and 73.4 ha, respectively.

A.1.4 Coefficient of Variation

For some purposes it is desirable to have a measure of variability that is dimensionless. Such a measure, called the *coefficient of variation*, is obtained by expressing the standard deviation as a percent of the mean. Thus the coefficients of variation for shoreline length and surface area, respectively, are:

$$\frac{\sigma_X}{\mu_X} \times 100 = \frac{1.43}{4.55} \times 100 = 31.4\%$$

and

$$\frac{\sigma_Y}{\mu_Y} \times 100 = \frac{73.4}{111.8} \times 100 = 65.7\%$$

A.1.5 Covariance

When the population contains more than one variate, it will normally be desirable to have a parameter expressing the strength of the relationship between any given pair of variates. A suitable parameter, called the *covariance*, is obtained by making an alteration in the definition formula for the variance. Note that we can write the definition formula for the variance as

$$\frac{\Sigma(X - \mu_X)^2}{N} = \frac{\Sigma(X - \mu_X)(X - \mu_X)}{N}$$

The covariance is obtained by replacing one of the parenthesized expressions in the numerator by the corresponding expression for the second variate. Thus

$$\text{covariance of } X \text{ and } Y = \frac{\Sigma(X - \mu_X)(Y - \mu_Y)}{N}$$

As with the variance, a more convenient formulation for computing is obtained by expanding the numerator as follows:

$$\begin{aligned}\Sigma(X - \mu_X)(Y - \mu_Y) &= \Sigma(XY - Y\mu_X - X\mu_Y + \mu_X\mu_Y) \\ &= \Sigma XY - \mu_X\Sigma Y - \mu_Y\Sigma X + N\mu_X\mu_Y\end{aligned}$$

$$= \Sigma XY - \frac{(\Sigma X)(\Sigma Y)}{N} - \frac{(\Sigma X)(\Sigma Y)}{N} + \frac{N(\Sigma X)(\Sigma Y)}{N^2}$$

$$= \Sigma XY - 2\frac{(\Sigma X)(\Sigma Y)}{N} + \frac{(\Sigma X)(\Sigma Y)}{N}$$

$$= \Sigma XY - \frac{(\Sigma X)(\Sigma Y)}{N}$$

With this formulation of the numerator the computing formula for the covariance becomes

$$\text{covariance of } X \text{ and } Y = \frac{\Sigma XY - (\Sigma X)(\Sigma Y)/N}{N}$$

The covariance of X and Y is usually denoted by σ_{XY}. The covariance of shoreline length and surface area is

$$\sigma_{XY} = \frac{\Sigma XY - (\Sigma X)(\Sigma Y)/N}{N} = \frac{5988.2 - (45.5)(1118)/10}{10} = 90.1$$

A.1.6 Correlation Coefficient

A dimensionless and more easily interpretable measure of relationship between two variates is obtained by dividing the covariance by the product of the standard deviations for the two variates. This is called the *correlation coefficient.* The correlation coefficient between shoreline length and surface area is

$$\text{correlation coefficient} = \frac{\sigma_{XY}}{\sigma_X \sigma_Y} = \frac{90.1}{(1.43)(73.4)} = .86$$

The correlation coefficient is usually denoted by the Greek letter rho (ρ). It can range between $+1$ and -1. A correlation coefficient of $+1$ indicates a perfect relationship between the two variates with all points falling on an upward sloping straight line when one variate is plotted against the other. A correlation coefficient of -1 indicates a perfect relationship between the two variates with all points falling on a downward sloping straight line when one variate is plotted against the other. A correlation coefficient of zero indicates lack of any linear relationship between the two variates. Intermediate correlation coefficients indicate a relation that is less than perfect, with the points clustering more closely about a straight line as the correlation coefficient approaches either $+1$ or -1.

Note, however, that the correlation coefficient must be interpreted in terms of a straight line relationship. Curvilinear relationships are not adequately measured by the correlation coefficient. For the data of Table A.1 the relationship is slightly curvilinear. The plot of squared shoreline length against surface area is more nearly linear, as reflected by the fact that the correlation coefficient between surface area and squared shoreline length is .87.

A.1.7 Mean and Variance of Combination Variates

We are now in a position to investigate how the mean and variance of a combination of variates will behave in relation to the mean and variance of the individual variates that make up the combination. Let a and b represent two constants, and let us examine the mean and variance of the composite variate

$$Z = aX + bY$$

The mean of the composite variate will be

$$\mu_Z = \frac{\Sigma Z}{N} = \frac{\Sigma(aX + bY)}{N} = \frac{\Sigma aX + \Sigma bY}{N} = \frac{a\Sigma X + b\Sigma Y}{N}$$

$$= a\frac{\Sigma X}{N} + b\frac{\Sigma Y}{N} = a\mu_X + b\mu_Y$$

The variance of the composite variate will be

$$\sigma_Z{}^2 = \frac{\Sigma(Z - \mu_Z)^2}{N}$$

For simplicity let us work with the numerator and then divide by N later. Then

$$\begin{aligned}
\Sigma(Z - \mu_Z)^2 &= \Sigma(Z^2 - 2Z\mu_Z + \mu_Z{}^2) \\
&= \Sigma Z^2 - 2\mu_Z \Sigma Z + N\mu_Z{}^2 \\
&= \Sigma(aX + bY)^2 - 2(a\mu_X + b\mu_Y)\Sigma(aX + bY) \\
&\quad + N(a\mu_X + b\mu_Y)^2 \\
&= \Sigma(a^2X^2 + 2abXY + b^2Y^2) - 2N(a\mu_X + b\mu_Y)^2 \\
&\quad + N(a\mu_X + b\mu_Y)^2 \\
&= \Sigma(a^2X^2 + 2abXY + b^2Y^2) - N(a\mu_X + b\mu_Y)^2 \\
&= \Sigma(a^2X^2 + 2abXY + b^2Y^2) \\
&\quad - N(a^2\mu_X{}^2 + 2ab\mu_X\mu_Y + b^2\mu_Y{}^2) \\
&= a^2\Sigma X^2 + 2ab\Sigma XY + b^2\Sigma Y^2 - Na^2\mu_X{}^2 \\
&\quad - 2Nab\mu_X\mu_Y - Nb^2\mu_Y{}^2
\end{aligned}$$

$$= (a^2\Sigma X^2 - Na^2\mu_X{}^2) + (b^2\Sigma Y^2 - Nb^2\mu_Y{}^2) + (2ab\Sigma XY - 2Nab\mu_X\mu_Y)$$

$$= a^2[\Sigma X^2 - (\Sigma X)^2/N] + b^2[\Sigma Y^2 - (\Sigma Y)^2/N] + 2ab[\Sigma XY - (\Sigma X)(\Sigma Y)/N]$$

Dividing by N in the denominator gives

$$\sigma_Z{}^2 = a^2\left[\frac{\Sigma X^2 - (\Sigma X)^2/N}{N}\right] + b^2\left[\frac{\Sigma Y^2 - (\Sigma Y)^2/N}{N}\right] + 2ab\left[\frac{\Sigma XY - (\Sigma X)(\Sigma Y)/N}{N}\right]$$

$$= a^2\sigma_X{}^2 + b^2\sigma_Y{}^2 + 2ab\sigma_{XY}$$

Bear in mind, however, that these parametric equations apply to populations. In other words the data set must include measurements on all observational units that could be considered as members in the population of interest. In most practical circumstances we will be faced with the necessity of estimating the parametric values from a subset or sample of the population. Probability provides a scientific basis for making such estimates, so we will delve briefly into probability before moving on to estimation from samples.

A.2 INTRODUCTION TO PROBABILITY

Probability, chance, or gambling figures heavily in surveys of environmental resources. Limitations on funds, time, and manpower usually make it impossible to measure everything for which information is needed. We are forced to take samples, and we gamble on getting good samples. When we win, we save time and money by sampling. When we lose, we are likely to make poor decisions in our management because our information is bad. Any gambler would like to have the odds in his favor; and the best way to ensure this is to have a good feel for the way chance works.

A.2.1 Randomization

The first thing to note and remember is that the laws of chance do not hold unless chance controls the game. When chance is in control of a process, we say that we have a *random process*. Any time the dice become loaded, so to speak, we won't know how to play the game and are very likely to lose. Therefore we go to some trouble to make sure that chance controls the selection of our samples. Chance, or randomness, can be introduced by mechanical methods such as shuffled cards, numbered chips in a hat, and so on. However, the easiest and best way is to use random numbers. Computers and calculators can be programmed to generate numbers that are, for all practical purposes, random. Tables of uniform random numbers are also available in statistical handbooks. These are tables of numbers in which the digits 0, . . . , 9 occur with approximately equal frequency and thoroughly mixed up.

There are several possible procedures in using a random number table. For the sake of our discussion assume that each of the possibilities we are considering for random selection has been given a unique serial or tag number. The simplest procedure is to poke a finger or pencil blindly at a page of the random number table and then take the nearest set of digits in any predetermined direction as the tag number of the next item for selection. Include as many digits as there are in the largest tag number. If there is no tag number corresponding to the random number obtained, reject the random number and get a new one by repeating the procedure.

A better method, however, is to use the numbers obtained from the blind poke as row and column numbers in the random number table. The random number used is the one located at the intersection of this row and column. Also the rejection of all numbers greater than available tag numbers may require an excessive number of entries in the table. The process can sometimes be modified so that fewer numbers are rejected. Suppose that the sampling units are numbered 1 through N, and that 10^K is the lowest power of 10 that is greater than or equal to N. Divide N into 10^K and discard the remainder. Suppose that the quotient is M. If the number drawn from the table is less than or equal to $N \times M$, divide it by N and use the *remainder* as the tag number. If the remainder is zero, use N as the tag number. Discard numbers from the table that are greater than $N \times M$. Unless the selection scheme is specifically designed to the contrary, each eligible tag number must have an equal chance of appearing.

A.2.2 Probability

The next thing is to pin down a little more securely the basic idea of probability. The probability of something occurring is essentially the fraction of times we can expect it to happen in a long series of occurrences. It may be more familiar to think in terms of percentages than fractions, and that is alright too. Thus something that is certain to occur has a probability of 1 or 100%. On the other hand a probability cannot be less than zero. Even a probability of zero, however, does not necessarily mean that the thing cannot happen. It only means that the chance of the happening is so small as to be neglible. For example, it has been known to rain occasionally under a weather forecast that gave precipitation probability as zero.

A.2.3 Independent Events

We need to distinguish between *dependent* and *independent* events. Two things being independent means that one does not affect the chances of the other, that is, the chances for the second event are the same regardless of how the first one turns out. The classic example of independent events is successive tosses of a fair coin. If the outcome of the first event in a series does change the chances for subsequent events, then the events are said to be dependent. It is usually easier to work with a series of independent events than with dependent events.

Suppose for example that we have two independent events, each with two possible outcomes, and the outcomes for each event are equally likely so that each outcome has probability of .5 or 50% chance of occurring. We expect the first outcome to show half the time for the first event. Look for a moment only at the times when the first event has the first outcome. Regardless of how the first event turned out, we still expect the second event to have its first outcome only half the time. The first event will have its first outcome only have the time, and in half of this half the second event will have its first outcome. Thus $1/2 \times 1/2 = 1/4$ is the fraction of times we would expect both events to have their first outcomes. This is an illustration of the law that the *probabilities of independent events are multiplicative.*

If you want a physical model for illustration, think in terms of two successive tosses of a fair coin. There are two possible outcomes (head and tail) at each toss, and these outcomes are equally likely (probability $= 1/2$). The probability that both tosses will show heads is

$$\frac{1}{2} \times \frac{1}{2} = \frac{1}{4}$$

as per the multiplicative law.

We can ensure independence in our samples by separately randomizing the selection of each item to be included in the sample.

A.2.4 Mutually Exclusive Events

If one outcome for an event precludes another outcome, the outcomes are said to be mutually exclusive. A good example here comes with cutting a poker deck minus the jokers. The card that is turned up cannot be both a diamond and a heart, so the outcomes "diamond" and "heart" are mutually exclusive. However, the card turned up could be the deuce of hearts, so the outcomes "deuce" and "heart" are not mutually exclusive. Again, mutually exclusive outcomes are easier to handle. For instance, there are four suits in the deck, with equal numbers of cards for each. Since $1/4$ of the cards are of each suit, each has a probability of $1/4$. Now how about the probability of getting a heart or diamond (red card). There are ¼ + ¼ = ½ of the cards in these two suits, so the probability of getting a red card is ½. This is an illustration of the rule that the *probabilities of mutually exclusive outcomes are additive.*

A.2.5 Probability Distribution

The additive and multiplicative laws serve to guide much of gambling. The first use we will make of them is to develop the concept of a *probability distribution.* Suppose that we are tossing a handful of three fair coins and counting the number of coins with heads up at each toss. An outcome, such as the number of heads, that has a numeric value is called a *random variable.* Obviously in this case we can get zero, one, two, or three heads. The present problem is to figure out the probabilities for all possible values of the random variable. The set of probabilities for all possible values of the random variable is called its *probability distribution.* In figuring out the probability distribution, we should keep in mind that the total probability for all possible outcomes of an event is 1, that is, something or other is certain to happen when an event takes place.

Let us first examine the possibilities for any given coin. The coin must either show heads or tails. These outcomes are mutually exclusive and equally likely. Therefore

$$1 = \text{probability heads} + \text{probability tails} = .5 + .5$$

Furthermore the three coins are independent, so we can take the probabilities for the individual coins and multiply them together. As shown in Table A.2, there are eight possible (and mutually exclusive) sequences of heads and tails assuming that we have the coins numbered in some way so that we can distinguish one from the other. Each of these sequences has probability

$$.5 \times .5 \times .5 = .125$$

However, there are three sequences with two heads and three with one head. Since these are mutually exclusive sequences, we can add probabilities for these sequences which give the same number of heads. This gives the probability distribution in Table A.3.

A.2.6 Binomial Distribution

Also notice that the last series of steps we went through in developing the probability distribution for our example can be condensed into the formula

$$(.5 + .5) \times (.5 + .5) \times (.5 + .5) = (.5 + .5)^3 = .125 + .375 + .375 + .125$$

Table A.2 Possible sequences of heads and tails in tossing three fair coins

Coin number		
1	2	3

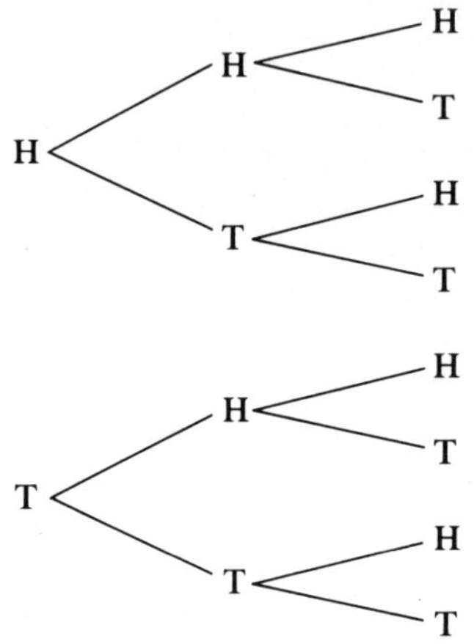

Table A.3 Probability distribution for number of heads in tossing three fair coins

Number of heads		Probability
0		.125
1	3 × .125 =	.375
2	3 × .125 =	.375
3		.125
Total for all outcomes		1.000

where the successive terms we get by working out the power are the probabilites of zero, one, two, and three heads, respectively. This sort of reasoning generalizes very easily to

$$(q + p)^n$$

where p is the probability of a head, q is the probability of a tail, and n is the number of coins. When the power expression is worked out and the terms with the same number of p's are added together, successive terms will give the probability of 0, 1, 2, . . . , n heads per toss. This is called the *binomial distribution* for sample size n.

As another example we can quickly work out the distribution for two coins instead of three. The formula is

$$(q + p)^2$$

with both q and $p = .5$. However, it will be easier to see which terms should be added together if we do not plug in the values of q and p until later. If we work out the power, we get

$$qq + pq + pq + pp$$

The next thing is to add the terms with the same number of p's and q's. This gives

$$qq + (pq + pq) + pp = qq + 2pq + pp$$

How we can plug in the values of p and q to obtain

$$(.5)(.5) + 2(.5)(.5) + (.5)(.5) = .25 + .50 + .25$$

where the first term is the probability of zero heads, the second is the probability of one head and one tail, and the third is the probability of two heads.

The binomial distribution applies to a large class of problems in which there are several independent events, each of which may occur in two ways. The probabilities p and q do not necessarily have to be equal. The usual terminology is to call one of the possible outcomes a "success" and the other a "failure." The probability that applies to a success is p, and that for a failure is q. In our case a head was a "success" and a tail was a "failure." Also it is customary to work out the algebra in $(q + p)^n$ ahead of time. The general form of the result is

$$q^n + \frac{n!}{(n-1)!} pq^{n-1} + \cdots + \frac{n!}{i!(n-i)!} p^i q^{n-i} + \cdots + p^n$$

where i is the number of successes that the term represents.

The exclamation mark in this formula warrants some explanation. This is factorial notation, and $n!$ is pronounced "n-factorial." The value of $n!$ is defined as

$$n! = n \times (n-1) \times (n-2) \times \cdots \times 1$$

with $0!$ defined to be 1. For instance,

$$4! = 4 \times 3 \times 2 \times 1 = 24$$

The multiplier

$$\frac{n!}{i!(n-i)!}$$

is of still further interest in that it gives the number of combinations of i things which can be formed from n things. There is still another bit of shorthand that one may see used in the formula for the binomial distribution. It is

$$\binom{n}{i} = \frac{n!}{i!(n-i)!}$$

A.2.7 Cumulative Probability Distribution

The cumulated form of a probability distribution is more directly useful in probability work than the simple probability distribution itself. The *cumulative probability distribution* is just the running sum of the probability distribution. It shows the probability that a random variable will take on a value *less than or equal to* the value given. The cumulative distribution for the three coin example as compared with the simple distribution is given in Table A.4. Notice that the last value of the cumulative probability distribution must always be 1.

Let us illustrate the use of a cumulative distribution by answering three types of questions about the three coin example. First what is the probability of getting two heads or less? To answer this question we need only consult the cumulative distribution for number of heads = 2. The answer is .875. Any question of this "less than or equal to" nature can be answered directly from the cumulative distribution.

A question of the "greater than" type can be answered by subtracting the value of the cumulative distribution from 1.0. For instance, the probability of getting more than one head is

$$1.0 - .5 = .5$$

Table A.4 Probability distribution and cumulative probability distribution for number of heads in tossing three fair coins

Number of heads	Probability	Cumulative probability
0	.125	.125
1	.375	.500
2	.375	.875
3	.125	1.000

Finally look at a question that involves middle values. What is the probability that the coins will not all show the same face? That is, what is the probability of getting one or two heads? We start with the largest applicable value from the cumulative distribution. We have already used the cumulative distribution to find that the probability of two or fewer heads is .875. From this we need to subtact the probability of getting less than one head (that is, zero heads). The distribution tells us that the probability of zero heads is .125. Therefore the probability of one or two heads is

$$.875 - .125 = .750$$

Much of statistical theory is concerned with developing probability distributions and analyzing the conditions under which they apply. However, one need not be a statistician to utilize the fruits of the statistician's labors. We will have occasion to use the normal distribution, t-distribution, F-distribution, and chi-square distribution later in this appendix.

A.3 DIRECT ESTIMATION FROM SIMPLE RANDOM SAMPLES

In most work with environmental resources it is economically impractical, if not physically impossible, to make observations or measurements on the whole population of interest. Furthermore the measurement process may itself be destructive, as when dissections are required to determine the presence or absence of disease or pesticide residues. With destructive measurements it is obviously undesirable to do a complete census since this would eliminate the resource. We usually must be content with examining only a portion of the population in which our interest lies. The portion actually examined is called a *sample.*

A.3.1 Nature of Simple Random Sampling

The manner in which the sample is selected is extremely important since it determines the information that can be gleaned from the sample. If we allow our selection to be governed by accessibility and ease of measurement, we will learn little or nothing about the less accessible parts of the population. If the sample is suitably selected, the measurements made on the sample can be used to estimate the parameters of the population. These estimates are called *statistics*, and the equations that are used to arrive at the statistics are called *estimation equations* or simply *estimators*.

If the sample has been randomly selected, it may be possible to state with some assurance that the true (but unknown) value of the population parameter lies between certain upper and lower limits. These limits are called *confidence limits*, and the interval between the limits is called a *confidence interval.*

The information gained from the sample will usually form the basis for choosing one of the possible courses of action open to us in managing the environmental resource. Our strategy in choosing or rejecting a possible course of action will usually be to accept the action if there is a good chance that the population parameter lies in the range where the action is beneficial. If the sample leads us to believe that there is relatively little chance of the population parameter being in the range of benefit, some alternative course of action will be chosen. The alternative procedure might well be to take no action at all. The use of information from samples to determine the choice of alternative courses of action is called *statistical decision making.*

The term *elementary unit* will be used to designate the individual unit in the population on which the actual measurement is made. If we try to determine the volume of usable wood in a forest tract, the elementary unit will be the individual tree. When we undertake to find the number of animals of a particular species in an area, the elementary unit will be the individual animal. If we are attempting to assess the extent of damage by a leaf-eating insect, the elementary unit might be the individual leaf. Elementary units for measurement, then, vary according to the material under investigation and what it is that we want to know about the material. It is important to be quite clear about what the elementary unit is in any sampling job, but the details that need to be considered depend on the nature of the resource.

For purposes of selecting a sample, the elementary units may be grouped into *sampling units.* For example, one ordinarily does not choose plants individually for measurement in vegetation surveys. Instead, plots are established and all plants growing on the plot are measured when the plot is visited. The choice of sampling unit should not be taken lightly since it can have a decided effect on the cost and accuracy of a survey. The structure of the population in terms of sampling units is called the *sampling frame.*

The final sample will usually contain several sampling units. The number of *sampling units* included in the sample is called the *sample size.* The sample size divided by the total number of sampling units in the population is called the *sampling fraction.* The sampling fraction may also be expressed in percentage form by multiplying by 100.

There are many possible ways of selecting a sample from a sampling frame. Some selection procedures will allow us to learn more about the population from a given size of sample than others. An ideal selection procedure should be efficient from an information standpoint, and yet not too costly in terms of field time, travel, personnel, and so on. In other words we are after the most information per dollar spent in collecting and analyzing the sample.

The theory of statistics dictates that chance should determine the units to be included in the sample, with the proviso that the probability of each unit being in the sample must be known. This is called *random sampling.* If we make the size of a random sample relatively large, we can be reasonably confident that the likelihood of obtaining an atypical sample is low (although not impossible). Random sampling does not require that the probability of obtaining the different possible samples of a given size be the same, only that the probability be known for each sample. Estimation becomes simpler, however, if we give each possible *sample* the same probability of occurrence. Such a sampling scheme is called an *equal probability selection method* (EPSEM). The situation is even further simplified if we also give each *sampling unit* the same chance of being selected at any point in the sample selection process. This is called *simple random sampling.*

Simple random sampling may be either with or without replacement. In sampling with replacement we allow a sampling unit to be measured several times in the same sample. When sampling is without replacement, a sampling unit cannot be measured more than once in a sample. Since measuring the same sampling unit more than once would normally be a waste of time, simple random sampling is understood to be without replacement unless otherwise specified. For the sake of brevity let us call the selection of the next sampling unit to be included in the sample a "draw." When sampling is with replacement, all units are eligible at every draw. When sampling is without replacement, units already drawn for inclusion in the sample are ineligible. Simple random sampling requires that the total number of sampling units in the population must be known ahead of time. Also each sampling unit must be assigned an identification number for use in drawing the sample. The need to know the population size and number each sampling unit can be a real drawback at times in natural resources work. In effect, a map or aerial photo of the tract to be surveyed must be available before the sample can be selected.

A.3.2 Estimating Mean and Total

When a simple random sample has been selected, the least we will want to do is estimate the mean, total, and variability in the population. The question then arises as to how these estimates should be calculated. The natural suggestion is to adapt the parametric formulas from Section A.1 so that they can be applied to sample data. The obvious modifications are to run summations over the sample items and to replace the population size N by the sample size n. The sample mean is thus

$$\text{sample mean} = \frac{\Sigma X}{n}$$

where the summation is over the items in the sample and n is the sample size. The sample mean is usually denoted by $\overline{X}$ or $\overline{Y}$.

Suppose that Long Mirage Lake, Middle Mirage Lake, and Little Mirage Lake were selected as a simple random sample of size 3 from the population of Mirage Lakes. Then the sample means for shoreline length and surface area, respectively, would be

$$\overline{X} = \frac{\Sigma X}{n} = \frac{6.8 + 5.0 + 3.0}{3} = 4.93 \text{ km}$$

and

$$\overline{Y} = \frac{\Sigma Y}{n} = \frac{202 + 77 + 42}{3} = 107.0 \text{ ha}$$

Before accepting the sample mean as a desirable way of estimating the population mean, we should at least inquire whether we might expect the average of a large number of sample means taken from the same population to approximate the parametric mean of the population. The average of all possible sample estimates with each possible estimate being weighted by its probability of occurrence is called the *expected value* of the estimator. If the expected value of the estimator is equal to the parametric value, then the estimator is said to be *unbiased.* Unbiased estimators are preferable, but a slight bias is often tolerable. Most textbooks on sampling methods contain proofs that the sample mean is an unbiased estimator of the population mean in simple random sampling either with or without replacement.

Given an unbiased estimator of the population mean, we can obtain an unbiased estimator of the population total simply by multiplying it by the population size. Thus we can estimate the total shoreline length and surface area of the Mirage Lakes as

$$\text{estimated population total} = N \times \overline{X} = 10 \times 4.93 = 49.3 \text{ km}$$

and

$$N \times \overline{Y} = 10 \times 107.0 = 1070 \text{ ha}$$

Conversely, given an estimate of the total, we can divide by N to estimate the mean.

A.3.3 Estimating Variance, Standard Deviation, and Coefficient of Variation

Encouraged by this approach we might attempt to estimate the population variance by the formula

$$\frac{\Sigma X^2 - (\Sigma X)^2/n}{n}$$

where the summations are over the sample and n is again the sample size. However, textbooks on sampling theory show that the expected value of this estimator is

$$\sigma^2 \times \frac{(n - 1)}{n}$$

instead of σ^2. The reason for the bias is that the population mean must first be estimated from the sample before we can proceed to calculate deviations from the mean. In effect, one unit of information in the sample is used up in estimating the population mean, leaving only $(n - 1)$ units of information available for estimating variability. However, the expected value of the estimator

$$\frac{\Sigma X^2 - (\Sigma X)^2/n}{n}$$

makes it apparent that we can correct for the bias by multiplying this estimator by the factor $n/(n - 1)$. This gives

$$\frac{\Sigma X^2 - (\Sigma X)/n}{n} \times \frac{n}{(n - 1)} = \frac{\Sigma X^2 - (\Sigma X)/n}{(n - 1)}$$

as the unbiased estimator of the population variance. This estimator is usually designated S^2. Also the quantity $(n - 1)$ in the denominator is known as the *degrees of freedom* for estimating variance.

For the sample of lakes we have

$$S_X^2 = \frac{\Sigma X^2 - (\Sigma X)^2/n}{(n - 1)} = \frac{80.24 - (14.8)^2/3}{2} = 3.61 \text{ sq km}$$

and

$$S_Y^2 = \frac{\Sigma Y^2 - (\Sigma Y)^2/n}{(n - 1)} = \frac{48{,}497 - (321)^2/3}{2} = 7075.0 \text{ sq ha}$$

The population standard deviation is normally estimated as

$$S = \sqrt{S^2}$$

even though a trivial bias is involved in this latter step. For our example then

$$S_X = \sqrt{3.61} = 1.90 \text{ km}$$

and

$$S_Y = \sqrt{7075} = 84.1 \text{ ha}$$

Likewise the coefficient of variation is normally estimated as

$$CV = \frac{S}{X} \times 100$$

giving

$$\frac{1.90}{4.93} \times 100 = 38.5\% \text{ for shoreline length}$$

and

$$\frac{84.1}{107.0} \times 100 = 78.6\% \text{ for surface area}$$

A.3.4 Estimating Covariance and Correlation Coefficient

In order to obtain an unbiased estimate of the covariance, we must again use the degrees of freedom in the denominator. Thus the covariance estimator is

$$S_{XY} = \frac{\Sigma XY - (\Sigma X)(\Sigma Y)/n}{(n-1)} = \frac{1884.6 - (14.8)(321)/3}{2} = 150.5$$

The estimator of the correlation coefficient, designated as r, is normally taken as

$$r = \frac{S_{XY}}{S_X S_Y} = \frac{150.5}{(1.90)(84.1)} = .94$$

The estimate that we actually obtain by taking a sample will depend on which one of the possible samples our random selection procedure happens to bring to our attention. If different samples give widely different estimates, there is a relatively high risk of getting a bad estimate from a sample. Therefore it is important to have a measure of variability to be expected in samples of size n from a population.

A.3.5 Mean and Variance of Sample Means

The estimated means from all possible samples of size n can be viewed as a new population which has been derived from the original population. This population of sample means has parameters in its own right. If we had several samples of size n at our disposal, we could estimate the mean from each sample and then proceed to estimate the mean and variance of sample means with the formulas already discussed. However, we will usually be able to afford only one sample. Therefore we need a way of estimating the mean and variance of *sample means* from a single sample.

Estimating the mean of sample means is not much of a problem since the average of an unbiased estimator such as the sample mean is the same as the mean of the original population. Thus the sample mean is itself an unbiased estimate of the mean of all sample means. However, estimating the variance of sample means requires a little more thought.

We saw in Section A.1.7 that the variance of a combination

$$Z = aX + bY$$

is

$$\sigma_Z{}^2 = a^2\sigma_X{}^2 + b^2\sigma_Y{}^2 + 2ab\sigma_{XY}$$

If the variables X and Y are independent, their covariance will be zero. Therefore a combination composed of a sum of variables with each variable multiplied by a constant should be the sum the variances with each variance multiplied by the square of the constant. Although this result was based on measuring pairs of variables on the same set of sampling units, it can serve as a guide for speculating on the nature of variance for sample means. Statistical texts contain the proof that our speculation is correct.

In calculating a sample mean it makes no difference whether we do the summation first or divide each value by n (the sample size) first and then do the summation, that is,

$$\frac{\Sigma X}{n} = \frac{X_1}{n} + \frac{X_2}{n} + \cdots + \frac{X_n}{n}$$

$$= \frac{1}{n}X_1 + \frac{1}{n}X_2 + \cdots + \frac{1}{n}X_n$$

Thus the sample mean is a sum of variables (the sample values) with each multiplied by the constant $a = 1/n$. Our simple random sampling procedure ensures independence between the X's that make up the sample, so they should have zero covariance. Furthermore the variance of individual X values is simply $\sigma_X{}^2$. Therefore we would expect the variance of a sample mean to be

$$\Sigma\left(\frac{1}{n}\right)^2 \sigma_X{}^2 = \frac{1}{n^2}\,\Sigma\sigma_X{}^2 = \frac{n}{n^2}\,\sigma_X{}^2 = \frac{\sigma_X{}^2}{n}$$

Since we already have a procedure for estimating $\sigma_X{}^2$, we can estimate the variance of the sample mean as

$$S_{\bar{X}}{}^2 = \frac{S_X{}^2}{n}$$

For our sample of three lakes, then, the estimate of the variance of means would be

$$S_{\bar{X}}{}^2 = \frac{S_X{}^2}{n} = \frac{3.61}{3} = 1.2 \text{ sq km}$$

and

$$S_{\bar{Y}}^2 = \frac{S_Y^2}{n} = \frac{7075}{3} = 2358 \text{ sq ha}$$

However, there is one small complication. If we took samples without replacement so large that they included the entire population, all samples would have the same mean, namely, the population mean. Whenever the sample is large enough that it constitutes an appreciable portion of the population in sampling without replacement, we should take account of this effect by multiplying the estimated variance of the mean by the fraction of the population that is *not* included in the sample. The modified formula is

$$S_{\bar{X}}^2 = \frac{S_X^2}{n}(1 - f)$$

where f is the sampling fraction. This fraction of the population which is not included in the sample is usually called the *finite population correction factor*. This correction is not applied in sampling with replacement. It can also be ignored when the sample size is small relative to the population size (say less than 5%).

Since the sampling fraction is $f = n/N$, there are several alternative ways of expressing the finite population correction factor. Two such formulations are

$$1 - \frac{n}{N} \quad \text{and} \quad \frac{N - n}{N}$$

If our sample of three lakes were drawn without replacement, the finite population correction factor should be applied. This gives

$$S_{\bar{X}}^2 = \frac{S_X^2}{n}(1 - f) = \frac{3.61}{3}(1 - 0.3) = 0.84 \text{ sq km}$$

and

$$S_{\bar{Y}}^2 = \frac{S_Y^2}{n}(1 - f) = \frac{7075}{3}(1 - 0.3) = 1650 \text{ sq ha}$$

as the estimated variance of the means for shoreline length and surface area, respectively.

A.3.6 Standard Error of Mean and Total

The standard deviation of the sample mean is estimated as

$$S_{\bar{X}} = \sqrt{S_{\bar{X}}^2}$$

giving

$$S_{\overline{X}} = \sqrt{0.84} = 0.92 \text{ km}$$

and

$$S_{\overline{Y}} = \sqrt{1650} = 40.6 \text{ ha}$$

for shoreline length and surface area, respectively. The standard deviation of the sample mean is also called the *standard error of the mean* since it is a measure of the error incurred by using the sample mean as an estimate of the population parameter. When the sample mean is given, it is usually accompanied by its estimated standard error to provide an indication of the variability associated with the estimate. A common form is

$$\overline{X} \pm S_{\overline{X}}$$

Estimates of other population parameters made from samples also have standard errors associated with them. For example, totals can be estimated by multiplying the corresponding estimate of the mean by the number of units in the population N. Therefore the estimate of the standard error for a total is also obtained by multiplying the estimate of the standard error of the mean by N, that is,

$$\text{estimated standard error of total} = N \times S_{\overline{X}}$$

A.3.7 Standard Errors of Combination Estimates

Freese (1962) gives a convenient summary of rules for estimating standard errors of combinations of sample estimates. Therefore we follow a similar format in summarizing these rules. Let U and V be sample estimates with standard errors S_U and S_V, respectively. Also let c and d be constants. Then the estimated standard error of the combination estimate

$$Z = cU$$

is

$$S_Z = \sqrt{S_Z{}^2}$$

where

$$S_Z{}^2 = c^2 S_U{}^2$$

Thus

$$S_Z = cS_U$$

The estimated standard error of the combination

$$Z = cU + dV$$

is

$$S_Z = \sqrt{S_Z^2}$$

where

$$S_Z^2 = c^2 S_U^2 + d^2 S_V^2$$

The estimated standard error of the combination

$$Z = cUV$$

is

$$S_Z = \sqrt{S_Z^2}$$

where

$$S_Z^2 = Z^2 \left(\frac{S_U^2}{U^2} + \frac{S_V^2}{V^2} + 2\,\frac{S_{UV}}{UV} \right)$$

with S_{UV} being the covariance of U and V.

The estimated standard error of the combination

$$Z = c\,\frac{U}{V}$$

is

$$S_Z = \sqrt{S_Z^2}$$

where

$$S_Z^2 = Z^2 \left(\frac{S_U^2}{U^2} + \frac{S_V^2}{V^2} - 2\,\frac{S_{UV}}{UV} \right)$$

Textbooks of sampling theory should be consulted for the derivation of these rules. However, it should be noted that the rules for combinations involving products and ratios (such as UV and U/V) are approximations in which some minor terms are omitted.

A.3.8 Interval Estimates

Up to this point when we turned the crank on an estimator it has dumped out a single number. When we place our bets on a single number as an estimate, we have a *point estimate.* The point estimate that a given estimator will produce depends on which one of the possible samples of size n is selected. The standard error of the sample mean is an expression of this variability between estimates, or sampling error. In view of the existence of the sampling error, point estimates can give an unwarranted impression of exactness. The estimate we make from the sample is only our best guess as to the value of the parameter we are trying to estimate. It would seem more realistic to give a low estimate and a high estimate which should bracket the true population value unless we have obtained a rather unusual sample.

This latter approach to estimation is known as *interval estimation*, because the high and low estimates form the ends of an interval which should encompass the population parameter unless the interval is determined from an unusual sample. The interval determined from the sample is called a *confidence interval* because the procedure for setting the interval must allow us some specified confidence that it will encompass the true value of the population parameter. The endpoints of the confidence interval are called *confidence limits.* Of course we cannot be absolutely sure that the confidence interval includes the true value unless the sample includes the whole population. There are two ways to increase our confidence in an interval estimate. One way is to make the interval larger (wider). The other way is to increase the sample size (make the number n of sampling units in the sample larger).

In order to set confidence limits on an estimate, we look to statistical theory and ask what sort of probability distribution the estimates will have if a large number of samples of the same size are taken from a given population. The *central limit theorem* of mathematical statistics states that the means of large random samples drawn from a population with finite variance will approximate a *normal distribution*, and that the larger the sample size the better will be the approximation.

A.3.9 Normal Probability Distribution

Normal probability distributions are bell shaped curves with the peak occurring at the mean, as shown in Figure A.1. At first the curve tails off from the peak rapidly, but the slope gradually becomes less steep as one moves away from the peak. Approximately 68% of the population lies within 1 standard deviation on either side of the peak (mean) as measured by the area under the curve. Roughly 95% lies within 2 standard deviations, and 99% within 2.56 standard deviations on either side of the mean. The exact shape of a normal distribution is determined by its mean and standard deviation. An increase in the population mean moves the curve bodily to the right, and a decrease moves the curve bodily to the left. An increase in the standard deviation causes the curve to have a wider spread, and a decrease in the standard deviation reduces the spread.

In order to facilitate making tables of probabilities and cumulative probabilities, we can standardize any given normal distribution by subtracting the

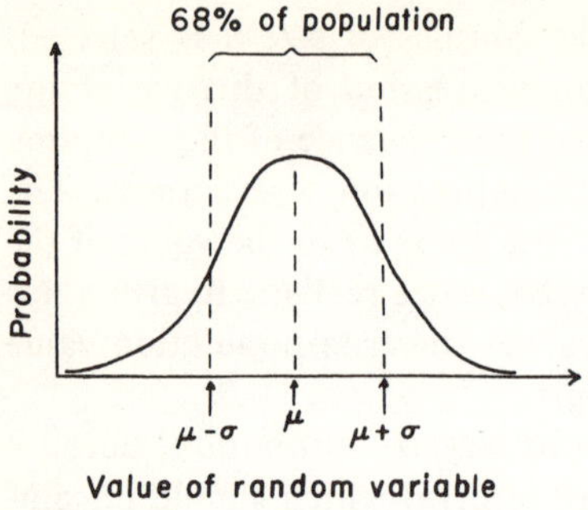

Figure A.1 Diagrammatic representation of a "normal" probability distribution.

mean from each value in the population and then dividing the result by the standard deviation, that is,

$$Z = \frac{X - \mu}{\sigma}$$

where Z is the standardized value (also called standard normal deviate), and X the value of the original variable.

In effect the Z transformation expresses everything as a difference from the mean in standard deviation units. The transformed or standardized Z values have a mean of 0 and a standard deviation of 1. Since the variance is the square of the standard deviation, it is also equal to 1. Through standardization the normal curve with mean 0 and variance 1 can be used as a reference curve for any other normal distribution.

Any book of statistical tables will contain tables of the probability distribution and cumulative probability distribution for the standard normal curve. Alternatively the (mean − 1 standard deviation), mean, and (mean + 1 standard deviation) can be plotted opposite the 34%, 50%, and 68% points on normal probability graph paper to define the cumulative probability distribution as a straight line through the three points.

A.3.10 Large Sample Confidence Limits for the Mean Based on Normal Distribution

To apply the normal distribution for setting confidence limits on an estimated mean, we need the mean and standard deviation for the population of sample means. Since the sample mean is an unbiased estimator of the population mean from which the sample was drawn, we know that the mean of sample means is μ. Likewise we know that the standard deviation of sample means is

$$\frac{\sigma}{\sqrt{n}}$$

Putting these facts together gives

$$\Pr\left(\mu - Z\frac{\sigma}{\sqrt{n}} \leq \overline{X} \leq \mu + Z\frac{\sigma}{\sqrt{n}}\right) = C$$

where Pr (. . .) = probability of the event in parentheses
μ = population mean
$\overline{X}$ = mean of sample of size n
σ = population standard deviation
C = desired confidence probability
Z = number of standard deviation units necessary to make $\pm Z$ encompass area C under the standard normal curve

The inequality inside the parentheses can be rearranged as follows:

$$\Pr\left(-Z\frac{\sigma}{\sqrt{n}} \leq \overline{X} - \mu \leq +Z\frac{\sigma}{\sqrt{n}}\right) = C$$

$$\Pr\left(-\overline{X} - Z\frac{\sigma}{\sqrt{n}} \leq -\mu \leq -\overline{X} + Z\frac{\sigma}{\sqrt{n}}\right) = C$$

$$\Pr\left(\overline{X} + Z\frac{\sigma}{\sqrt{n}} \geq \mu \geq \overline{X} - Z\frac{\sigma}{\sqrt{n}}\right) = C$$

$$\Pr\left(\overline{X} - Z\frac{\sigma}{\sqrt{n}} \leq \mu \leq \overline{X} + Z\frac{\sigma}{\sqrt{n}}\right) = C$$

Setting $Z = 1$ gives a confidence probability of approximately 68%, $Z = 2$ gives approximately 95%, and $Z = 2.56$ gives approximately 99% confidence. The only difficulty here is that we do not know the population standard deviation σ required to compute our confidence limits. In a large sample, however, the sample estimate of the standard deviation should be a good approximation to σ, so the confidence limits should not be distributed greatly by making the substitution

$$S_{\overline{X}} = \frac{\sigma}{\sqrt{n}}$$

For example, suppose that we have taken a sample of $n = 36$ for which

$$\overline{X} = 10 \qquad \text{and} \qquad S_{\overline{X}} = 1.5$$

To make a 95% confidence interval estimate of the population mean, we use

$$\Pr\,(\overline{X} - 2S_{\overline{X}} \leq \mu \leq \overline{X} + 2S_{\overline{X}}) \simeq .95 \text{ or } 95\%$$

to calculate the interval

$$10 - 2(1.5) \le \mu \le 10 + 2(1.5)$$
$$10 - 3 \le \mu \le 10 + 3$$
$$7 \le \mu \le 13$$

Another way of writing this is

$$\overline{X} = 10 \pm 3 \text{ with } 95\% \text{ confidence}$$

Confidence limits based on the normal distribution will usually be acceptable for samples of size $n = 30$ or larger.

A.3.11 Small Sample Confidence Limits for the Mean Based on *t*-Distribution

The central limit theorem does not apply to small samples, so we cannot count on the means of small samples to be normally distributed. However, many types of measurements produce a line that is approximately straight when cumulated percentages are plotted against the size of measurement on probability paper. By cumulated percentages we mean the percentage of the measurements that are less than or equal to any given value. If the plot is approximately straight, it indicates that the measurements themselves follow a normal distribution.

If samples are taken from a population that is normally distributed, statistical theory tells us that the sample means will follow a probability distribution known as a t-distribution. Even after standardization, t-distributions depend on the degrees of freedom $(n - 1)$ for the sample. If the assumption of normality for the original population seems reasonable, the following confidence limits are available on the basis of the t-distribution:

$$\Pr(\overline{X} - t\,S_{\overline{X}} \le \mu \le \overline{X} + t\,S_{\overline{X}}) = C$$

where $\overline{X}$ = mean of sample of size n
$S_{\overline{X}}$ = estimated standard error of sample mean
C = desired confidence expressed as a decimal
t = t value necessary to make $\pm t$ encompass area C under the t-distribution curve for $n - 1$ degrees of freedom as determined from a table of t available in statistical handbooks. The t value for a t-distribution is analogous to the Z value for a normal distribution. For degrees of freedom exceeding 30 there is very little difference between t values and Z values.

As an example suppose that we are willing to regard the population of Mirage Lakes as being normally distributed for shoreline length. Then the

95% confidence limits for the mean shoreline length as determined from our sample of three lakes would be

$$\Pr(\overline{X} - t\,S_{\overline{X}} \le \mu_X \le \overline{X} + t\,S_{\overline{X}}) = C$$

$$\Pr[4.93 - (4.3)(0.92) \le \mu_X \le 4.93 + (4.3)(0.92)] = .95$$

$$4.93 - 3.96 \le \mu_X \le 4.93 + 3.96$$

$$0.97 \text{ km} \le \mu_X \le 8.89 \text{ km with 95\% confidence}$$

There are very wide limits, which only goes to show that means from samples of size $n = 3$ are highly variable. Unlike the large sample limits, the t-distribution limits take into account the fact that the standard error must be estimated from the sample.

A.3.12 Sample Size

The specifications for a survey will usually give the allowable standard error, and we will have to decide how large a sample is needed to satisfy the requirements. This is accomplished by inverting the formula for the standard error of the mean.

If sampling is with replacement, the inverted formula is

$$\sigma_{\overline{X}} = \frac{\sigma}{\sqrt{n}}$$

$$\sqrt{n} = \frac{\sigma}{\sigma_{\overline{X}}}$$

$$n = \frac{\sigma^2}{\sigma_{\overline{X}}^2}$$

If sampling is without replacement, the inverted formula is

$$\sigma_{\overline{X}} = \frac{\sigma}{\sqrt{n}}\sqrt{\frac{N-n}{N}}$$

$$\sigma_{\overline{X}}^2 = \frac{\sigma^2 N - \sigma^2 n}{nN}$$

$$nN\sigma_{\overline{X}}^2 = \sigma^2 N - \sigma^2 n$$

$$nN\sigma_{\overline{X}}^2 + \sigma^2 n = \sigma^2 N$$

$$n = \frac{\sigma^2 N}{N\sigma_{\overline{X}}^2 + \sigma^2}$$

In either case it will be necessary to have some prior information on the standard deviation of the population. We usually will not know the exact value of the standard deviation ahead of time, but may be able to estimate a reasonable upper limit for the standard deviation on the basis of prior experience with sampling in similar populations. Alternatively we can take a few preliminary samples in order to obtain a conservative estimate of the standard deviation.

To illustrate the use of the formulas, suppose that a standard error of 2 units is desired in sampling with replacement from a population having σ of about 10 units. The necessary sample size is

$$n = \frac{\sigma^2}{\sigma_{\bar{X}}^2} = \frac{10^2}{2^2} = \frac{100}{4} = 25$$

The desired standard error of the mean can be stated as a percentage of the mean if σ is replaced by the coefficient of variation.

If the specifications are stated in terms of a confidence interval, the fact that

$$Z\sigma_{\bar{X}} = \text{half-width of confidence interval}$$

means that $\sigma_{\bar{X}}$ in the preceding formulas should be replaced by

$$\frac{\text{half-width of confidence interval}}{Z}$$

If the sample size is likely to be small, t should be used in place of Z. However, this creates a little problem in that the sample size must be known to find the degrees of freedom. Therefore it is necessary to start with a guessed value of t, estimate n, revise the value of t, revise the estimate of n, and so on until n no longer changes.

A.4 INDIRECT ESTIMATION BY REGRESSION

It is not at all uncommon in survey work for important measurements to be expensive, difficult, or to have undesirable side effects. In cases like these, discretion is often the better part of valor, and we are well advised to seek something that is more amenable to measurement but still closely related to our original variable of interest. For instance, a tree's diameter, height, and form are the primary determinants of its volume and much easier to measure. Thus we are in a situation that calls for measuring one thing indirectly by actually measuring another. It is customary to call the predictor variables *independent variables* and the variable to be predicted the *dependent variable.*

The whole idea of indirect measurement sounds pretty good, but we must first determine the relationship between the variable we want to know about and the variables we plan to measure as its alter ego. This means taking *both* measurements on a sample of individuals in order to determine the nature of the relationship. In other words we want a prediction equation to use in estimating the dependent variable from the independent variables. This prediction equation is usually developed by regression analysis.

A.4.1 Regression Model and Residuals

Let the independent variables be denoted by $X_1, X_2, \ldots, X_m$ and the dependent variable by Y. Also let us place a hat on the $\hat{Y}$ when its value is estimated instead of actually measured. Regression analysis is appropriate for determining the coefficients (b's) of any prediction equation (or model) that takes the form

$$\hat{Y} = b_0 + b_1X_1 + b_2X_2 + \cdots + b_mX_m$$

Differences between the actual and predicted values are called *residuals.* Thus a residual takes the form

$$(Y - \hat{Y})$$

where Y is the actual measured value for an observational unit, and $\hat{Y}$ is the predicted value for that unit.

The regression procedure is designed to minimize the sum of squared residuals over the sample from which the prediction equation is developed, that is, the predictive equation is designed to minimize

$$\Sigma(Y - \hat{Y})^2$$

A.4.2 Normal Equations

The first step in the regression procedure is to calculate ΣX and ΣX^2 for each variable, and also the sum of the cross products for each pair of variables. Theses calculations are performed for the dependent variable as well as for the independent variables. These sums, sums of squares, and sums of cross products are used to construct a set of simultaneous linear equations called the *normal equations.* The normal equations are then solved for the b coefficients that determine the prediction equation.

The normal equations for five independent variables are given in Table A.5. Note particularly the diagonal terms that are underlined. If there is no

Table A.5 Normal equations for five independent variables in regression analysis

$$\underline{b_0 n} + b_1\Sigma X_1 + b_2\Sigma X_2 + b_3\Sigma X_3 + b_4\Sigma X_4 + b_5\Sigma X_5 = \Sigma Y$$

$$b_0\Sigma X_1 + \underline{b_1\Sigma X_1^2} + b_2\Sigma X_1X_2 + b_3\Sigma X_1X_3 + b_4\Sigma X_1X_4 + b_5\Sigma X_1X_5 = \Sigma X_1Y$$

$$b_0\Sigma X_2 + b_1\Sigma X_1X_2 + \underline{b_2\Sigma X_2^2} + b_3\Sigma X_2X_3 + b_4\Sigma X_2X_4 + b_5\Sigma X_2X_5 = \Sigma X_2Y$$

$$b_0\Sigma X_3 + b_1\Sigma X_1X_3 + b_2\Sigma X_2X_3 + \underline{b_3\Sigma X_3^2} + b_4\Sigma X_3X_4 + b_5\Sigma X_3X_5 = \Sigma X_3Y$$

$$b_0\Sigma X_4 + b_1\Sigma X_1X_4 + b_2\Sigma X_2X_4 + b_3\Sigma X_3X_4 + \underline{b_4\Sigma X_4^2} + b_5\Sigma X_4X_5 = \Sigma X_4Y$$

$$b_0\Sigma X_5 + b_1\Sigma X_1X_5 + b_2\Sigma X_2X_5 + b_3\Sigma X_3X_5 + b_4\Sigma X_4X_5 + \underline{b_5\Sigma X_5^2} = \Sigma X_5Y$$

term in the prediction equation involving a given b coefficient, the entire row and column crossing at the underlined b term are eliminated from the normal equations.

The normal equations can be expanded to encompass more than five independent variables, but problems of this magnitude are usually handled by computer programs. Hand held calculators are available with a preprogrammed facility for solving systems of six simultaneous linear equations.

A.4.3 Regression Example

The regression procedure can be applied to populations as well as samples. We will illustrate the procedure by developing a regression equation to predict the surface area of Mirage Lakes from the shoreline length and the square of the shoreline length. To make this prediction equation fit the general form of the regression model, we let

Y = surface area,
X_1 = shoreline length,
X_2 = square of shoreline length

Then our prediction equation takes the form

$$\hat{Y} = b_0 + b_1X_1 + b_2X_2$$

The sums, sums of squres, and sums of cross products are

$$\Sigma X_1 = 45.5$$

$$\Sigma X_2 = 227.45$$

$$\Sigma Y = 1118.0$$

$$\Sigma X_1^2 = 227.45$$
$$\Sigma X_2^2 = 6840.4613$$
$$\Sigma Y^2 = 178{,}920$$
$$\Sigma X_1 X_2 = 1218.245$$
$$\Sigma X_1 Y = 5988.2$$
$$\Sigma X_2 Y = 33{,}706.76$$

There are no terms in our model involving b_3, b_4, or b_5 so we eliminate three rows and columns from the normal equations leaving

$$\begin{aligned} b_0 n + b_1 \Sigma X_1 \quad + b_2 \Sigma X_2 \quad &= \Sigma Y \\ b_0 \Sigma X_1 + b_1 \Sigma X_1^2 \quad + b_2 \Sigma X_1 X_2 &= \Sigma X_1 Y \\ b_0 \Sigma X_2 + b_1 \Sigma X_1 X_2 + b_2 \Sigma X_2^2 \quad &= \Sigma X_2 Y \end{aligned}$$

Substituting the calculated values, we have

$$\begin{aligned} 10 b_0 + \quad 45.5 b_1 + \quad 227.45 b_2 &= 1118.0 \\ 45.5 b_0 + \quad 227.45 b_1 + \quad 1218.245 b_2 &= 5988.2 \\ 227.45 b_0 + 1218.245 b_1 + 6840.4613 b_2 &= 33{,}706.76 \end{aligned}$$

as the set of simultaneous equations to be solved.

The solution to these equations is

$$b_0 = 70.327697 \qquad b_1 = -34.886297 \qquad b_2 = 8.802152$$

as can be verified by substituting these b coefficients back into the normal equations.

A.4.4 Analysis of Variance Table

Having developed a regression equation from a sample, we will need an index of how well it serves the purpose. We might also wish to explore whether all the terms in the regression equation are really necessary. In either case a first step is to assemble an *analysis of variance table.* For purposes of illustration we will treat the 10 lakes as being a sample from a larger population of lakes.

The format of the analysis of variance table is shown in Table A.6. The column heading SS, DF, and MS denote sum of squares, degrees of freedom, and mean square, respectively.

Table A.6 Format for analysis of variance table

Source	SS	DF	MS	F ratio
Regression	II	IV	VI	VIII
Residual	III	V	VII	
Total	I			

Item I, the total sum of squares, is calculated first. This is simply the numerator of the variance for the dependent or Y variable. Therefore the formula is

$$\text{total SS} = \Sigma Y^2 - (\Sigma Y)^2/n$$

For our example this is

$$\text{total SS} = \Sigma Y^2 - (\Sigma Y)^2/n = 178{,}920 - (1118)^2/10 = 53{,}927.6$$

Item II, the regression sum of squares, is calculated next from the formula

$$\begin{aligned}\text{regression SS} = {} & (b_0 \Sigma Y) + (b_1 \Sigma X_1 Y) + (b_2 \Sigma X_2 Y) \\ & + (b_3 \Sigma X_3 Y) + (b_4 \Sigma X_4 Y) + (b_5 \Sigma X_5 Y) - (\Sigma Y)^2/n\end{aligned}$$

omitting any parenthesized term that contains a b coefficient not present in the regression equation. Since the example does not contain b_3, b_4, or b_5, the appropriate formula is

$$\begin{aligned}\text{regression SS} &= (b_0 \Sigma Y) + (b_1 \Sigma X_1 Y) + (b_2 \Sigma X_2 Y) - (\Sigma Y)^2/n \\ &= (70.327697)(1118) + (-34.886297)(5988.2) \\ &\quad + (8.802152)(33{,}706.76) - (1118)^2/10 \\ &= 41{,}419.87\end{aligned}$$

Item III, the residual sum of squares, is calculated by subtracting the regression sum of squares from the total sum of squares. The residual sum of squares is the quantity that the regression procedure is designed to minimize. For example, we have

$$\begin{aligned}\text{residual SS} &= \text{total SS} - \text{regression SS} \\ &= 53{,}927.6 - 41{,}419.87 = 12{,}507.73\end{aligned}$$

Item IV, the regression degrees of freedom, is the number of b coefficients in the regression equation *excluding* b_0. For our case, this is 2.

Item V, the residual degrees of freedom, is the number of observations minus the number of b coefficients in the equation. In this case b_0 is *included* in the count of b coefficients. For our case this is $10 - 3 = 7$.

Item VI, the regression mean square, is calculated by dividing item II by item IV. For the present example

$$\text{regression MS} = \frac{\text{regression SS}}{\text{regression DF}} = \frac{41{,}419.87}{2} = 20{,}709.9$$

Item VII, the residual mean square, is calculated by dividing item III by item V. For the present example

$$\text{residual MS} = \frac{\text{residual SS}}{\text{residual DF}} = \frac{12{,}507.73}{7} = 1786.8$$

Item VIII, the F ratio, is calculated by dividing item VI by item VII. For the example

$$F \text{ ratio} = \frac{\text{regression MS}}{\text{residual MS}} = \frac{20{,}709.9}{1786.6} = 11.6$$

The completed analysis of variance table for the example is shown in Table A.7.

A.4.5 Determining whether the Regression is Adequate

When a regression equation is computed from a sample, we must keep in mind the fact that the b coefficients are only estimates of the true values for the population. It is possible that the true values of the b coefficients for the independent variables in the population as a whole are zero. In this case we would be deluding ourselves in thinking that the equation derived from the sample is of any value for prediction.

The F ratio in the analysis of variance table provides a means of checking the possibility that $b_1, b_2, \ldots, b_m$ are all zero for the population as a whole. We must first decide what chance we are willing to take of declaring that

Table A.7 Analysis of variance table for regression example

Source	SS	DF	MS	F ratio
Regression	41,419.87	2	20,709.9	11.6
Residual	12,507.73	7	1,786.8	
Total	53,927.60			

there are nonzero coefficients for the population when, in fact, they are all zero. For example, a 1 in 20 chance of being wrong is .05 probability. Next consult a table of the F-distribution in any statistical handbook. The table entries will be significance level, numerator degrees of freedom, and denominator degrees of freedom. Significance level is the probability of being wrong as just explained. Tables usually offer a choice of .05, .01, or .001. Use the regression degrees of freedom as the numerator DF and the residual degrees of freedom as the denominator DF. The table will show a critical F value corresponding to these arguments. If the calculated F ratio from the analysis of variance exceeds the critical value, we say that the regression is significant and reject the possibility that the population values of $b_1, b_2, \ldots, b_m$ might all be zero. For the example, an F table will show a critical value of 4.74 corresponding to significance level of .05, 2 numerator DF, and 7 denominator DF. Since the calculated F ratio of 11.6 exceeds this critical value, we would reject the idea that b_1 and b_2 for the larger population might both be zero.

Even though the b coefficients are different from zero, the regression equation may not do a particularly good job of prediction. The *coefficient of determination,* denoted by R^2, is a convenient index in this regard since it can be interpreted as the fraction of the variability in the dependent variable that is "explained" by the independent variables. The formula is

$$R^2 = \frac{\text{regression SS}}{\text{total SS}}$$

For the example we have

$$R^2 = \frac{41{,}419.87}{53{,}927.60} = .77$$

The decision is then a subjective one whether or not this fraction is large enough to satisfy the practical purpose of indirect estimation.

Except for b_0, the F ratio test from the analysis of variance table examines the entire set of b coefficients collectively. It does not address the possibility that some subset of the b coefficients might be zero with others being nonzero. There is a different F ratio test for dealing with this question. The first step in performing such a test is to calculate a new regression equation from which the subset of b coefficients in question have been eliminated. This is usually called a "reduced model." For example, we might wonder whether the shoreline length contributes anything over and above the square of the shoreline length. In other words, could b_1 be considered zero? The reduced model for testing this question is

$$\hat{Y} = b_0 + b_2 X_2$$

Note that we remove the terms we want to test in setting up the reduced model. The normal equations for this model are

$$b_0 n + b_2 \Sigma X_2 = \Sigma Y$$

$$b_0 \Sigma X_2 + b_2 \Sigma X_2^2 = \Sigma X_2 Y$$

Substituting in sums, squares, and cross products gives

$$10b_0 + \quad 227.45b_2 = 1118$$

$$227.45b_0 + 6840.4613b_2 = 33{,}706.76$$

The solution to this reduced set of normal equations is

$$b_0 = -1.137706 \qquad b_2 = 4.965386$$

The analysis of variance table for this reduced model is given in Table A.8. In order to test the components dropped in forming the reduced model, we construct an F ratio as follows.

First calculate the difference between the residual SS for the reduced and full models. This is

$$\begin{aligned} \text{reduction SS} &= \text{residual SS for reduced model} \\ &\quad - \text{residual SS for full model} \\ &= 12{,}824.88 - 12{,}507.73 = 317.15 \end{aligned}$$

Next calculate the corresponding difference between residual DF for the two models. This is

$$\begin{aligned} \text{reduction DF} &= \text{residual DF for reduced model} \\ &\quad - \text{residual DF for full model} \\ &= 8 - 7 = 1 \end{aligned}$$

Table A.8 Analysis of variance table for reduced model

Source	SS	DF	MS	F ratio
Regression	41,102.72	1	41,102.7	25.6
Residual	12,824.88	8	1,603.1	
Total	53,927.60			

Then compute the reduction MS as

$$\text{reduction MS} = \frac{\text{reduction SS}}{\text{reduction DF}} = \frac{317.15}{1} = 317.15$$

Now form an F ratio as

$$F \text{ ratio} = \frac{\text{reduction MS}}{\text{residual MS for full model}} = \frac{317.15}{1786.80} = .18$$

This is compared against a critical F value using the reduction DF as numerator DF and the residual DF for the full model as denominator DF. For a significance level of .05 in this example the critical F value is 5.59. If the calculated F ratio exceeds the critical value, the b coefficients dropped in forming the reduced model are significantly different from zero and should probably be retained in the prediction equation.

In our case the calculated F ratio is much less than the critical value. Therefore we would drop the shoreline length as an independent variable and use the reduced model

$$\hat{Y} = -1.137706 + 4.965386X_2$$

as our predictor. Since X_2 is actually the square of the shoreline length, and we had previously denoted the shoreline length by X, this is

$$\hat{Y} = -1.137706 + 4.965386X^2$$

In order to place standard errors on estimates from regression equations, one must make use of so-called C multipliers developed by matrix inversion. Space limitations preclude developing the necessary background in matrix algebra, so the reader is referred to Freese (1964) or Draper and Smith (1966) for this aspect of regression analysis.

A.5 NOMOGRAMS

Calculating predicted values from regression equations can be cumbersome when there are several independent variables. Tabulations are also awkward in such circumstances. Nomograms or alignment charts provide a graphical means of evaluating such equations. A nomogram is a set of three graduated lines positioned in such a way that a mathematical function of two variables can be read from one of the lines by laying a straight edge across the other two lines. A nomogram for the simple function

$$Y = 0.5X_1 + 2.0X_2$$

is shown in Figure A.2.

Each nomogram will serve to reduce two terms of an equation to a single term. Regressions with more than two independent variables will thus require a series of nomograms, with the value read from one nomogram being used as an entry for another nomogram. Nomograms for complicated equations may have nonuniform scales and curved lines.

A.6 STRATIFIED SAMPLING

Simple random sampling is a reasonable procedure to use when we know little about the population before we start sampling. Should we happen to have some prior knowledge about the population, however, we would not usually want to use simple random sampling. Simple random sampling does not allow us to use our previous knowledge to improve the sample, except for deciding on the sample size.

In many cases we will be able to divide the population ahead of time into groups of sampling units that we already know are different in some respect. We need a selection system that permits us to take advantage of our ability to subdivide the population into relatively homogeneous parts.

In a forest setting we might have a forest composed of several stands which have different species, ages, site qualities, or stockings. The stands are much more uniform than the forest as a whole. The fact that stands are different is obvious from a cursory examination before a survey is started. We would like to use the differences between stands to improve the survey we are planning.

The selection method known as *stratified random sampling* is useful in these circumstances. In sampling jargon the groups of relatively uniform sampling units are called *strata*. The general idea is to take separate samples in the respective strata, and then pool the information from the individual

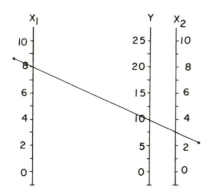

Figure A.2 Nomogram for $Y = 0.5X_1 + 2X_2$ with cross line representing $10 = (0.5)(8) + (2)(3)$.

strata that make up the population. In this way the rather large differences *between* strata do not get mixed up in the sampling error. The stratification of the sample leaves only the relatively small variations *within* strata to be reflected in the sampling error.

A.6.1 Sample Selection

If we make the sample size for a stratum proportional to the number of possible sampling units in the stratum, the individual sampling units still have an equal probability of getting into a sample (EPSEM). However, sampling units from different strata cannot get into *the same* sample. This way of allocating the samples to strata simplifies the calculations involved in making estimates.

However, there is nothing that says we have to put more samples in the larger strata. If it suits our purpose, we can allocate more samples to the more valuable strata, or perhaps to the more variable strata. An application of calculus shows that the sampling error and cost can be minimized simultaneously by making the allocation proportional to the factor.

$$\frac{\text{stratum size} \times \text{standard deviation for stratum}}{\sqrt{\text{cost per sampling unit for stratum}}}$$

As an illustration suppose that we are concerned with campground occupancy in three (hypothetical) parks. The first park has 30 tent sites, the second has 50 tent sites, and the third has 100 tent sites. We select at random and without replacement 5 tent sites in each of the first two parks and 10 tent sites from the third park. Although this allocation is not optimum, we might nevertheless see it as being desirable. Continuous records of occupancy are kept for the sample tent sites through the season, yielding the data shown in Table A.9. Each park constitutes a separate stratum.

A.6.2 Estimation

The rules for combining estimates given in Section A.3 are used in pooling estimates from the several strata. An independent sampling operation for each stratum ensures that the covariance terms are zero.

Estimates for the individual strata (parks) are shown in Table A.10, as calculated from the estimators of Section A.3.

The estimate of the total visitation for all parks (strata) is simply the sum of the estimated totals for the individual parks, namely,

$$\begin{aligned} \text{estimated total for population} &= 1260 + 1050 + 7900 \\ &= 10{,}210 \text{ tent site days} \end{aligned}$$

Table A.9 Number of days occupied for sample tent sites

Sample tent site	Park 1	Park 2	Park 3
1	60	10	75
2	50	20	90
3	30	35	85
4	40	25	100
5	30	15	65
6			70
7			95
8			60
9			80
10			70

Table A.10 Estimated tent site days by park (stratum)

Type of estimate	Park 1	Park 2	Park 3
Mean per tent site	42	21	79
Standard error of mean	5.3	4.1	4.0
Park total	1260	1050	7900
Standard error of total	159	205	400

The estimated mean per tent site for the population of three parks is the estimated total for the population divided by the total number of tent sites in the population, or

$$\text{estimated mean per tent site} = \frac{10{,}210}{180} = 56.7 \text{ occupied days}$$

The same result is obtained by weighting the means for the strata by the size of the stratum, that is,

$$\frac{(30)(42) + (50)(21) + (100)(79)}{30 + 50 + 100} = \frac{10{,}210}{180} = 56.7$$

Applying the rules for standard errors of combined estimates given in Section A.3, the standard error of the estimated population total is the square root of the sum of the squared standard errors for stratum totals, namely,

$$\sqrt{159^2 + 205^2 + 400^2} = 476.8$$

The standard error of the estimated mean per tent site across the entire population is obtained by dividing the standard error of the estimated total by the total number of tent sites in the entire population, namely,

$$\frac{476.8}{180} = 2.65$$

A.7 INTRODUCTION TO MULTISTAGE SAMPLING

Stratification reduces the sampling error if we are successful in setting up the strata so that each one contains quite uniform sampling units. However, stratification does nothing to reduce the travel time involved in taking the sample since each stratum must be visited and sampled individually. We may wish to subdivide the population with an eye toward reducing travel costs rather than sampling error. The procedure for accomplishing this is known as *multistage sampling*, *subsampling*, or sometimes also *cluster sampling*.

A.7.1 Sample Selection

The idea here is to divide the population into groups of adjacent sampling units in such a way that one group is as nearly like any other group as possible. Field work is thus reduced by selecting only a sample of groups, with the remaining groups not being visited at all. A sample of sampling units is then chosen from each of the chosen groups. This puts us in the position of taking samples from samples. The original groups are usually called *primary* or *first-stage units.* The individual sampling units within primary units are called *secondary* or *second-stage units.*

In fact there is no reason to stop at two stages. We can draw a sample of first-stage units; then draw a sample of second-stage units from the selected first-stage units; then draw a sample of third-stage units from the selected second-stage units; and so on. This allows us to localize our final sample instead of running all over the map taking samples scattered through the entire population.

A.7.2 Estimation

Let us take a different view of the campground data in Tables A.9 and A.10 in order to illustrate the concept of multistage sampling. Suppose now that the three parks are a random sample of 15 parks under our jurisdiction.

Suppose further that the total number of tent sites in all 15 parks is 900. The estimation problem is simplified by selecting first-stage units with replacement, and we will assume that this is the case for our sample of 3 parks from the 15 parks. If a first-stage unit appears in the sample more than once, a separate subsample is taken each time.

We are again interested in the number of days that the tent sites are occupied. Our first concern will be to estimate the total number of tent site days for all 15 parks. We have already made estimates of the total for each of the sample parks. Since there are 3 parks in the sample and 15 parks in the population, the total for the population should be approximately $^{15}/_{3} = 5$ times larger than the total for 3 sample parks. Therefore our estimate of the total for the population of 15 parks is

$$(1260 + 1050 + 7900) \times \frac{15}{3} = 10{,}210 \times 5 = 51{,}050 \text{ tent site days}$$

In general, estimation of the population total in this simple form of multistage sampling involves a stepwise series of expansions with the expansion factor for the ith stage being

$$\frac{M_i}{m_i}$$

where M_i is the number of units in the population at the ith stage, and m_i is the number of units in the sample at the ith stage. Thus we first expanded tent site (second-stage unit) figures to the park level (first-stage unit), and then expanded park level (first-stage unit) estimates to the population level.

After estimating the population total, we can proceed to estimate means for the various levels. Thus we can estimate the mean per tent site as

$$\text{estimated mean per tent site} = \frac{\text{estimated population total}}{\text{number of tent sites in population}}$$

$$= \frac{51{,}050}{900} = 56.7 \text{ days/tent site}$$

and the mean per park as

$$\frac{\text{estimated population total}}{\text{number of parks in population}} = \frac{51{,}050}{15} = 3403 \text{ days/park}$$

One approach to placing a standard error on the estimated population total is to note that we obtain an estimate of the population total by multiplying the estimated total for any first-stage unit in the sample by the number of first-stage units in the population. Thus

estimated total for sample park $\times$ 15 parks
$= 1260 \times 15 = 18{,}900$ as estimated from first sample park
$= 1050 \times 15 = 15{,}750$ as estimated from second sample park
$= 7900 \times 15 = 118{,}500$ as estimated from third sample park

The estimate 51,050 is simply the mean of these three estimates. We can treat the separate estimates from first-stage units as random variables and calculate the standard error of their mean. This procedure gives 33,737 as the standard error of the estimated population total. However, there are more efficient but more complicated methods based on the analysis of variance which will not be covered here.

There may also be situations where we do not know the total number of second-stage, third-stage, etc. units in the population. In these cases we can estimate the total number of such units in the population from the average number of subunits per next larger size unit in the sample. The price we pay for doing so is a bias in the estimates. This bias need not trouble us too much provided that a reasonably large number of first-stage units are included in the sample.

The topic of multistage sampling is a large one with many variations on the basic theme. The reader is advised to consult a textbook on sampling, and probably also a statistician, before undertaking large multistage sampling projects.

A.8 INTRODUCTION TO DOUBLE SAMPLING

Double sampling schemes involve making a relatively easy and/or inexpensive measurement on a large number of sampling units in the first phase, followed by a smaller number of more difficult and/or expensive measurements in the second phase. Three common types of double sampling will be introduced via illustrations of their possible applications.

A.8.1 Double Sampling with Stratification

When a survey project involves a mail questionnaire, the questionnaire will serve to divide the mailing list into respondent and nonrespondent categories. The proportions of these two categories in the mailing list provide estimates of the corresponding proportions of potential respondents and nonrespondents if the questionnaire were distributed to the entire population. Such a full mailing would allow stratification of the population into respondent and nonrespondent strata. The sample mailing provides an estimate of the population proportion in each stratum.

When sorted according to response/nonresponse, the sample mailing list is similar to a stratified sample drawn with the sample size proportional to the size of stratum. Characteristics of the respondent category (stratum) would be estimated from the information in the completed questionnaires. Characteristics of the nonrespondent stratum would probably be estimated by interviewing a sample of nonrespondents.

Since a random sample of a random sample is itself a random sample, the final sample resembles a stratified sample but is no longer of the proportional type. The sampling operation has involved two phases. The first phase of sample mailing and recording response or nonresponse provides estimates of stratum sizes. The second phase provides information on the actual variables of interest within the strata, and involves different kinds of measurement operations in the two strata. This type of two-phase sampling system is called *double sampling with stratification.*

Procedures for estimating means and totals are fairly apparent by analogy with stratified sampling. However, the stratum sizes now have a sampling error associated with them, which complicates the calculation of standard errors. A textbook on sampling methods should be consulted for procedural details.

A.8.2 Double Sampling with Ratio Estimation

As a setting for introducing another type of two-phase sampling, consider the problem of updating survey information from permanent sample plots. The establishment and original measurement of the plots constitutes the first phase of sampling. However, changes taking place in the field will eventually place the currency of the original survey information in doubt. To update the information base we might choose to remeasure a sample of the original plots. The ratio of remeasurement to original measurement provides an estimate of the percent change taking place since the original survey. This type of two-phase sampling is called *double sampling with ratio estimation.*

Again the procedures for making estimates of totals and averages are fairly obvious. However, standard errors of these estimates involve contributions from both phases of sampling and can be somewhat complicated.

A.8.3 Double Sampling with Regression Estimation

Indirect estimation by regression has already been introduced in Section A.4. A common procedure for implementing such indirect estimation is to take measurements of independent variables on a large sample, with measurements of both dependent and independent variables on a smaller sample serving as a basis for developing the regression equation. The large sample of independent variables can be considered as one phase of sampling, and the small sample for regression analysis as a second phase. Thus we have *double sampling with regression estimation.* As with the other types of double sampling, the estimation of means and totals will be reasonably straightforward, but standard errors must take account of sampling variation in both phases.

One can also envision survey designs involving three or more phases of sampling. These designs can become rather complex, and should be used under the guidance of a professional statistician.

A.9 SAMPLING WITH UNEQUAL PROBABILITIES

When some sampling units are larger or more important than others, one may question the wisdom of giving all sampling units an equal chance of appearing in the sample. If sampling units are given different probabilities of appearing in the sample, one must take account of this fact in making estimates from the sample data. We will consider here only the simplest case of sampling with unequal probabilities. This is the case of sampling with replacement in which the ith sampling unit has the probability p_i of appearing each time a sampling unit is drawn for inclusion in the sample. Thus the probabilities differ between sampling units, but the probability for a given sampling unit does not change from one draw to the next.

In this case we must use the reciprocals of the probabilities as weights in making estimates. Each sampling unit in the sample provides an estimate of the population total by the formula

$$\text{estimated total} = \frac{X_j}{p_j}$$

where X_j is the jth unit in the sample, and p_j is the probability of that sampling unit being selected on any given draw. The average of these estimates over the sampling units appearing in a sample of size n is

$$\text{estimated total} = \frac{1}{n} \sum^{n} \left(\frac{X_j}{p_j}\right)$$

Since the mean is the total divided by the number of units N in the population, the estimated mean is

$$\text{estimated mean} = \frac{1}{nN} \sum^{n} \left(\frac{X_j}{p_j}\right)$$

Let e_j be the estimate of the population total from the jth unit in the sample. The estimated variance of these estimates is

$$S_e^2 = \frac{\sum^n e_j^2 - \left(\sum^n e_j\right)^2 \Big/ n}{n - 1}$$

$$= \frac{\sum^n \left(\frac{X_j}{p_j}\right)^2 - \left[\sum^n \left(\frac{X_j}{p_j}\right)\right]^2 \Big/ n}{n - 1}$$

and the estimated variance of the mean of n such estimates from the sample is

$$S_{\text{total}}^2 = \frac{\sum^n \left(\frac{X_j}{p_j}\right)^2 - \left[\sum^n \left(\frac{X_j}{p_j}\right)\right]^2 \Big/ n}{n(n - 1)}$$

Therefore the standard error of the estimated total as computed from the sample is

$$S_{\text{total}} = \sqrt{S_{\text{total}}^2}$$

and the standard error of the estimated mean is

$$S_{\overline{X}} = \frac{S_{\text{total}}}{N}$$

A.10 CHI-SQUARE ANALYSIS OF CROSS TABULATIONS

Many of the data from social and economic surveys take the form of cross tabulations. For instance suppose that one question in a questionnaire called for respondents to place themselves in one of three categories with respect to variable A and another question called for choice of one of four categories of variable B. A cross tabulation or contingency table could then be prepared showing the numbers of respondents in the various possible combinations of categories for A and B. Assume for purposes of illustration that the results were as shown in Table A.11.

A.10.1 Expected Frequencies for Independent Categories

The question then arises regarding a possible relationship between A and B. If we divide each of the column totals by the total number of respondents,

Table A.11 Hypothetical cross tabulation of variables A and B showing number of respondents in each cell

Categories for variable B	Categories for variable A 1	2	3	Total for B
1	76	93	98	267
2	51	72	74	197
3	24	33	23	80
4	9	6	7	22
Total for A	160	204	202	$n = 566$

we have an estimate of the probabilities of a respondent falling in each of the categories of A. These are

$$\frac{160}{566} = .28 \qquad \frac{204}{566} = .36 \qquad \frac{202}{566} = .36$$

rounded to two decimal places. Likewise dividing each of the row totals by the total number of respondents gives estimates of the probabilities that a respondent will fall in the respective categories of B as follows:

$$\frac{267}{566} = .47 \qquad \frac{197}{566} = .35 \qquad \frac{80}{566} = .14 \qquad \frac{22}{566} = .04$$

In Section A.2 it was stated that the probability of two independent events occurring simultaneously is the product of their probabilities. If there is no relationship between variables A and B, the responses to the two questions will be independent. If the responses are independent, the expected number of respondents in any given cell is

$$\text{probability for row} \times \text{probability for column} \times n$$
$$= \frac{\text{row total}}{n} \times \frac{\text{column total}}{n} \times n$$
$$= \frac{\text{row total} \times \text{column total}}{n}$$

The expected number of respondents in each cell under the hypothesis of independence is given in Table A.12.

Table A.12 Expected number of respondents by cell if variables A and B of Table A.11 are independent

Categories for variable B	Categories for variable A 1	2	3	Total for B
1	75.5	96.2	95.3	267
2	55.7	71.0	70.3	197
3	22.6	28.8	28.6	80
4	6.2	8.0	7.8	22
Total for A	160.0	204.0	202.0	$n = 566$

A.10.2 Chi-Square Test of Independence

The Chi-square (χ^2) distribution provides a means of determining whether the differences between observed and expected cell frequencies can be attributed to chance variation. The chi-square criterion is

$$\chi^2 = \Sigma \frac{(O - E)^2}{E}$$

where O denotes an observed cell frequency, E denotes an expected cell frequency, and the summation is over the cells in the table excluding the totals. For the example,

$$\chi^2 = \Sigma \frac{(O - E)^2}{E} = 4.43$$

The entries for finding a critical value in a table of χ^2 are the probability that one is willing to accept of being wrong and the degrees of freedom for χ^2. The probability of being wrong is the probability of declaring that the variation between observed and expected values is not due to chance when in fact it is due to chance. For purposes of illustration we will accept a 1 in 20 chance or .05 probability of error. The degrees of freedom for χ^2 is $(R - 1) \times (C - 1)$ where R is the number of rows and C is the number of columns. For the example this is

$$\text{DF for } \chi^2 = (4 - 1) \times (3 - 1) = 3 \times 2 = 6$$

A χ^2 table will show the critical value for the probability of .05 and 6 degrees of freedom to be 12.6. If the computed χ^2 exceeds the critical χ^2, we declare that the variation is not due to chance, thus implying that there is a relationship between the variables. However, the computed χ^2 for the example is less than the critical value of 12.6, so we conclude that there is no convincing evidence of a relationship between variables.

Chi-square analysis is meant for use with relatively large samples, so it should not be applied to cross tabulations having cells with an expected frequency of less than 5. Small expected frequencies may require combining categories to obtain a table with fewer cells having larger expected frequencies. Some statisticians also recommend reducing each difference between observed and expected values by 0.5 before squaring, which gives a smaller value for the computed χ^2. This is called Yates' correction for continuity. Yates' correction can usually be ignored unless the sample size is small and there are only two categories for each variable.

A.11 BIBLIOGRAPHY FOR APPENDIX

Bliss, C. *Statistics in Biology,* Vol 1. New York: McGraw-Hill, 1967.

Bliss, C. *Statistics in Biology,* Vol. 2. New York: McGraw-Hill, 1970.

Bruce, D., and F. Schumacher. *Forest Mensuration.* New York: McGraw-Hill, 1950.

Cochran, W. *Sampling Techniques,* 2nd ed. New York: Wiley, 1963.

Draper, N., and H. Smith. *Applied Regression Analysis.* New York: Wiley, 1966.

Freese, F. *Elementary Forest Sampling.* Washington, D.C.: U.S. Department of Agriculture, Agricultural Handbook 232, 1962.

Freese, F. *Linear Regression Methods for Forest Research.* Madison, Wis.: U.S. Department of Agriculture, Forest Service, Forest Products Laboratory, Research Paper FPL-17, 1964.

Freese, F. *Elementary Statistical Methods for Foresters.* Washington, D.C.: U.S. Department of Agriculture, Agricultural Handbook 317, 1967.

Freund, J. *Mathematical Statistics,* 2nd ed. Englewood Cliffs, N.J.: Prentice-Hall, 1971.

Mood, A., and F. Graybill. *Introduction to the Theory of Statistics,* 2nd ed. New York: McGraw-Hill, 1963.

Snedecor, G., and W. Cochran. *Statistical Methods,* 5th ed. Ames, Iowa: Iowa State University Press, 1967.

Sokal, R., and F. Rohlf. *Biometry.* San Francisco: Freeman, 1969.

Spiegel, M. *Schaum's College Outline Series: Theory and Problems of Statistics.* New York: McGraw-Hill, 1961.

Sukhatme, P., and B. Sukhatme. *Sampling Theory of Surveys with Applications,* 2nd ed. Ames, Iowa: Iowa State University Press, 1970.

INDEX